Project Management in Product Development

Leadership Skills and Management Techniques to Deliver Great Products

Project Management in Product Development

Leadership Skills and Management Techniques to Deliver Great Products

George Ellis
VP Engineering and Chief Engineer, Kollmorgen IA-EU,
Radford, VA, USA

AMSTERDAM • BOSTON • HEIDELBERG • LONDON
NEW YORK • OXFORD • PARIS • SAN DIEGO
SAN FRANCISCO • SINGAPORE • SYDNEY • TOKYO

Butterworth-Heinemann is an imprint of Elsevier

Butterworth-Heinemann is an imprint of Elsevier
The Boulevard, Langford Lane, Kidlington, Oxford, OX5 1GB, UK
225 Wyman Street, Waltham, MA 02451, USA

Notices

Knowledge and best practice in this field are constantly changing. As new research and experience broaden our understanding, changes in research methods, professional practices, or medical treatment may become necessary.

Practitioners and researchers must always rely on their own experience and knowledge in evaluating and using any information, methods, compounds, or experiments described herein. In using such information or methods they should be mindful of their own safety and the safety of others, including parties for whom they have a professional responsibility.

To the fullest extent of the law, neither the Publisher nor the authors, contributors, or editors, assume any liability for any injury and/or damage to persons or property as a matter of products liability, negligence or otherwise, or from any use or operation of any methods, products, instructions, or ideas contained in the material herein.

ISBN: 978-0-12-802322-8

Library of Congress Cataloging-in-Publication Data
A catalogue record for this book is available from the Library of Congress

British Library Cataloguing-in-Publication Data
A catalogue record for this book is available from the British Library

For information on all Butterworth-Heinemann publications visit our website at http://store.elsevier.com/

Transferred to Digital Printing in 2016

Endorsements for *Project Management in Product Development*

Everyone recognizes the importance of sustainable innovation for any growing business. Less well understood is the vital role outstanding project management plays in innovative businesses. Our approach to project management at Danaher evolved greatly over the last two decades and played no small part in our success. George Ellis nicely captures the "state of the art" and demonstrates how process alone is not enough—true "Total Leadership" in project management differentiates the winners from the runners-up. If you want to win the innovation game, read his book.

—Lawrence Culp, Jr., Former CEO, Danaher Corporation

Tools for project management have expanded over the years, most recently agile methods as applied to software development, and lean methods adopted from manufacturing; earlier additions include critical-chain and phase-gate methodologies. Many fine books cover each of these, but none covers them all. Ellis does, as well as traditional waterfall/critical path methods. Especially valuable is Ellis' comparison of the methods, their relative strengths and weaknesses, where each applies and doesn't apply. Along the way he reminds us of the importance of leadership and interpersonal skills in project management by way of interesting side comments and bits of advice for dealing with technical specialists, customers, and bosses. All of this separates Ellis' book from and puts it above the rest in project management.

—John Nicholas, Professor, Quinlan School of Business, Loyola University, Chicago

Based on his many years of practical experience, George Ellis tackles an application area of project management that is rarely covered. This book is a very down-to-earth and thorough exposition, rather than being theoretical and academic. It is well illustrated with lists, tables, charts and explanatory diagrams. As well as techniques such as critical path management, it delves into Phase-gate management, Agile project management, "Lean Product Development,"

Risk, and "Patents" issues, all in the context of product development. Ellis also emphasizes the importance of looking upon project management as a leadership responsibility rather than just an administrative position. This book is a valuable addition to any product development manager's reference library.

—R. Max Wideman, FPMI, Richmond Hill, Ontario, Canada

The section on patent law is at the right level of detail to help project managers get up to speed. Clear, step-by-step explanations take the mystery out of reading a patent and searching for prior art. The reader will learn how to avoid serious pitfalls, and will acquire the understanding necessary to discuss patent issues with the development team. The patent material alone makes George Ellis' *Project Management in Product Development* an excellent investment.

—Alan L. Durham, Judge Robert S. Vance Professor of Law,
University of Alabama School of Law, and author of
Patent Law Essentials: A Concise Guide

Dedication

To Rachel and Regina, my advisors in this and so many other endeavors

Contents

Part III - Advanced Topics

Preface

Project management has undergone a revolution over the last two decades, especially for product development. Effective new methods and powerful tools abound. The results are in: they work. But in the face of so many advances, schedule and quality problems still dog product development projects in every industry. Project managers have better tools and are working as hard as ever, but too often the results are disappointing. What's happening?

For one thing, the role of leadership in project management is widely misunderstood. Many companies treat a project manager (PM) almost as administrator: "If projects are not going well, we need the PM to follow process better." There's no doubt that managing to process is essential for success: stay organized, follow up, work the issues, report regularly. This is *transactional* leadership and every PM must master these skills. This book has ample material to help readers grow in this area. But this is only half the story.

PMs must also grow as *transformational* leaders. They must build a vision for the product and create common purpose within the team. They must also be connected with their team, understanding the needs, abilities, and goals of each member. And they must display the kind of character people can follow. These are the skills that are required for a PM to inspire their team to care, to want to win, to do their best. Transactional and transformation skills work together to create *total leadership*. Too many writers present PMs with a false choice: "be a great manager or be a visionary." The outstanding PM will be both and this book accentuates that point throughout.

Projects that develop new products are fundamentally different from other project types. First, PMs are typically dealing with complex technology, partially understood customer needs, and team members that don't always have the best interpersonal skills. The PM is not always the technical expert, but must understand enough to make tough decisions when the team is not fully aligned. They also need knowledge of multiple project management techniques if they are to create the optimal plan for each project. Also, PMs need to know enough about patents to understand where there's an opportunity to protect an invention and when to seek legal counsel. This book addresses those topics, presenting numerous project management methods as different tools in the toolbox. There's even a chapter dedicated to patents for PMs. And every chapter focuses on product development. None of this says building an office park or running a marketing campaign is easier than developing a product. But they are different and product developers need a book that focuses on their domain.

The recent growth of project management methods has been exciting. The critical path method (CPM) was developed in the 1950s; even so, it's still the method of choice in many industries. In the 1980s, Phase–Gate project management (PGPM) was added as the need for cross-functional teams and process became clear. PGPM set up standards that every project in an organization had to meet and, in doing so, became the foundation for project portfolio management, another important advance. But even with these improvements, project results remained largely disappointing, with launch delay being the most common complaint. A number of alternatives to CPM were introduced starting with critical chain project management (CCPM) in the late 1990s, a method with intense focus on schedule. A few years later, lean product development (LPD) brought thinking from lean manufacturing to the engineering department: increase value, reduce waste. Today, both of these methods are flourishing, improving product development across all industries.

Around 2000, as software projects grew more common, it became clear that the existing project management methods focused on "high cost of iteration" projects—generally, hardware projects that demand a great deal of planning. Large investments in factory equipment, supplier tooling, and regulatory certification demand that most hardware designs be fully validated before release. CPM, PGPM, and CCPM are built to serve this dynamic. By contrast, software projects often have a low cost of iteration—new versions can be built in days. So, highly detailed planning at the outset is not as important. Accordingly, a family of Agile methods (Scrum, eXtreme Programming, and Scrumban, for example) were created and have become the dominant methods in many software industries. These methods build the product up, starting with the most basic functioning version and then iterating again and again until the full-featured version is released. That's not practical when you're tooling $1M worth of castings, but it works well for many software projects.

So, with all these project management methods, which one should a PM learn? The answer is, to some degree, all of them. Accordingly, this book will present these methods as alternatives, describing where each fits best. Each will be compared to the others across a dozen characteristics. Further, each will be deconstructed so PMs can pick and choose the components that fit their needs. Want to mix Kanban boards from LPD into your CPM project? No problem. Want to build the *minimum viable product* into your CCPM project? Also, not a problem. When you understand the methods, you can mix and match with ease. This book treats each method like a set of tools. None are better or worse; it's just that each does some things better than the others.

Finally, PMs must be versed in visual workflow management. This is the skill of building a simple, credible "picture" that can be used to drive good action. Visual workflow management can be (and should be) applied across a wide range of issues from a 2-week cost-reduction effort to a *key performance indicator* (KPI) that lasts the life of the project. Visual workflow management has three primary requirements:

- *Simple*: it fits on one page with minimum text and a person outside the core project team can understand it quickly.
- *Credible*: people trust the data. They understand it, they believe the source, and they know the picture is updated regularly.
- *Drives good action*: the information is there to allow the person to make the decision that their role requires, be that a team member or the company president.

Visual workflow management is taught throughout this book. These skills are needed to present status to the project sponsor and steering committee. They are needed to communicate with the team, ensuring consensus concerning the major issues. Most importantly, they are needed for the PM to know he or she truly understands the project issues throughout its life.

Recent advances in project management are broad and vibrant. There are a host of new methods that can be tailored to every project. The principles that build good leaders who can leverage those methods are known. The tools to maximize the efficiency of these methods are also widely available. It is the goal of this book to bring all of these factors together so you and your organization can reliably launch great products on time and on budget.

This book is separated into three parts:

- **Part I: The Fundamentals**
  The first three chapters provide an overview of projects and their use in product development along with the basic techniques you can use right now to get started. This part is written assuming you bought this book because you had an immediate need and that you shouldn't have to read through hundreds of pages to start managing projects. Accordingly, the critical path method, the foundation of project management, is presented in Chapter 2 (Planning) and Chapter 3 (Execution).

- **Part II: Leadership Skills and Management Methods**
  Chapter 4 explains the value of leadership skills and then provides material to help PMs build those skills. Chapters 5–8 present four competing project management methods, including critical path management with Phase–Gate (Chapter 5), critical chain project management (Chapter 6), lean product development (Chapter 7), and Agile methods (Chapter 8). The goal is to provide a balanced view of each without advocating any.

- **Part III: Advanced Topics**
  Three advanced topics are presented in this last part:
  - Chapter 9 Risks and Issues: Preparing for and Responding to the Unexpected
    Describes risk management using two lines of defense. The first line is diligent preparation—adequate planning, choosing a strong team, and having the right processes in place. The second is competent leadership in resolving issues since some risks will survive even the most diligent preparation.

- Chapter 10: Patents for Project Manager
  Discusses patents and patent law as they relate to project management for product development. It is likely that as PM, you will at some point come into contact with patent law, especially when developing innovative products. Unfortunately, patents are often misunderstood by development teams. PMs are in a unique position to help the company accomplish its goals related to patents so you will want to be familiar with the basics.
- Chapter 11: Reporting
  Provides detailed discussion on reporting with a focus on reporting up—to the sponsor or steering committee. The first section focuses on oral presentations; the remainder discusses the use of quantification in project management, beginning with metrics and then developing KPIs, and finally creating a project dashboard.

Acknowledgments

Writing a book is a large task and requires support from many people. First, I want to thank my mother, who, when the facts should have dissuaded her, was sure I would grow into someone she would be proud of. And thanks to my father for his unending insistence that I obtain a college education; that privilege was denied to him, an intelligent man born into a family of modest means. Thanks to my beautiful wife Regina and our dear daughter Rachel, who tolerated my absences on countless weekends, evenings, and vacations as I labored on the manuscript. Both were also skillful and tolerant advisors.

Many thanks go out to the people of Danaher Corporation, my long-time employer. I'm grateful to the management team, especially John Boyland and Dan St. Martin, who supported me in this task from the beginning. Special thanks go to my colleagues who reviewed various chapters. Scott Davis and Josiah Bogue helped me get off to a good start, advising me on early material. Scott Davis also used his considerable skill and experience in Agile to bring balance to the material; he also was a valued advisor on leadership and several other chapters. I appreciate the advice George Yundt provided, applying his skill in patents to Chapter 10. Thanks also to Andrew McCauley for his sage advice on topics relating to communication.

Special thanks go to Mark Woeppel, President and CEO of Pinnacle Technologies, whose reviews of critical chain project management were invaluable. I met Mark after watching his presentation "Deconstructing Critical Chain: A failed Implementation Leads to an Evolution" (referenced in Chapter 6). I was impressed by how he combined a deep knowledge of the critical chain method with great flexibility to meet his client's needs. I subsequently contacted him asking for his advice, which he gave in generous doses. His hours reviewing the material, correcting errors, and bringing balance will benefit every reader.

I want also to thank Tony Ricci for his guidance on how to gently start lean conversion. Finally, I want to thank those reviewers Elsevier engaged on my behalf, especially John Nicholas, whose diligent and knowledgeable review of each chapter combined with his constant encouragement caused me to look forward to his comments. Thanks also to Prof. Alan Durham for the many improvements he brought to the chapter on patents.

Part I

The Fundamentals

Part I will present the fundamentals of project management. This part is designed to get the you off to a quick start: presenting the why's and how's of project management as it applies to product development, and then quickly jumping into the planning and execution of a simple project.

CHAPTER 1 INTRODUCTION

Describes the reasons projects are used to develop product and the make-up of a typical project team. Discusses how projects for product development are different from other project types. Discusses reasons to be a project manager (PM) and what makes a person a good fit to the role.

CHAPTER 2 THE CRITICAL PATH METHOD: PLANNING PHASE

Presents a step-by-step planning process appropriate for a basic new-product development project. Discusses each step in detail, giving the largest coverage to schedule development.

CHAPTER 3 THE CRITICAL PATH METHOD: EXECUTION PHASE

Provides details at multiple levels on how to execute the project planned in Chapter 2.

Chapter 1

An Introduction to Project Management for Product Development

1.1 THE PROJECT: FLEXIBILITY, COMMUNICATION, AND ACCOUNTABILITY

A project is a sequence of activities undertaken to accomplish a specified outcome at a defined time using a defined set of resources. The project team is made up of the people who work on the various tasks of the project including the project manager (PM) who leads the team and the project sponsor, typically a member of the senior staff, who provides oversight and approvals.

The project structure rests on a complex chain of commitments made among the project team members and between the organization and the project team. The most basic of all those commitments is the "Iron Triangle" shown in Figure 1.1 [1]. In product development, the triangle represents that the team will be given time and resources (people, expenses, and support) in exchange for delivering a product that meets the specification. The PM can breech that commitment by missing all or part of a deliverable, delaying the completion date, or exceeding the budget. The project sponsor can also breech the commitment by expanding the deliverables (*scope creep*), pulling committed resources from the team, or shortening the schedule. These are breeches whether the sponsor or PM takes the actions directly or fails to prevent others from doing so. This triangle is deceptively simple—in a single project, it may

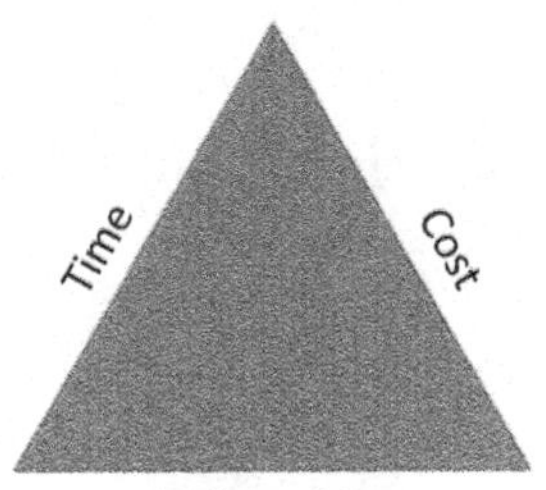

FIGURE 1.1 The Iron Triangle of project management.

Project Management in Product Development. http://dx.doi.org/10.1016/B978-0-12-802322-8.00001-2

take on a 100 forms, each of which may need to be negotiated, documented, and ultimately executed. It may be said that all of project management—the techniques, processes, and tools—rests on this most basic principle.

Modern business has come to rely on projects because of their flexibility—they can be right-sized for almost any activity from planning an offsite meeting to building a product platform to completing a business acquisition. The team size can vary from two to hundreds. They can be focused on a single function, like having a team of three programmers develop a firmware upgrade for an existing product. Or they can be the most international, cross-functional group in the organization; for example, creating a new product platform can require people from marketing, multiple disciplines of design engineers, manufacturing engineers, process engineers, and members from sourcing, quality assurance, regulatory compliance, finance, and other departments. The project can last weeks, months, or years. Team members can drop off when the bulk of their contribution is finished; new members can be added as new functions receive more focus.

The supreme flexibility of the project is one reason why today's businesses have come to rely on them so heavily. Another is their ability to break down the silos that exist in most companies as shown in Figure 1.2. Attempting to manage a cross-functional activity with the thick walls created by functional organizations usually results in disappointment. The teams don't work to a common goal because the different functional departments have different objectives. When a quality problem appears on the factory floor, fingers can start pointing. The manufacturing engineer may think poor manufacturability demands product redesign while the design engineer may think the assembly process needs to be improved. Both may be right and the silo organization incentivizes each to blame the other—"if I can get her to solve this, I can work on what my boss really wants." When problems like this come up from time to time, organizations can manage their way through them. But this type of conflict can arise

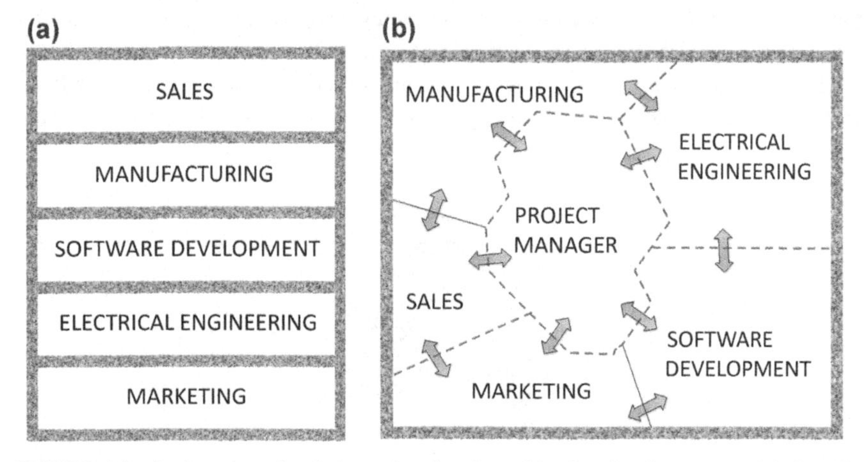

FIGURE 1.2 Project teams break down the silos formed by functional structures. (a) The silo structure of functional structures. (b) Projects connect people across functions.

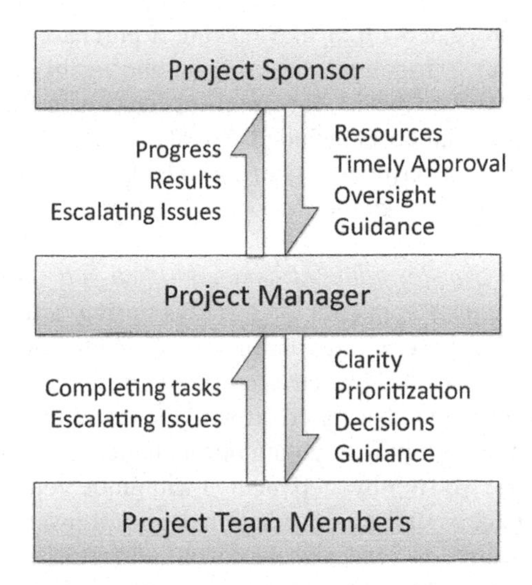

FIGURE 1.3 The chain of accountability created by the project structure.

daily when a cross-functional team works on a complex project—without the thin walls and constant communication fostered by the project structure, progress can grind to a halt.

Another reason the project structure is so effective is the transparency it creates for accountability. The PM has responsibility to the sponsor for progress and reporting. In addition, projects have a series of tasks called the *work breakdown structure*, each with a clear owner who is responsible for that task. At the same time, the sponsor has duties to the PM such as timely approvals and guarding project resources. The PM also has responsibilities to the team such as organizing project work, keeping prioritization clear, and making decisions. So a chain of accountability is created from the sponsor through the PM to the team as shown in Figure 1.3.

Without the project structure, accountability is unclear. Consider this simple case: your boss asks you and a coworker to do something today. The agreement is that you will get things started and your coworker will take it from there. Suppose you do your part and give it to your coworker, but progress stops there. Is it your responsibility to ensure your colleague finishes or will your boss look after it? If the boss's ask is informal, who knows? If she sets up a project with a manager, it's clear. The lack of clarity from even this simple example can create problems, and those problems get exponentially worse as larger teams work together on more complex goals.

1.2 PROJECT MANAGEMENT AND LEADERSHIP

Most texts on project management focus heavily on *hard skills* such as ability to follow a process, stay organized, and report regularly on progress. These skills are often referred to as *transactional leadership* because they relate to actions that are repeated often like making a purchase at a store. You're expected to carry out transactional tasks again and again, accurately and on time. File weekly reports. Check your team's actions against a process. Ensure the budget has all material costs included. Sweat the details, keep order. Strong transactional leadership is critical for PMs. Think of transactional leadership as working within the system—doing the same thing again and again, striving to improve over time.

Compare this to *transformational leadership*, the *soft skills* that deal with events that occur irregularly and are too complex to fit a process. They include building a common purpose within your team and displaying the kind of character a team will want to follow. They require you to genuinely care for the best interests of your team: sharing credit with them, helping them with career choices, coaching them on how to communicate better, and resolving interpersonal conflicts fairly. No step-by-step method will guide you in these areas.

Strong PMs have solid transformational and transactional leadership skills. Both are required to lead a project well. Unfortunately, many authors emphasize one set of skills at the expense of the other. Many books on project management will dedicate the great majority of the material to transactional skills—showing detail after detail on Gantt charts and resource allocation techniques. They may mention leadership skills, but the passion for transactional skills comes through clearly. It's as if transactional skills are necessary but transformational ones are "nice-to-haves." Perhaps the most respected guide to traditional project management is the *Project Management Body of Knowledge* (PMBOK®) from the Project Management Institute (PMI). PMI is the largest certifying organization for PMs in the United States. Curlee and Gordon endorse the PMBOK® Guide, calling it an "essential document for any project manager" [2]. But they also point out that the PMBOK® Guide "offers some guidance with regard to [transformational[1]] leadership…but [offers] little direction in this area" [3].

At the other end of the spectrum, some authors espouse transformational leadership skills while denigrating transactional leadership. They present the reader with a choice: be a transformational leader (a "visionary") or be a transactional leader (just a "manager"), as if you can't be both. A common refrain is "managers manage things, but leaders lead people." Don't believe it. While vision and connection are necessary for success, if you master them without management skills, you'll be able to imagine great things and infect others with your enthusiasm, but unable to get much done; you won't inspire anyone for very long. Management skills such as the ability to build a plan, execute the steps correctly and according to the schedule, and timely reporting to all concerned are also necessary for success.

1. Note that Curlee, like many authors, substitutes "leadership" for transformation and "management" for transaction.

Transactional skills without transformation produce dull bureaucrats that push a bored team through checklists to develop mediocre products. Transformational leadership without transaction produces lofty goals and enthusiasm without the ability to finish. The outstanding leader will master both—he or she will be able to define a worthy destination and build common purpose with their team to go there; then they will have the discipline to build and execute a plan so the team arrives at that destination. Chapter 4 will focus entirely on this topic; throughout this book, the need for both types of leadership skills will be highlighted.

1.3 PRODUCT DEVELOPMENT PROJECTS

This text is focused on projects for product development. Product development projects share many characteristics with other project types such as building construction, event planning, marketing initiatives, and election campaigns. Each of these has a project structure with customer goals, a list of tasks, a budget of some sort, and a target completion date. They generally have the same organizational structure, which is to say a temporary project team led by a PM and resources who likely don't report directly to the PM. So, all project types have much in common, but product development projects are different, requiring: (1) a large effort to manage innovation, (2) an unusually high level of collaboration, (3) a low level of determinism at the outset, and (4) high reliance on technical experts.

1.3.1 Innovation Management

Because many new products require new technology in order to be competitive, development projects often require innovation management. Of course, all project types require a certain level of innovative thinking. However, product development (along with its sibling, technology development) is unique in that often the innovation is the primary source of value added by the project. If the product doesn't do something different from what's already available, it may not be successful. Further, if the company can't protect the innovation over time, any advantage may be short lived. Innovation management is a large topic and what follows is a brief overview.

Innovation can be defined as the use of inventions to solve unmet customer or market needs [4]. When the bulk of the value of the project derives from innovation, the PM must understand the inventions: how the customer values them, how they improve upon competitors' offerings, what development risks are associated with them, and what intellectual property will need protection. The PM must understand them because it's common for the innovative components of the project to create unexpected barriers. Development may stall because the team lacks the technical ability to resolve a problem or because a supplier cannot deliver on a critical commitment. If that time comes, the PM will be called on to help decide among delaying launch, cutting a feature, or increasing resources. Only a keen understanding of the innovation will enable the PM to guide the team. Of course, the PM will receive input from team members: marketing should understand best how the innovation brings value to the marketplace. Technical experts should have the best understanding of the development risks. However, team members will often bring conflicting advice—the better the PM understands the issues, the better decisions he will make.

The PM must also understand intellection property (IP) law well enough to guide prudent action. The primary questions are which inventions should be reviewed for (1) pursuing a patent, (2) avoiding patent infringement, or (3) protecting trade secrets. This is a complex process of reviewing patents and other literature. It usually requires collaboration with an IP expert, typically a lawyer. Ignore it at your peril. Miss an opportunity to patent, and the product may be copied quickly. On the other hand, if the new product infringes someone else's technology, it can be a serious problem. The product's lifetime profitability can be significantly reduced by a single patent infringement case.

Again, the PM will be advised by team members, but team members do not reliably drive good action around patents—inventors often regard their inventions as "obvious," which, if true, would bar patentability. But the legal meaning of *obviousness* is complex and should be determined by an expert. The organization should be able to rely on the PM to ensure each invention gets a fair hearing. And if the company elects to purse a patent, the PM must manage the tasks related to completing patent applications in a timely way. Finally, the PM must manage information during the patenting process—an advertisement, a demonstration at a trade show, or even the wrong conversation with a customer can eliminate the possibility of patenting. The PM must guide the team to avoid these risks. Patents and the related responsibilities of a PM are the topic of Chapter 10.

1.3.2 Managing Extreme Collaboration

Product development projects are among the most cross-functional activities in any organization. A development project requires collaboration with almost every part of an organization: R&D, manufacturing, application engineering, customer support, sourcing, finance, service, sales, and marketing, among others. You may have team members spread out around the world and from many and varied cultures. All projects require collaboration, but the depth and breadth of collaboration for product development are as large as any project type.

This level of collaboration often leads to resource allocation conflicts. For example, suppose a PM is leading a project that requires a new X-ray sensor. If the company has only one expert in X-ray technology and there are several active projects related to this field, the PM will have to share this resource. This means the health of the PM's project depends in part on the health of competing projects. So the PM must keep up with competing projects and, when a conflict arises, be ready to make the case for her project.

1.3.3 Lack of Determinism at the Outset

All projects have risks at the start, but product development projects are subject to more unknown risks than other project types. For example, market requirements are normally known only partially. Moreover, it's often unclear exactly what the customer will value and by how much. This is especially true of the most innovative products because customers won't understand them fully until they use them; so, they may find it difficult to imagine how those products will benefit them. No one will argue that a project to build an airport is complex, but before any earth is moved, a great deal about what the customer wants must be clear. Technologies are often only partially understood at the outset of a project. There may be proof-of-concept units, but many questions remain about reliability, manufacturability, and production costs. New suppliers are often required and they bring risks, especially when they are critical for innovative components. In such cases, patents or know-how may force the company to rely on a firm whose capability will be demonstrated only as the project proceeds. During the project, measuring a supplier's progress can be difficult, especially if they fall behind since they may perceive incentive to conceal the potential for delay. And the most innovative products often bring regulatory compliance uncertainty because the product is doing something new that may not have been contemplated by the regulatory agency.

1.3.4 Reliance on Technical Expertise

Most project types rely on technical expertise at some level, but for product development that reliance is high. A single project might need an electronic hardware design guru, a team of extraordinary programmers, an outstanding material scientist, and a mechanical genius. And because they all might be doing something new, they may be unable to predict when or even if they will be able to deliver their tasks. And experts are not always the easiest personalities to work with: "you can expect the interaction to test your interpersonal behavior skills" [5].

1.3.5 Summing It Up

Taken together, these factors make fundamental differences in how product development projects are managed (see Table 1.1). Innovation must be managed, which requires technical knowledge, market/customer understanding, and, at a minimum, enough familiarity with IP law to know when to seek assistance from an attorney. The complexity of the collaboration will have PMs making decisions in a broad range of areas, many of which will be unfamiliar.

TABLE 1.1 Examples of How Product Development Varies from Other Project Types

Area	Extra Demand Placed on PM
Innovation management	Understanding the inventions and how they bring value to the customer; understanding how the competition is likely to respond.
	Understanding IP law well enough to avoid risks of (1) losing the ability to patent, (2) patent infringement, and (3) losing ownership of know-how.
Extreme collaboration	Being able to engage every member in a team of unusual technical and cultural diversity.
	Managing conflicts that result from sharing a finite pool of resources.
Lack of determinism at the start	Remaining open minded and curious as the project proceeds. Being willing to reset project priorities as data becomes available.
Reliance on technical experts	Being able to make decisions in areas where the PM is not fully knowledgeable.
	Dealing with the difficult personalities that are overrepresented in teams of technical experts.

The lack of determinism requires an open and curious mind, always taking in new information and ready to change[2] project plans when the data demands it. This can be especially challenging for the personalities that value order and stability most—the very type of person often drawn to project management. The reliance on technical expertise can be difficult since the PM is likely to encounter more than the average number of difficult personalities; this will test the PM's patience and leadership skills. The elements of Table 1.1 work together to demand more flexibility on the part of the PM than most other project types.

Returning to the PMBOK® Guide, Curlee and Gordon find weaknesses in the "linear thinking" it teaches when dealing with projects that have too many unpredictable factors,[3] a category product development projects commonly fall into. They write that PMBOK® does not adequately address issues such as communication and working in a highly distributed team [6].

---

2. In this text, we'll refer to such changes as *project innovation*, the application of invention to solve a problem in project planning and execution.
3. The authors use the term *complexity*.

None of this is to say product development projects are more difficult than other project types. Building a 120-story building is extremely challenging. So is planning a championship football game, or producing a large-budget movie. But it's difficult to think of any other project type that includes the four elements above to the degree that new product development does. To be sure, we can learn from PMs of every stripe. Nevertheless, product development is a class all its own.

1.4 WHY ORGANIZATIONS NEED PMs

Today's organizations rely on projects to accomplish their most important goals and almost every product of any complexity is developed by a project team. So, what value does the PM bring? Without proper project management, the smartest, hardest working, and best-intentioned engineers will quickly find themselves mired in confusion and conflict. The capable PM will inspire the team to deliver their best effort, focus them on satisfying the customer, make decisions, manage the many details of a complex project, identify and mitigate risks early, manage the customers and the suppliers, empower the team, resolve the inevitable conflicts within the team, and protect the team from distractions while ensuring they are recognized for their contribution. Success in a project depends on many factors ranging from company core competences to good luck, but strong project management is one of the most important elements.

Success in product development is complicated to measure. According to Kerzner, there are seven primary measures for projects, the first three of which come from the Iron Triangle [7]:

1. Was the project completed in the allocated time period?
2. Was the project completed within the budget?
3. Does the product exhibit all the features and performance requirements?
4. Did the customer want the new product?
5. Were the scope changes well managed?
6. Was the project completed without disturbing other work in the organization?
7. Was the project able to be completed in the corporate culture?

This set of goals is not exhaustive: we could add product manufacturability, IP protection, team member satisfaction, and team member professional growth along with many other metrics. A strong PM is required for a team to be successful in the many dimensions that are important to an organization.

Unfortunately, the current state of project success is somewhat disappointing. According to a study done by the PMI, fewer than 70% of projects in large organizations meet stated project goals. Another study, this time of large engineering projects by Miller and Lessard, determined that less than half of projects met most of their stated objectives—20% had to be fully abandoned [8]. The 2009 Standish's Chaos Report estimates 44% of IT projects were challenged and 24% were failures, and Curlee extrapolates this to estimate that more than 30% of all projects end in failure [9]. However, these statistics might not be quite as discouraging as they seem. Reinertsen [10] points out that the failure rate should be above zero because if every project succeeds, the company is probably being too conservative. Nevertheless, these failure rates are too high. Apparently, industry has yet to fully understand the discipline of creating and managing projects. So, the opportunity for skilled PMs seems large and enduring.

1.5 DO YOU WANT TO BE A PM?

Perhaps you picked up this text because you are an "accidental" PM—you came to work one day and discovered you were a PM. Perhaps you aspire to be a PM and you want to understand how well it fits you and what opportunity it brings. Perhaps you're a leader in a company looking to improve your project management team and you want to better understand the discipline. Whatever your situation, you should find these questions interesting: (1) What are the skills that a PM should possess and (2) What are the incentives for a person to become a PM?

1.5.1 What Skills Should a PM Possess?

The first reason to become a PM is because you want to help a team become more successful. If your first goal is to help others—increase their job satisfaction today and improve their career prospects tomorrow—you'll be able to withstand the ups and downs that come with this career. On my best and worst days, recalling those cases where I have been able to help people is an enduring source of satisfaction.

Another is the drive to help your company reach its goals through serving customer and market needs. As a PM, you will expend so much energy in the service of your company's goals and those of its customers that you should derive satisfaction knowing you increased the value of your organization. And companies value this greatly, for example, "A research undertaken by McManus and Wood–Harper highlights that companies look for a number of potential skill sets in their project managers, the most important being a solid understanding of the business objectives" [11].

You should also want to lead people. Project management is the first leadership role for most engineers and developers. The PM role is structured to build your ability to lead through inspiration. (Normally your team members will not report directly to you, so your ability to lead through "coercive incentives" such as disciplinary action is limited.) This is by intention—if you can't lead people that don't report to you, you won't be able to lead those that do. In my career, I've had many cases of people I've led coming into my organization or moving out of it and my ability (or inability!) to lead those people remained nearly unchanged. We'll talk more about this in Chapter 4, Total Leadership for Project Managers.

Your prospects of leading in your organization will almost always develop faster through project management than through mastering technology. The technology path is long because it takes years to master technology (10,000 hours, generally 5–10 years, according to Gladwell [12]). Certainly you can take on leadership roles as a technical expert, but those opportunities normally will come more slowly. None of this should push you to project management if technology is your first love. All things being equal, you will excel at what you enjoy, so don't leave your passion for the imagined rewards of a career in leadership. However, if your ultimate goal is to lead people, good project management can get you there fast.

Table 1.2 provides a sample of characteristics of strong PMs; it's not exhaustive (see Ref. [13] for more examples).

TABLE 1.2 Reasons to Pursue or Avoid Project Management

You May Be a Good Fit as a PM If...	Be Concerned about Becoming a PM If...
You want to help others to succeed.	You like working alone.
You want to serve by helping your company and its customers accomplish their goals.	You don't share the vision of your organization. You don't find their goals appealing.
You want to lead others through inspiring them. You are willing to funnel recognition to your team.	You thrive on recognition for your personal contributions such as solving tough technical problems or mastering technology.
You are highly organized. You sweat the details. You're driven to finish what you start.	You're a big-picture person—you like to create and innovate, but you'd prefer to let others work through the details.
You get satisfaction when you help resolve personal conflicts. You get along with people and you have patience for difficult personalities.	People that don't get along frustrate you—you don't enjoy helping them find consensus. You struggle getting along with a number of people yourself.
You like working closely with people from different cultures even though it can be difficult at times.	You find the challenges of bridging cultural differences uninteresting.
You like variety in your daily work. You can tolerate changing priorities.	You value a predictable job. You like to focus on lengthy tasks until you're finished.
You can lead (or want to learn how to lead) a team relying mostly on inspiration.	You might not mind directing others, but you have little interest in inspiring others to action.
You have enough self-confidence to lead even when you are regularly in a position having less technical understanding than some of those you lead.	You become defensive easily when others indicate you are not as smart as they are.

If, after reading this, you're not sure about project management, you might consider being a "working" PM. The focus of this text is the full PM: someone whose primary role in the project is to plan, manage, and report while the great majority of the project work is done by other team members. This is a model that fits medium and large projects best with teams of, say, four or more people. However, when the project team drops to two or three people, it's common to use a simpler model: a working PM's primary role is to complete project work; her secondary role is to execute the modest amount of project management work these smaller projects require. The role is a compromise between project management and project engineering and so it can be an ideal step for those uncertain if they want to commit to project management as a career. As a working PM, you'll be exposed to the primary functions of a PM—scheduling, following process, reporting—but all on a smaller scale and you'll be able to retain your technical work. Some people use the role of working PM as a stepping stone to full project management; others remain as working PMs indefinitely because the mix of management and engineering is appealing. See Table 1.3 for an overview.

1.5.2 Project Management Support Structure

If you have decided you'd like to be a PM, next consider the implications of taking on that role at a given company. Project management is a complex job and requires a strong support structure from the organization. Here are a few questions to ask yourself as you consider being a PM in a particular organization:

- Does the company value project management? There's a wide range of acceptance of this discipline, with some companies seeing the role as primarily administrative (filling in forms, holding meetings, but making few decisions) and others seeing it as a leadership role critical to project success. Obviously,

TABLE 1.3 Comparing Full and Working PMs

	Full PM	Working PM
Time spent on project management	Up to 100%	Typically <25%
Type of work	Project management	Mostly technical with some project management
Size of project managed	Medium to large	Small
Opportunity for leadership in the project	High	Limited
Requirements for organization skills	High	Medium

if the role is viewed as primarily administrative, your opportunities for growth will be limited.

- Does the company have an established group of PMs? If so, meet the team and see how you would fit in and how the position would fulfill your aspirations. If there are no PMs currently, is it because the company doesn't sustain the role well (many people leave the position, the company is slow to replace them) or has the importance of the role only recently been recognized? In either case, it's probably wise to be concerned—if the commitment to this discipline is recent or varies over time, it may not survive the ups and downs of the organization's business cycle. Try to find out why the role is receiving newfound attention and if it is broadly supported by the senior leadership.
- Are the business opportunities solid? Would you be working in an existing business with sustained revenue? If so, is that business healthy and growing? If not, are you part of a group embarking on a new and exciting journey? New and exciting journeys can bring opportunity for growth. However, the company's commitment can flag if growth in the new area is disappointing. The stronger the business you will be serving, the more opportunities you are likely to encounter.
- Is the project team strong? Do they have the technical skills to execute the anticipated projects? Do they work together well with no more than ordinary conflicts? If you're leading a weak team there will be a limit to what you can accomplish and to what you'll learn.
- Does the company have a commitment to the processes to support project management? These processes are among the most collaborative in any organization and you'll need support from every corner of the company. If they don't exist, a large effort must be expended to create them. Is the company motivated to do this? If they do exist, find out if people use them and if they are kept up to date. Processes that are developed and left in a drawer have little value. Ask if the company regularly revises them—that's usually a good sign.
- How does the organization view the workload? In some companies, PMs can be expected to put in whatever labor is necessary to meet project requirements. Be sure your expectations are aligned to those of the organization.

If these questions don't have clear answers, bear in mind that a move to project management is almost always going to bring some level of risk. The important thing is to understand the risk and act reasonably.

1.6 PROJECT STAKEHOLDERS

Project *stakeholder* refers to anyone that has a substantial interest in the project. These include six major groups interacting with the PM as shown in Figure 1.4 [14,15].

- The customer or customers for the products being developed. This can range from one company to millions of consumers.
- Senior staff—the leadership of the company that steers the product development projects.

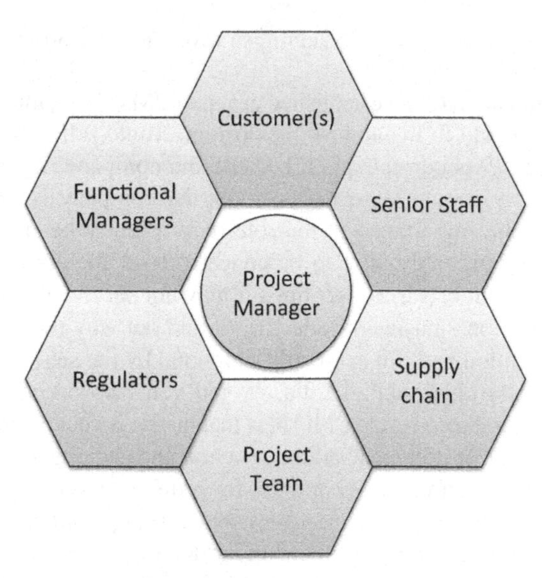

FIGURE 1.4 The project stakeholders.

- Function managers such as director of engineering or vice president operations.
- Regulators such as UL (Underwriters Laboratories) and FDA (Food and Drug Administration).
- The supply chain—your suppliers and the suppliers they rely on.
- The project team members.

The PM must serve each stakeholder, meeting their needs where possible and, where not, renegotiating [16].

Finally, the project sponsor is typically a member of senior staff or a functional manager that has high-level responsibility for the project within the organization. The sponsor has several roles [17]:

- Champion the project at the highest levels of the company.
- Remove barriers in the organization.
- Provide guidance for the PM.
- Ensure functional managers keep their commitments to the projects.

Often the sponsor is an informal role and sometimes it can be a combination of more than one person. Some projects lack a sponsor and this can make managing the project more difficult.

1.7 CERTIFICATION

One question that comes up often for new PMs is: Should I get certified? There are a number of agencies that certify PMs, the largest in the United States being the PMI, which offers the widely popular Project Management Professional (PMP)®

certification. Appendix A shows a number of certifying organizations for different regions of the world.

Certification is valued differently in different industries. Highly regulated industries like defense, pharmaceuticals, and financial services demand certification. These industries must comply with complex regulations in the planning and execution of their projects.

On the other hand, the requirement for certification in most product development organizations is lower. Naturally, many of the things you would learn would be valuable and all things being equal, certification would benefit you; those benefits arise not simply because of the knowledge you gained but also because it shows a high level of motivation, which is likely to be appreciated by prospective employers. Certainly, it can open doors to join a new firm or move to a new position in your company.

However, the question is not whether certification has value, but rather what is the most valuable way to invest the many hours required to attain it? Many companies developing new products will place a high value on *domain* knowledge—knowledge about their customers, markets, and technologies. Remember, PMs in product development must make decisions when experts on the team are not in full alignment. Also, they are in a position to spot gaps in project plans that team members may have missed. Domain knowledge is highly valuable in such situations.

It's also important to understand that competence as a PM is a combination of knowledge, skills, and ability to perform. Professional certification measures competence. Dinamore compares certification to a written driver's test, which is only one part of proving you can drive; the road test fills many of the gaps the written test leaves. He points out what certification does not provide: "Lacking are skill-related evaluations, indicating that task can be skillfully carried out…and that those skills can be applied together with other skills in project settings" [18].

In the end, these types of questions are complex because there are many unknowns. You may invest a large effort in a PMP® certification only to work for a company that does not value it. Or you may invest a lot of energy learning domain knowledge in a specific area only to land a position in another area. Such risks are difficult to quantify. In these cases, the best answer may be to follow your passion, understanding that if you're doing what you love, you're more likely to be good at it and so success is more likely to follow, whether at your next position or one that follows.

1.8 REFERENCES

[1] Lewis JP. Project planning, scheduling, and control. 3rd ed. McGraw-Hill; 2001. p. 14.

[2] Curlee W, Gordon RL. Complexity theory and project management. John Wiley and Sons; 2011. p. 16.

[3] Curlee W, Gordon RL. Complexity theory and project management. John Wiley and Sons; 2011. p. 38.

[4] Dougherty D. Innovation management at Rutgers business school. Available at: https://www.youtube.com/watch?v=rZU0tv6OMI4.
[5] Heerkens GR. Project management. McGraw-Hill; 2002. p. 35.
[6] Curlee W, Gordon RL. Complexity theory and project management. John Wiley and Sons; 2011. p. 26–27.
[7] Kerzner H. Project management. A systems approach to planning, scheduling, and controlling. John Wily & Sons; 2006. p. 7.
[8] Chapman C, Ward S. Project risk management: processes, techniques, and insights. Chichester, UK: John Wiley & Sons; 1997. p. 10.
[9] Curlee W, Gordon RL. Complexity theory and project management. John Wiley and Sons; 2011. p. 17.
[10] Reinertsen DG. The principles of product development flow: second generation lean product development. Celeritas Publishing; 2009. p. 49.
[11] McManus J. Leadership. Project and human capital management. Butterworth-Heinemann; 2006. p. 7.
[12] Gladwell M. Outliers. The story of success. Little, Brown, and Company; 2008. p. 35.
[13] Heerkens GR. Project management. McGraw-Hill; 2002. p. 39.
[14] Rosenau MD, Githens GD. Successful project management. A step-by-step approach with practical examples. John Wiley and Sons Inc.; 2005. p. 15.
[15] Managing projects large and small. The fundamental skills to deliver on budget and on time. Harvard Business School Publishing Corporation; 2004. p. 6.
[16] Rosenau MD, Githens GD. Successful project management. A step-by-step approach with practical examples. John Wiley and Sons Inc.; 2005. p. 32.
[17] Kerzner H. Project management best practices. John Wiley and Sons, Inc.; 2006. p. 298.
[18] Dinamore PC. The right projects done right. From business strategy to successful project implementation. Jossey-Bass; 2006. p. 89.

Chapter 2

The Critical Path Method: Planning Phase

In this chapter we'll discuss best practices for planning a new project using the critical path method (CPM). CPM is the traditional project management method used in every project type in every industry starting in the 1930s. It sets a baseline for all project management methods. The concept is intuitive: build a plan like a recipe, with one task feeding the next. Each task has an owner; if every owner finishes their task on time, the project will finish on time. In projects of any complexity, several tasks will be active simultaneously throughout the development. But at any given time, one will be the *critical task*. The critical task is the one that if delayed one day will delay the project one day. The path of execution through all the critical tasks is the *critical path*. According to CPM, project managers (PMs) should identify and focus on the critical path.

The planning process will follow the flow chart of Figure 2.1. It starts from a clear understanding of the *value proposition*, the reasons customers will purchase the products being developed; it ends with approval and funding. Project execution can then begin, as will be discussed in Chapter 3.

2.1 ENGINEERING PROCESS FLOW CHARTS

Figure 2.1 is one development process. It includes the major steps required for a project of modest complexity. There is considerable variation across industry because good processes are tailor-made by and for the organizations they serve. Whatever process an organization uses, it should be documented, followed, and continuously improved. Often, less mature organizations will plan and execute projects ad hoc, leading to high variation among projects in execution quality and speed. Further, continuous improvement is difficult—without a defined current state, how can you move to a better state? At the other end of the spectrum are companies with onerous and lifeless processes, full of requirements that may have made sense at one time but bring little value today. These processes may sit nearly unused except when an ISO auditor asks for a review. Strong processes are living documents, adapting to changing business needs.

Processes are often depicted in a simplified form such as the flow chart of Figure 2.1. Steps are shown like a cooking recipe where the next step begins when the previous step is finished: add flour, add water, mix, and so on. However,

Project Management in Product Development. http://dx.doi.org/10.1016/B978-0-12-802322-8.00002-4

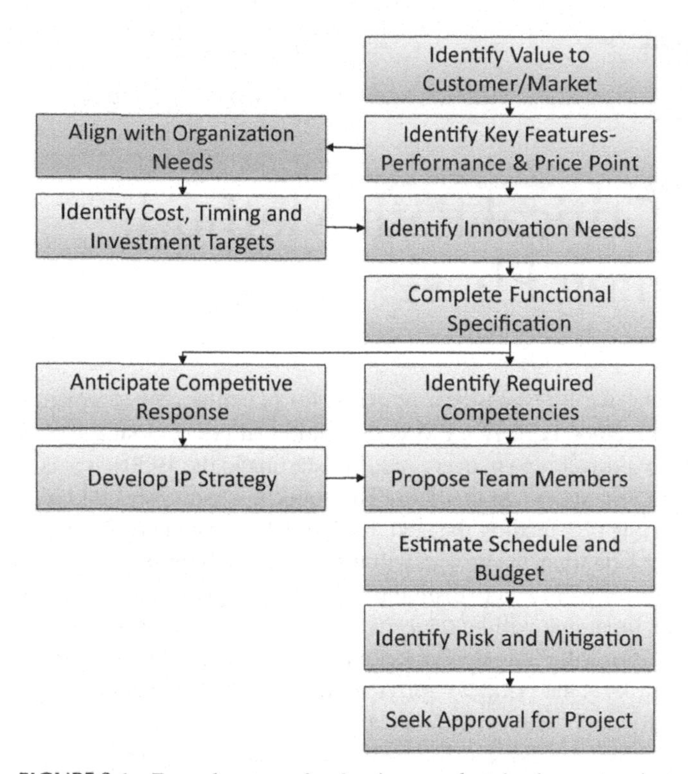

FIGURE 2.1 Example process for planning a product development project.

in product development there is almost always interdependency and iteration, neither of which are normally shown in process flow charts. For example, in Figure 2.1, "Identify Innovation Needs" and "Identify Risk and Mitigation" are shown as independent steps when, in fact, a great deal of risk identification will occur when innovation needs are being defined. Iteration in the steps is common. You may define the innovation needs before you finish the functional specification; however, during the diligence of completing the functional spec, you are likely to review competitive literature and there you may discover new innovation needs. If a process showed all such iteration paths, it would become a jumble of interconnecting arrows since, to some degree or other, learning in almost every step can reveal the need to rework many other steps. So, process charts like Figure 2.1 add value because they make the process easier to understand; they are not meant to be lock-step sequences.

Another issue that arises in any engineering process is defining what "done" means. Consider "Complete functional specification." When is this complete? When a document is published internally? When the marketing manager has signed off? When senior management has reviewed it? It may be those things and more. Whether each step is done is a combination of meeting specified requirements and good judgment. In mature organizations, standard work is created to define as much of each step as is practical—there may be a standard template, a formal review meeting, and a defined approval process. In younger organizations, the functional spec may be free form. More standard work reduces variation from one project to the next; it sets a sort of floor that every team must meet. But more standard work also makes the process heavier, inadvertently pushing teams toward a *check the box* mentality because there can be so many boxes to check. An organization must right-size its processes, balancing the benefits of defining a floor against creating overly demanding requirements.

Finally, each step of this process demands a mix of transactional and transformational leadership. A pure transactional approach may lead a PM to execute steps rapidly in order to meet the letter of formal requirements without meeting the spirit. A pure transformational approach may lack diligence in completing the details. The outstanding PM will be diligent in completing the details while recognizing each step has opportunities for interaction that may lead to the discovery of new ways to improve project performance.

2.2 CRITICAL PATH PROJECT MANAGEMENT: PROCESS OVERVIEW

Chapters 2 and 3 show the many processes and skills sets of project management working together. The goal is to present a system of practices and tools working together; so, it will be necessary to limit detail. The approach here is to present each topic thoroughly enough to use on a project of modest complexity. Later chapters will expand upon selected topics. For example, we'll discuss risk identification in enough detail here for some projects; Chapter 9 is provided for those readers that want to delve deeper into that topic.

The project management method in this chapter is often called *traditional PM*. It is the most intuitive project management method and new product development has relied on it for more than half a century. The method works by breaking the project into tasks and then allocating time and resources to each task. Tasks are then interconnected in a *project network* as shown in Figure 2.1, with one task starting after all its predecessors are done (*predecessors* tasks must be completed before *successor* tasks can begin).

In almost all cases, there will be multiple paths; for the example of Figure 2.1 there are two paths from "Complete functional specification" to "Propose team members": on the left, "Anticipate competitive response" and "Develop IP strategy" and on the right, "Identify required competencies." Whichever of the two takes the longest would be on the critical path. The critical path is then the set of tasks that sets the schedule for the project. CPM teaches that if each task is completed on time, on budget, and with the necessary quality, the project will be successful. However, if the project encounters delays, as product development projects often do, the team should focus on the critical path in order to maintain schedule.

2.2.1 Identify Value to the Customer/Market

Identify Value to Customer/Market

A *value proposition* is a statement of the value a new product family will bring to its customers. In this context, *value* can be defined as performance and features the customers are willing to pay for. The whole purpose of a company is sometimes distilled down to a single function: to create value. Created value or *value-add* can be defined as the difference between the price a customer pays for the product and the cost of the materials in it. There are many ways a company can increase the value they add, for example, by reducing the material cost of existing products, by expanding into new geographies, and, the subject of this text, by producing new products.

For new products, understanding the value your company provides requires a thorough understanding of customer needs. That can be simple if a customer wants a modification of a product your company already produces because the ways value can be added are limited to that modification. But understanding value becomes more complex as the difference between the new product and existing products increases. At the far end of the scale, when a company develops a new product platform that includes numerous innovations, understanding value from the customer perspective is complex indeed.

At the core of the problem, few customers can understand the value of a highly innovative product at the outset of a project. Imagine asking a consumer to evaluate a smartphone a few years before they were introduced. How many would have understood the value of capturing and then posting a video to social media seconds later? Today, people do this so often that nearly everyone understands. But, before the innovation was marketed, it must have been challenging to understand what people would pay for this. On the other hand, the Segway personal transporter "never came close to early expectations" [1]; the company seems to have grossly overestimated how much the customer would value the device. You can't simply ask a potential customer—they may not know until they get some road time (literally, with a Segway) and that won't happen on a wide scale until the product is largely developed. Unfortunately, that's not a good time to find out you've been on the wrong track because, by that point, most of the investment will have been spent.

Understanding customer value is a discipline all to itself and developing expertise in it is beyond the scope of a PM's role. Your project team should include members from sales and/or product marketing[1] [2], to lead this effort. However, the PM should resist the urge to compartmentalize this function. Understanding customer value is so central to the function of every team member, it should be thoroughly understood by all; the PM must lead this collaborative effort.

1. In many organizations, sales leads this activity for customer-driven projects and product marketing leads for market-driven products. In others, the marketing function is divided between sales and senior management. There are many models in industry. According to Deebs [2] "Most B2B [business-to-business] companies are sales-driven organizations, and most [business-to-consumer] companies are marketing-driven organizations, with numerous examples of companies overlapping between the two."

Here are examples of the problems that occur when the entire team is not well versed in the value of the new product:

- The team may expend considerable effort adding features or performance the customer is not willing to pay for. This often happens in a new product where there are many new features.
- Customer feedback is a sampling process. Your sales/marketing lead will select a limited number of people from a small number of customers (for *B2B* businesses–businesses that sell to businesses) or a sample set of consumers (for *B2C* businesses—those that sell to consumers). The sample might not be representative. A few examples for B2B: the group might include too many working engineers who often overestimate value of highly technical features and downplay the barriers created by adding cost. Or, the group might include too many commercial leaders who don't understand how a feature will improve the experience of the end user. For B2C businesses, the demographics of the sample may not represent the intended market.
- A feature may be added simply because it is part of a competitor's offering. An off-hand comment by a customer or two can cement the belief that a feature is critical. It's easy to overestimate the value of features in such a case.
- The importance of a feature may be missed because your team is unfamiliar with it. It can be easy to dismiss customer comments about features the team misunderstands.
- Technical barriers may arise through the course of the project that force compromises to the original specification. Cost may increase, schedule may extend, or features/performance may need to be downgraded midway through the project. The less the team understands what the customer values, the less likely team consensus will lead to the optimal decision.
- Team members may discover better ways to deliver features and benefits through the course of the project. The less informed each team member is about the value proposition, the more their creativity will be encumbered.

All of these problems can be mitigated if the entire team thoroughly understands the value the new product will offer. Here are a few steps you can take as a PM to facilitate:

- Sit in when marketing or sales conducts customer interviews and take team members with you. Listen as they explain how they will use new features and what benefits they anticipate.
- Watch the end user working with the current solution and, when possible, perform the observed functions yourself in the customer setting. Try to experience the strengths and shortcomings of the current solution.
- Ensure the customer sample includes decision makers. If you're designing a child's toy, you'd probably want to understand how the adult will value that toy too. If you're selling to another company, ensure decision makers—senior staff and purchasing, for example—are included.

- Attend trade shows where products similar to yours are displayed; bring team members with you. You're likely to see products before they are advertised; perhaps some will demonstrate features planned for (or that ought to be added to) your product. As a second benefit, you'll get to make a more thorough estimation of a competitor's response to your product.
- Show a little skepticism when reviewing customer requests. The "wow" factor of seeing a new product can cause a customer to overestimate value. The question is whether the customer will be willing to pay for the product after the initial excitement has faded.
- In the B2B world, be aware that customers sometimes inflate potential volumes to get more attention to their projects; others overstate interest in your product because their primary intention is to use information from your company to leverage pricing with existing suppliers. In such cases, you can spend a lot of energy with little benefit. If you have any doubts, explore the option of nonrecoverable engineering (NRE) for customer-driven projects. Even a small amount of NRE can require management approval from the customer's company, thereby increasing confidence that the company is aligned.

2.2.2 Identify Key Features, Performance, and Price Point

Identify Key Features-
Performance & Price Point

The next step is to identify the features and performance the customer will value and to estimate the price the customer will be willing to pay. Features and performance cannot normally be taken in isolation. For example, a new smartphone might bring the largest screen on the market. The value of the screen is its ability to support other features such as clearer maps (implying navigation), easier web surfing (implying high-speed browsing), and so on; you can't simply ask a customer "what is the value of a big screen?" Features and performance are usually merged into various groups that are compared one against the other. The engineering team would take the lead in understanding the costs of those groups so the sales/marketing team can weigh the cost versus value to find the optimal solution: that set of features where the value/cost difference is high enough to justify their inclusion. For some companies, this is a formal process where feature sets are compared one to another using customer feedback and detailed financial analysis; in others the process is informal, relying on the intuition of key decision makers. As with the previous step, this activity is normally led by sales/marketing, but the entire team should understand how these features connect to customer value.

2.2.3 Align with Organizational Needs

Align with Organization
Needs

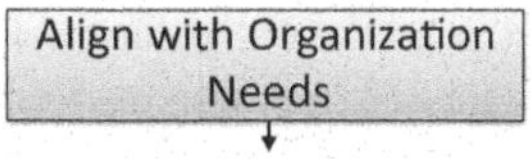

The team needs to understand how well this product aligns with organizational needs. Is this product strategic? In other words, is it a natural step along the path the company has planned for the coming years? A few questions to ask:

- Will this project allow us to build more competence in design and manufacturing technologies that we will need in future planned products?
- Will this project help us obtain patents that allow us to protect technologies important to our future?
- Will this product allow us to expand into new markets or geographies that we are planning to expand into?

The stronger the "yes" to these types of questions, the more strategic value this project brings. So, the project can be partly justified by the revenue it generates and partly because it moves your company closer to its vision. At the other end of the spectrum are tactical projects—those projects that are based almost completely on the opportunity the resulting product brings. Such projects can still make sense if the revenue they bring justifies the investment and risk.

2.2.4 Identify Cost, Timing, and Investment Targets

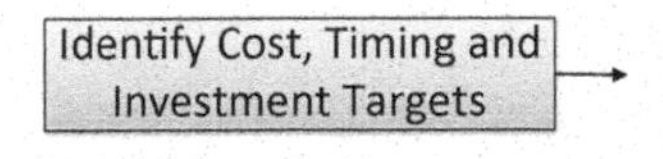

At this point, you can scope the project targets. The three critical targets are: cost of producing the product, time to launch, and total investment.

Maximum Cost Target

Cost target is a combination of the price point (from Section 2.2.2) and the margin required by the company. The *margin* of a product is defined as:

$$\text{Margin} = (\text{Price} - \text{Cost}) / \text{Price}$$

Standard margin uses standard cost (variable and fixed costs) where variable margin includes only variable costs. An example of a fixed cost is rent—it is independent of how much a given product is produced (assuming there is capacity). An example of a variable cost is the material needed to produce a component. Producing twice as many of a given product requires twice the material. Many organizations base most of their decisions on variable margin and others on standard margin. In general, the more innovative a product, the more margin the company will expect. There are two reasons: first, a customer should be willing to pay more for a more innovative product assuming those innovations solve problems important to that customer. Second, developing more innovative products normally brings more risks and higher cost—more investments from engineering, marketing, and manufacturing are required. So, more margin is needed to offset the investment and cover the risk.

The margin required depends on many things: company culture, market needs, and whether the product is strategic or tactical (normally, the more strategic, the lower margin the company will accept). Senior staff will normally make this determination, again with some organizations having a formal process and others more informal. However it's derived, the project team should use price and margin targets to derive a maximum cost:

$$\text{CostTarget} \leq (1 - \text{MinimalMargin}) \times \text{Price}$$

For example, if the product cost target is $1000 and the company demands 60% margin for this project, the target cost must be $400 or less.

Target Timing

The target timing of the project will depend on competing opportunities. Apart from direct costs, longer projects bring more risk (e.g., competitor beating you to market or team resources changing) and more hidden costs (e.g., more mind-share of senior staff for project reviews). A strategic project with large potential volumes may be allowed to take a large team 2 or 3 years where a small tactical project may be allowed only a few months. Again, this is often determined by senior staff.

Target Investment

The total investment—people-time plus expenses—must be evaluated to ensure the project pays back the development costs. A simple method for this is "payback period," which simply divides total investment by annual margin (in $, not %) after the product is at its full production rate. For example, let's imagine a project that costs $150k to develop. It's anticipated that the result of the project will be 100 units sold each year at $4000 each with a margin of 45%.

Total revenue:	100 × $4000 = $400k
Total margin $:	45% × $400k = $180k
Total investment:	$150k
Payback period:	$150k/$180k = 0.83 years = 10 months

There are many variations on this type of calculation. They may account for the time-varying value of money ($1 now is worth more than $1 next year), the expected revenue variation by year (e.g., due to lengthy customer adoption periods), and anticipated cost reductions over time (e.g., due to increasing volumes). For example, internal rate of return, net present value, and breakeven time are commonly used and so will be presented in Section 5.3.1. However, payback period is used often because it's intuitive. In the Excel sheet we'll use later in this chapter, payback period will be calculated in this manner.

2.2.5 Identify Innovation Needs

Identify Innovation Needs

At this point, we can identify the key needs for innovation: we understand the customer's needs, we know the cost targets, and we have a good understanding of the company's appetite to invest. In general, more strategic projects will encourage more investments in innovation and justify more risk. So, we should be able to identify the areas most likely to benefit from innovation. Ask a few questions:

- What are the technology needs voiced by the customers? Look beyond higher performance and high-end features. Consider also making the product easier to use or service, more reliable, and easier to deliver quickly. And look at how newer technologies could reduce cost.
- How are competitive solutions meeting those needs today? What technologies would allow your company to satisfy those needs in a better way?
- What are the core competencies of your organization? How can you apply those competences to solve customer problems?
- What technologies are available to you from other sources—other divisions of your company, or outside sources of innovation such as universities, other companies, or consultants?

One question comes up often: can the PM plan innovation? The answer is yes, but normally it can't be scheduled precisely. Innovation requires invention and without doubt, the process of human invention is mysterious. So the issue here is not controlling the moment of invention, but using schedule and team understanding of customer needs to create focus. With the right focus and a creative environment—the right people, teamwork, and tools—the team is more likely to invent during the project, and that invention is more likely to be in the areas that bring the customer the most value.

2.2.6 Complete Functional Specification

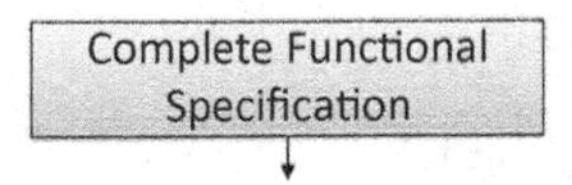

This is the point to ensure the functional specification, a document defining what the product needs to do, is complete. Presumably that document was started long before this point, probably shortly after sales/marketing discovered this opportunity. If the questions in the above sections are answered, you should be ready to complete the document, at least well enough to kick off the project. A successful project requires a thorough functional specification to define clearly what the project team will deliver. Without this document, the project will have a different meaning to every stakeholder. It is crucial to ensure this document is fixed and the team is in consensus.

One of the largest problems in project management for product development is *scope creep*, the phenomenon where the functional requirements expand as the project matures. Many people try to solve this problem by "freezing" the specification. This doesn't usually work very well. If new data arises that indicates the specification does not meet customer needs, it's imprudent to ignore that data. So, the functional spec probably will change through the life of the project; the problem is when it changes without recognizing that the other bars of the Iron Triangle (Figure 1.1) change accordingly.

Later in the project, when new data reveals the function needs to change, then time and cost must be adjusted to accommodate. The PM must manage the process to gain approval for these changes once the project starts. We'll discuss that more in Chapter 3 The Critical Path Method: Execution Phase, but for the moment, understand that the more complete and accurate functional specification is at the start, the easier the change management process will be later.

2.2.7 Anticipate Competitive Response

Anticipate Competitive Response

The next step is to anticipate what your competitors are likely to do when you release your new product [3]. Their response may be directly to your new offering or it may be indirect, coming about by them acting on the market factors that are driving your project. Start by asking what their most likely response will be and then take steps to reduce the negative effects. How can you anticipate their response? There are many options. Your team can study their literature and use your collective knowledge of the market and technology. You can interview their customers. Also, you can stop by their booth at trade shows—you might be surprised how much you can learn just talking to a competitor. I've never found a need to conceal my affiliation at trade shows; people are usually excited to talk about their new products, even to a competitor. Table 2.1 offers a few examples of actions you can take in reaction to a competitor's reaction to your product.

TABLE 2.1 Example Responses to Competitors

Competitor's Anticipated Response	Action(s) You Can Take
Reduce price of their competitive product.	Focus early on product cost. Ensure you can earn acceptable margins even if you must follow a price reduction from the competition.
Develop a similar product.	Seek patent protection on key technologies; accelerate development speed so you are harder to follow.
Add features your product lacks to their existing product.	Understand how the customer will value the features that your competitors may add. Consider adding those features to your product at the outset.
Expand their similar product into other geographies.	Evaluate if you can start a program to introduce your product quickly into those geographies.

2.2.8 Develop IP Strategy

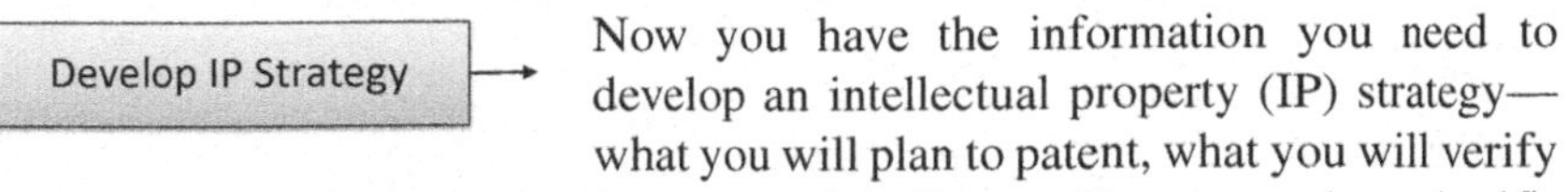

Now you have the information you need to develop an intellectual property (IP) strategy—what you will plan to patent, what you will verify is free to use, and what technology you might license. Patents can be a significant advantage in the market place for whoever develops technology first. Chapter 10 will discuss IP protection in more detail, but here are a few tips you can use to direct you in simple cases:

- If you developed a new way to do something and it's genuinely valuable, it may be patentable. Do you see anyone else in the industry using what you've developed? If not and if it does add significant value, it may be you discovered something new. (After all, if it adds value and your competitors knew about it, wouldn't they be using it?) Seek counsel either from a patent attorney or a patent agent.
- Don't rely on the inventor to determine patentability. Many inventors are unfamiliar with patent law and so may not be the best advisors for their own inventions. One well-known barrier to patentability is "obviousness," but the legal meaning of "obvious" is quite different from its use in ordinary language. And often the inventor will say something is obvious because she's spent the last 7 months thinking about it day in and day out. What's obvious to her and what's obvious according patent law can be two very different things. Again, seek legal counsel.
- If you are considering patenting an invention, protect the information. Tell only the people who need to know. Patentability is easily injured when the invention is disclosed before the patent application is filed. Disclosure is a complex topic, so seek legal advice. For the moment, be aware that the team can end the ability to patent an invention by publishing too much information about it (even on a blog), explaining it to someone outside the company, or showing the invention to an outsider in the office or at a trade show.
- Look at competitor's patents to reduce the chances that you are using technology your competitor owns. This is a complex area and requires legal counsel, but you can use sites like Google Patents (www.google.com/patents) to start a search. Also, you can prepare yourself by purchasing a book on patents for laypeople. Patent searches are discussed in detail in Chapter 10.

Again, patent law is complex, so seek legal advice early in the project. This is the point where the most opportunity to patent exists and where it's normally easiest to avoid infringing on the patented technology of others.

2.2.9 Identify Required Competencies

Identify Required Competencies

There are many disciplines you may need on the team. Some may work on your project full time, others may contribute a few hours a week, and

TABLE 2.2 Example Disciplines You May Need on Your Team

Design engineers	Software developers	Manufacturing engineers
Scientists	Lab technicians	Buyers
Quality assurance	Sourcing	Sourcing quality assurance
Chemists	Designers	Marketing communications
Documentation specialists	Assembly technicians	Testers
Test technicians	Finance	Product marketing
Sales	Application engineering	Customer service
Regulatory compliance	Safety	Legal

others may be part of a support team like a test lab or machine shop. However they come into the team, include all the disciplines needed to complete the work on time. Examples of disciplines to consider are shown in Table 2.2.

Compare the requirements in the functional specification to capability and capacity of your team and fill any gaps that you discover.

2.2.10 Propose Team Members

Propose Team Members

You now have the information to select the team: an initial understanding of what new technology is needed, the scope of the project, and the rough time frame. It's time to select the number of people and their competencies. Balance the team according to the needs. If this will be a complex product to manufacture, be sure that manufacturing and quality are well represented. If the product testing will be onerous, be sure to have the expertise available to manage and execute the tests. If there is complex technology development, include people with the capability to do the work.

The RACI matrix is sometimes used to formalize team membership and roles. In the RACI matrix each function is listed in the columns with major responsibilities shown in the rows. Then the role for each function/responsibility combination is defined as one of:

R **Responsible** to ensure timely completion and compliance to requirements.
A **Accountable** for the results.
C **Consulted** during the execution of the task.
I **Informed** during the execution of the task.

Any organization that struggles with role clarity should consider the RACI chart. The RACI chart is discussed in more detail in Section 9.3.3.

Your success as a PM will be largely determined by the quality of the team. An outstanding team with strong leadership will usually be able to make up ground when something goes wrong (and something almost always goes wrong at some point!). On the other hand, if the team isn't capable, the project won't move forward as expected. Remember, when the project goes well, there's lots of credit to share among the team members. When it doesn't—even for things that are out of your control—often you'll likely feel like you're carrying the blame because you are the PM.

2.2.11 Estimate Schedule and Budget

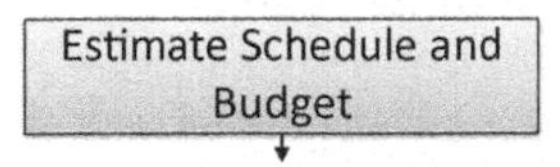

Now it's time to build an initial schedule and budget. Sit with key prospective team members and map out the approach. Ideally, the company will have a defined process for product development. If so, use the process as a checklist; estimate the resources (people and expense) needed for each step, and the time each will take. Of course, no process will capture 100% of a project's tasks—pay special attention to the items that are new or occur so rarely that they are not part of the company's processes. But a strong process will guide you through a majority of the planning.

Develop a Schedule

Begin the schedule by developing the *work breakdown structure* or WBS. This is a list of each task that must be accomplished during the project [4]. Figure 2.2 is an example of a WBS for a simple project to create a proof-of-concept (POC) unit (or "technology demonstrator"), followed up by building 10 pieces. The total project ("Create Product") is broken down into four major steps: Project Definition, Develop POC, Produce 10 Units, and Launch.

The WBS can be used to create a schedule: you need to understand dependencies (which task must be completed before the next starts) and the length of each step. There are many software tools that can support your efforts to build and execute schedules while following your budget. However, to keep things simple, we'll use an Excel tool developed for this book as shown in Figure 2.3, the schedule for the WBS of Figure 2.2.

As simple as this tool is, it has the features necessary to build a small- or medium-sized project using traditional project management techniques. The methods that use a project network like critical path and critical chain (Chapter 6) cascade tasks—when one task is done the next can start. Sometimes it's called *waterfall* because the Gantt chart (bottom right of Figure 2.3) looks something like a waterfall.

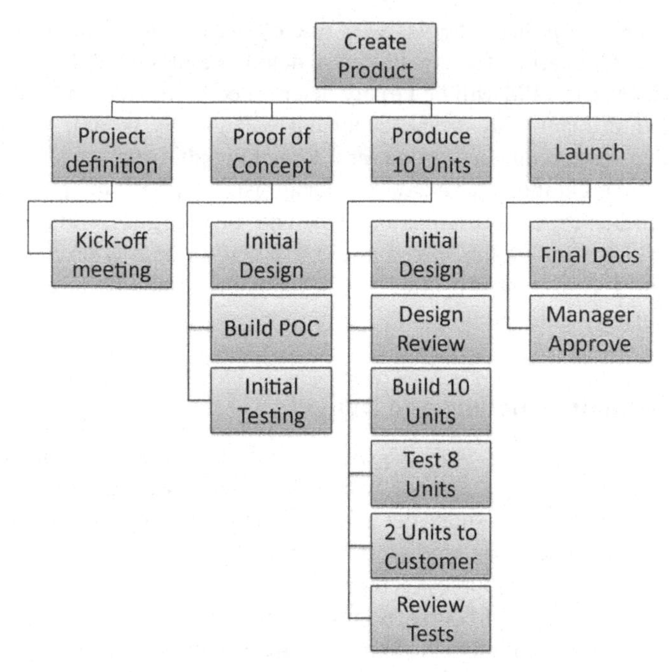

FIGURE 2.2 Example work breakdown structure (WBS).

Figure 2.3 contains detail on each task in the WBS. The key components of the WBS are:

A. The task number.
B. The task name.
C. If the task has a fixed start date, what that date is. Usually, tasks start as soon as its predecessors are complete, but some tasks may be fixed in time by events like a scheduled site visit from a regulator or a trade show.
D. If the task does not have a fixed start date, list one predecessor: a task that must finish before this task can start.
E. Optional second predecessor.
F. Optional third predecessor.
G. Duration of the task in days (7 days = 1 week).
H. A planned "done" date—this is set at project start and never changes (column H can be copied from column J at the project start). This column provides a history that can be used to estimate project health during execution.
I. A calculated start date (based on fixed start date or completion of predecessors).
J. A calculated done date (the start date plus the duration).
K. A "Yes" if the task is done.

The remaining columns are a simple Gantt chart, giving a picture of how the steps interconnect. The chart is automatically calculated from columns A–K.

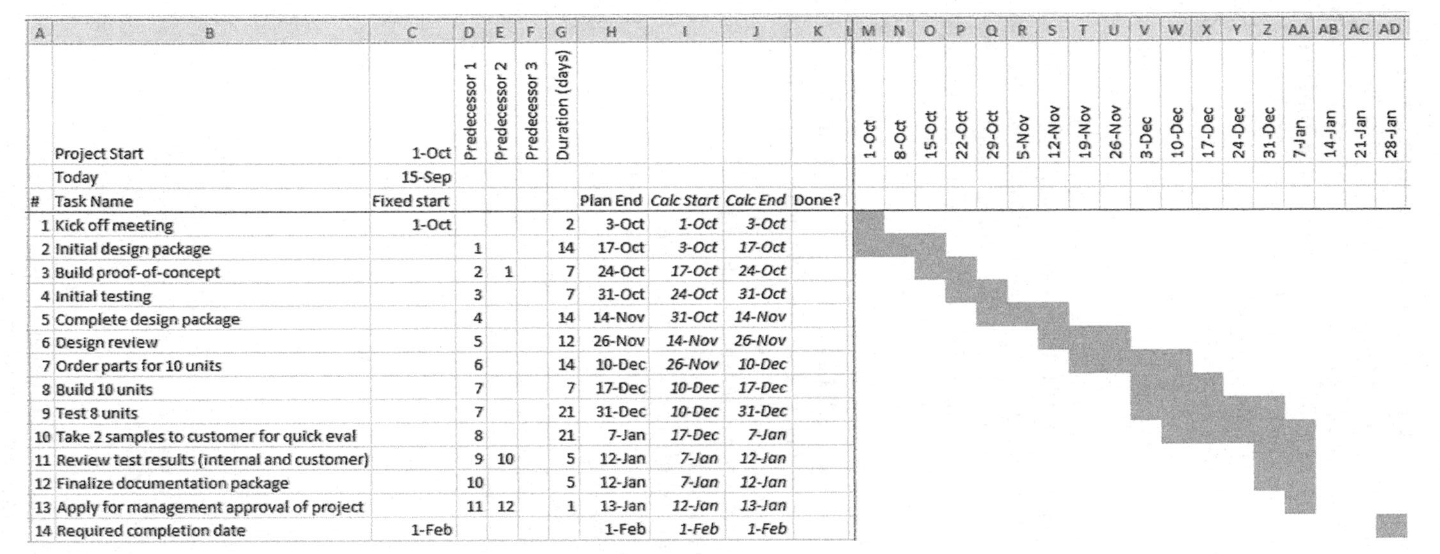

A	B	C	D	E	F	G	H	I	J	K
	Project Start	1-Oct	Predecessor 1	Predecessor 2	Predecessor 3	Duration (days)				
	Today	15-Sep								
#	Task Name	Fixed start					Plan End	*Calc Start*	*Calc End*	Done?
1	Kick off meeting	1-Oct				2	3-Oct	*1-Oct*	*3-Oct*	
2	Initial design package		1			14	17-Oct	*3-Oct*	*17-Oct*	
3	Build proof-of-concept		2	1		7	24-Oct	*17-Oct*	*24-Oct*	
4	Initial testing		3			7	31-Oct	*24-Oct*	*31-Oct*	
5	Complete design package		4			14	14-Nov	*31-Oct*	*14-Nov*	
6	Design review		5			12	26-Nov	*14-Nov*	*26-Nov*	
7	Order parts for 10 units		6			14	10-Dec	*26-Nov*	*10-Dec*	
8	Build 10 units		7			7	17-Dec	*10-Dec*	*17-Dec*	
9	Test 8 units		7			21	31-Dec	*10-Dec*	*31-Dec*	
10	Take 2 samples to customer for quick eval		8			21	7-Jan	*17-Dec*	*7-Jan*	
11	Review test results (internal and customer)		9	10		5	12-Jan	*7-Jan*	*12-Jan*	
12	Finalize documentation package		10			5	12-Jan	*7-Jan*	*12-Jan*	
13	Apply for management approval of project		11	12		1	13-Jan	*12-Jan*	*13-Jan*	
14	Required completion date	1-Feb					1-Feb	*1-Feb*	*1-Feb*	

M	N	O	P	Q	R	S	T	U	V	W	X	Y	Z	AA	AB	AC	AD
1-Oct	8-Oct	15-Oct	22-Oct	29-Oct	5-Nov	12-Nov	19-Nov	26-Nov	3-Dec	10-Dec	17-Dec	24-Dec	31-Dec	7-Jan	14-Jan	21-Jan	28-Jan

FIGURE 2.3 Simple Excel® tool to show project schedule.

Resource level the task list: if one person is working on multiple tasks at the same time, the schedule implies multitasking. This can work if the tasks take a small share of the person's time. But if more than one task requires most of the person's time, the schedule will put an undue burden on the resource. In such cases, you can make one task the predecessor to the other or lengthen all the tasks to account for the person being assigned only part time to each.

If there is an engineering subprocess that you're unfamiliar with, you'll want to get help from someone who knows it—an experienced PM or an engineering manager. If you've been through a few DFMEAs (design failure modes and effect analysis), you'll have a good estimate of how long one will take; if you never heard of a DFMEA, your estimates will be unreliable.

You might have noticed that the required completion date is 1 month after the calculated date of the final task. This is because the project includes a buffer. It's generally wise to include a buffer. Projects have risks, both known and unknown, and a buffer can allow the team to deal with ordinary problems without having to renegotiate the project schedule.

Task Commitments and Schedule

When you build a schedule, it's based on a series of commitments from the team. The Iron Triangle from Figure 1.1 comes up again and again. Take any of the tasks in the WBS as an example. The project schedule will allow a certain amount of time and a budget (for equipment or material), and the owner of that task will deliver a defined deliverable—say, a product feature or performance enhancement. These many small iron triangles lock together to build the large iron triangle that represents the whole project. Each commitment must be managed. Here are some steps you can take during the planning process to help manage commitments:

- All tasks should be clear and measurable. Avoid unmeasurable tasks like "improve performance" or "read about competitor." Include a measurement: "speed up cycle time by 25%" or "provide report analyzing competitor's product model RHR."
- A task should have a single owner, at least at any one time. Of course, many people can contribute to a task, but someone needs to be responsible for the result. Some tasks shift ownership among team members—that can work well so long as the handoff is clear to all involved.
- Always have buy-in from the task owner. Don't allow a team member to take responsibility for a task when they believe the task goal is unachievable. You may let them stretch the time so they can commit or you may have to find another owner. At the same time, set a reasonable tone when someone misses a commitment. If you berate team members in meetings, they'll be less likely to commit later. Strike the right balance—if the person is doing something they've never done before, you can't expect high estimation accuracy. What you can expect is team members to put forth a strong effort and, if that fails, to inform the team at the first sign of trouble so you can take quick action.
- Don't change a task without the owner accepting the change.

- Don't assign a task to someone who is not at the meeting. If you need to assign it to someone who is not present, you can assign it to yourself and then meet with the person later to transfer it.
- Tasks should be recorded where everyone can see them—use a shared folder on your intranet, a SharePoint site, or a project management system in the cloud. Avoid the emailed Excel sheet. There should be one place where the project is defined and that all stakeholders can access at any time.
- Plan tasks at the proper granularity—neither too small nor too large. Coarse granularity hides problems because delays are often unnoticed until the planned date elapses. Overly fine granularity makes the team feel "micromanaged" and can make the WBS unwieldy. If you're meeting weekly, a good granularity is a day or two for things in the immediate future. For tasks in the more distant future, use coarser granularity—tasks 2 months out should normally be a few days or a week long; tasks 6 months out are normally even coarser. As time moves on, those coarse blocks will move nearer and can be divided down to finer granularity. The benefit is flexibility: as the project matures, you'll learn more and your WBS will improve. If there are hundreds of tiny tasks 3 months out, adjusting the WBS will be challenging. Figure 2.4 shows a suggested granularity for a product development project.

Courser granularity provides greater flexibility to adjust to new information, which is frequent in new product development projects. The disadvantage is the accuracy of planning can be lower. So, the PM must strike a balance between the two. As shown in Figure 2.5, product development projects are frequently higher risk than other projects, especially when one or more difficult technical issues must be resolved during the project. On the other hand, they generally have a higher tolerance for delay than say a major construction project.

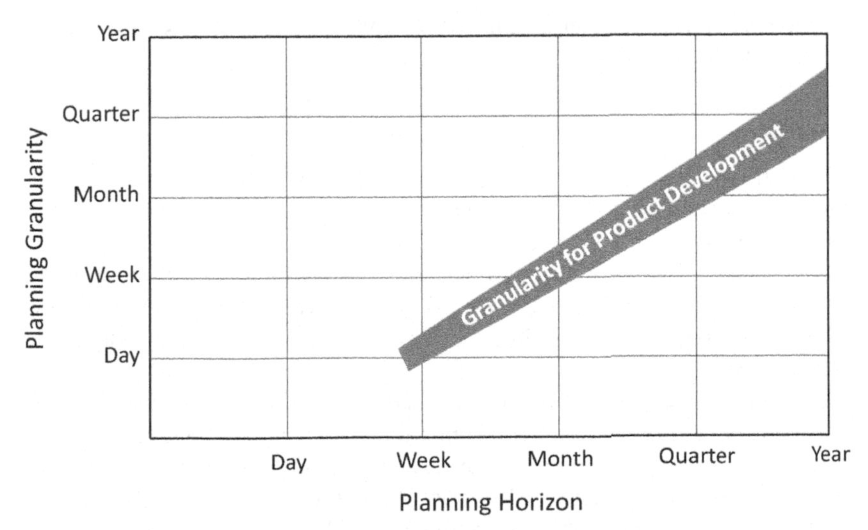

FIGURE 2.4 Suggested granularity for product development projects.

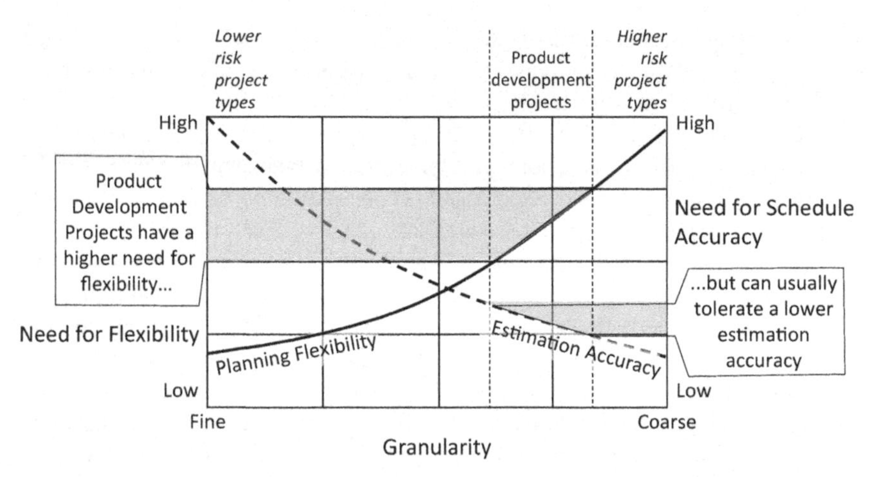

FIGURE 2.5 The inverse relationship of flexibility and accuracy for product development projects.

Develop a Budget

The next step is to develop a budget. Normally, this is done in two parts. First, you'll need to capture people time, usually the number of hours a person is expected to work on the project times a standard hourly rate for a given function set by your company. (Normally you don't use people's actual salary since that information is not widely shared.) For example, you might estimate the primary mechanical engineer will be 50% dedicated to the project. If the estimated project length is 4 months, the hours for that engineer would be:

$$4\text{ months} \times (52/12)\text{ weeks/month} \times 40\text{ hours/week} \times 50\% = 347\text{ hours.}$$

If a mechanical engineer is charged at $75/hour, the estimated cost for that engineer would be:

$$347\text{ hours} \times \$75\text{/hour} = \$26\text{k.}$$

Usually, the cost will include overhead and so will be far larger than the engineer's salary. That should be expected; salary is only part of the cost of each employee.

This chapter also uses a simple budgeting tool, which is another tab in the Excel file with the simple scheduling tool from above. Figure 2.6 shows a sample output. It has 11 separate lines, one for each team member; people time totals about $100k.

Most companies track employee time and expenses separately; usually capital expenses are tracked separately from ordinary expenses. The finance department will help you determine if an expense is a capital item; normally it must cost several thousand dollars and be part of a durable asset. Table 2.3 shows a few differences among the three expenses.

	A	B	C	D	E	F	G	H
1	**Revenue**							
2		Target Unit Sales (annual)	100					
3		Unit Price	$4,000					
4		Unit Cost	$2,200					
5		Total revenue (annual)	$400,000					
6		Margin	45%					
7		Total margin (annual)	$180,000					$180,000
9	**Investement**							
10		Team members	Role	Rate/hr	Hours	Cost		
11		Rachel	PM	$75	300	$22,500		
12		Christian	Design EE	$75	240	$18,000		
13		Sanjay	Design ME	$75	160	$12,000		
14		Kenneth	Design ME	$75	200	$15,000		
15		Ethan	Design Tech	$35	80	$2,800		
16		Aziz	Doc Specialist	$35	24	$840		
17		Betty	Design Docs	$35	200	$7,000		
18		Brandon	Manuf ME	$75	160	$12,000		
19		Lisa	Qual ME	$75	40	$3,000		
20		Annie	Marketing	$75	40	$3,000		
21		Toby	Communications	$45	120	$5,400		
22					1264		$101,540	
24		Expense					$42,500	
25		Flights and expense for trade show (3 people)				$3,000		
26		Customer visits (2 people, 3 x)				$6,000		
27		New heater station in factory cell				$6,500		
28		Test lab				$9,500		
29		Add product to web site				$5,500		
30		Patent search and application				$12,000		
32	**Total Investment**							$144,040
33		ROI (Months)						9.6

FIGURE 2.6 Simple Excel budgeting tool.

TABLE 2.3 Three Common Types of Expenses Tracked in Projects

	Employee Time	Ordinary Expenses	Capital Expenses
Purpose	Cover salary, benefits, and overhead.	Cover most expenses. Usually includes consultant charges.	Cover larger investments in assets with a long life.
How estimated	Based on estimates of how much time people will spend on a project.	Line-by-line estimate of items to purchase, often with buffer added.	Line-by-line estimate of items to purchase.
How tracked	Sometimes based on estimates of how much time people spent on a project. Other times tracked daily with a time card system.	Through purchase order and receiving system.	Through purchase order and receiving system.

You might have noticed that Figure 2.6 also includes ROI (return on investment) in cell H33. This is simply the initial investment ($144k) divided by the annual margin ($180k) and then multiplied by 12 to get units of months. This project pays its investment back in 9.6 months, assuming the project goes to full revenue immediately.

2.2.12 Identify Risk and Mitigation

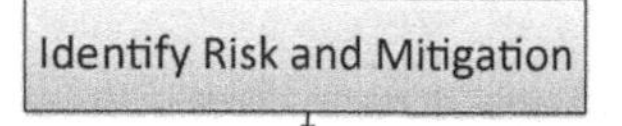

The next step is to identify and deal with known risks. These risks, sometimes called "known unknowns," represent the things the team can anticipate going wrong—late delivery from a supplier that has missed a few dates in the past, a team member that might take a few extra weeks to solve a problem, or a customer might decrease their actual purchases below what they predicted at the project start. These risks usually make up the majority of risks that the projects will encounter and they are the only risks the team can plan for. The more skilled the team and the more diligent process they follow to root out hidden risks, the more known risks the team will find.

However, for projects developing highly innovative products, no matter how long the team labors there will always be risks that are missed: a competitor surprises the team with an announcement of a patent for technology they relied on, a new technology emerges during the project and obsoletes the initial approach, or a critical supplier goes out of business without warning. These are examples of unknown risks, which cannot be fully planned for. In this chapter, we'll focus on the traditional approach of planning for known risks. In Chapter 9, Risks and Issues, we'll look at more advanced techniques to plan for risk. However, the techniques here are the most common and they will be sufficient for many projects.

Let's look at the standard risk avoidance techniques (see Figure 2.7):

1. Identify as many risks as possible
   a. Describe the risk.
   b. Define the risk type.
2. Assess and prioritize risks
   a. Likelihood (1–10): How likely is it this risk will become an issue?
   b. Severity (1–10): If it becomes an issue, how severe is that issue?
   c. Assessment: Product of Likelihood and Severity; 45 is set as "high."
3. Mitigate risks
   a. Mitigation strategy: How will the team reduce the likelihood or severity?
   b. Confidence: Does the team believe the strategy will work?

The processes to identify, prioritize, and mitigate risks vary widely. Many companies have no formal process. Others rely on the PM working for an hour to two to capture all risks he is aware of. More diligent companies might call an 8-hour 10-person meeting. Sometimes the scales are carefully defined and other times the assessment is simply low, medium, and high. A single mitigation

	A	B	C	D	E	F	G
1	Description	Type	Likelihood	Severity	Assessment (calc)	Mitigation strategy	Confidence
2	POC shows firmware runs too slow	Technical	6	9	54	Change to faster processor after POC	High
3	Consultant unable to find low-cost cover that resists salt spray	Technical	5	8	40	Accelerate salt spray test program allow testing of many materials	Medium
4	Competitive threat at primary customer	Competitor	6	8	48	Technical team to visit customer often to ensure full understanding of their needs.	Medium
5	Web site not ready at launch	Technical	8	3	24	No mitigation--launch wihtout web if necessary	Not applicable
6	Production supplier unable to meet delivery schedule	Sujpplier	8	7	56	Identify supplier for small samples	High
7							
8							

FIGURE 2.7 Sample risk identification and mitigation plan.

strategy may be listed for each item or the team may pursue multiple routes until confidence is high in all known risks.

It can be challenging to define when the initial risk identification process is "done." One way is to continue until the team believes the remaining risks that can be identified at the outset are small compared to those that cannot. It is the nature of product development, especially for highly innovative products, that there will almost always be a set of risks that are unknowable at the project start. If experience indicates that the remaining issues the team is able to identify are probably much smaller than those it cannot, it may be time to stop.

The team may be able to use history. If the company executes a retrospective process for tracking what went well in a project, this would be a good place to identify how many times known risks interrupted a project versus unknown risks. If history shows too many problems came from risks that ought to have been better identified at the start more diligence in risk management is called for. At the same time, avoid the fallacy pointed out in *The Drunkard's Walk* [5], a phenomenon where a wholly unpredictable effect appears predictable in hindsight. There is not always a bright line; mix some empathy with your best judgment when determining what a team could have reasonably been expected to have identified.

2.2.13 Seek Approval for Project

Seek Approval for Project

The final step for project planning is gaining approval from senior management to start the project in earnest. At this point, you should get validation for the plan, allocation of resources necessary to move to the next step, and an approved budget.

Your job as PM is to present the plan openly, highlighting the opportunities and the risks. Try not to "advocate" for the project: emphasizing the positives and obscuring the negatives. Be enthusiastic, but give an honest assessment of the weaknesses of the project as you understand them. Use standard forms where you can—the less people have to adapt to your new slide show format, the more they can be thinking about the issues.

When presenting complicated topics such as a challenging risk or needing outside expertise where the team is weak, work hard to make the issue "visual." The main three components of good problem visualization are:

- Concise.
  Fits on a single page or screen shot, has limited text, and lets pictures tell most of the story.
- Credible.
  Data supports the story. The team is in consensus. You've presented a full picture, not just the information that backs up your view.
- Drives good action.
  The decision that needs to be made is clear. The alternatives are laid out like a menu and you're not vested in any particular decision.

Finally, be respectful to nontechnical people. Treat them like they are smart, but untrained in science and engineering. Think of how a good doctor can explain a medical problem—they can tell you what you need to know and even give a basic explanation of what's going on even though you didn't attend medical school. Do the same for nontechnical managers. Present things at a business level—an overview of the issue, the level of risk exposure, when it can be contained, what expertise is needed to solve it—all the things that help commercial people make the right call.

2.3 WHEN YOU NEED HELP

This chapter has presented the process for planning a project with many interlocking pieces. If you find this a little overwhelming, you might want to get some help. The first place to look is your boss; perhaps she can coach you through a project or two. She could attend team meetings or you could walk her through weekly reports. Or, you might want to search for a mentor—a person you don't report to who has expertise you lack. It could be a more experienced PM or a department leader. You might be surprised by how much you could learn by having someone sit in on your kick-off meeting and two or three team meetings. Don't be shy about asking for help—you'll learn faster and make fewer mistakes. You'll also be more confident if you can bounce ideas off someone more familiar with project management.

2.4 REFERENCES

[1] Caulfield B. Steve jobs explains why we're not all riding a segway. Forbes.com Tech; September 27, 2010. http://www.forbes.com/sites/briancaulfield/2010/09/27/steve-jobs-explains-why-were-not-all-riding-a-segway/.

[2] Deebs G. Sales vs marketing for startups? depends if you are B2B or B2C. Forbes.com Entrepreneurs; May 14, 2014. http://www.forbes.com/sites/georgedeeb/2014/05/14/sales-vs-marketing-for-startups-depends-if-you-are-b2b-or-b2c/.

[3] Coyne KP, Horn J. Predicting your competitor's reaction. Harv Bus Rev April, 2009;87(4): 90–7. The Magazine.

[4] Managing projects large and small. The fundamental skills to deliver on budget and on time. Harvard Business School Publishing Corporation; 2004. p. 72.

[5] Mlodinow L. The Drunkard's walk, how randomness rules our lives. Pantheon Books; 2008. p. 167.

Chapter 3

The Critical Path Method: Execution Phase

In this chapter, we'll discuss how a project manager (PM) could manage the project planned in Chapter 2. The process is shown in the flow chart of Figure 3.1. It's based on the PM contributing in three different roles:

1. Working alone or with one or two team members.
2. Leading team meetings.
3. Representing the team in senior staff reviews.

These activities are all repeating, but with different time periods ranging from daily (upper left corner) to months (along the bottom and right side). Figure 3.1 is one example—the PM's roles will vary from one company to another. However, the principles described here should be relatively constant in those organizations that view the PM as a leader.

3.1 WORKING ALONE AND WITH ONE OR TWO TEAM MEMBERS

PM WORKING
Alone & in small groups

The PM will spend the majority of her time working alone and with one or two team members. Demands placed on the PM by these tasks require both transactional and transformational leadership.

3.1.1 Following Up on Issues Daily

Follow up on key issues

Following up is probably the single most identified trait of good project management during execution. These activities can demonstrate transactional leadership at its best. Track down a resource who has been stolen from the project and get them back on task. Review the life test preparation and make sure it's on schedule. Check to see if critical equipment has arrived; if not, get on the phone with the supplier. Figure out how to fill in for a colleague who is ill. The list goes on. Every day, project managers negotiate, remind, pester, demand, rearrange, and plead their cases in order to hold the schedule, budget, and product requirements together. Daily follow-up helps identify problems quickly, which provides more time to react.

Project Management in Product Development. http://dx.doi.org/10.1016/B978-0-12-802322-8.00003-6

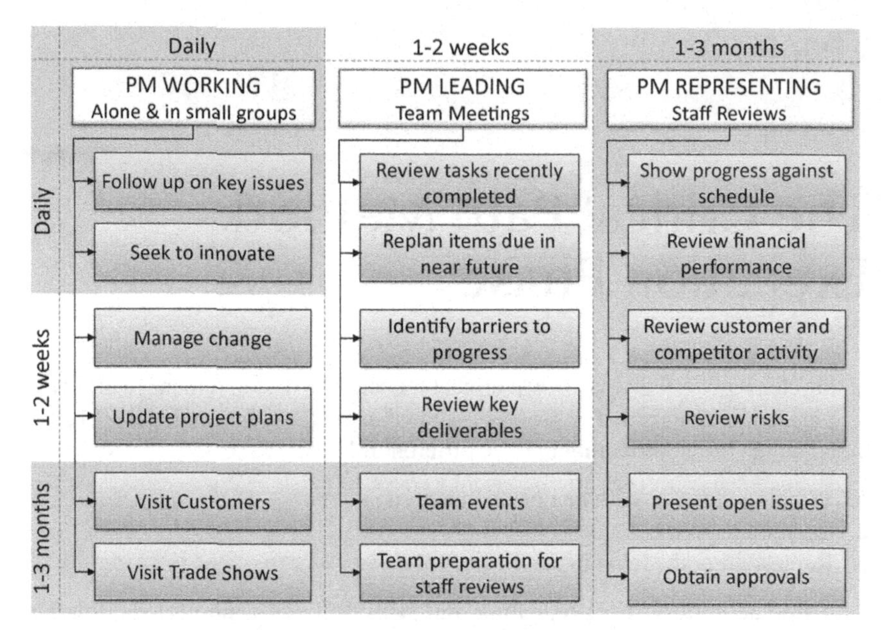

FIGURE 3.1 Executing a product development project.

3.1.2 Seek Opportunities for Project Innovation

Seek to innovate

The balance to follow up is *project innovation* (sometimes "Iterate and learn" [1], a *pivot*, or *nonlinear* thinking [2]). To quote Curlee: "Consider where nonlinear thinking might help…Once you find a place…proceed with this change in order to offer people with a creative option."

While follow-up seeks to check the box, project innovation asks why the box is there in the first place. Examples of project innovation are:

- You see a team member struggling with a challenging design problem. It occurs to you, he has not got the ability to solve the problem, but he doesn't know it. You quietly find him some help.
- You're fighting with getting a three-step test complete. It's a problem because the third step must be completed before you can order material for prototypes. But the team has lost 2 weeks because the test lab can't finish Step 2. It occurs to you to change the order—skip Step 2 for now because you can manage Step 3 without it. Four hours later, you're ordering material.
- A supplier has promised a new type of glass and they just informed you of an 8-week delay to deliver the production quality samples. You realize you can find a small shop that can provide prototype quantities to keep the project moving during those 8 weeks.
- The lead engineer on your team can't reach agreement with the lead engineer at a key customer. This has been going on for 4 weeks with no sign of progress. You realize you need to make a visit to the customer with your engineer to

break the logjam. At the meeting, you see the two engineers don't communicate well—no one is at fault; it's just not good chemistry. When you return, you assign someone else to manage the technical communication.

- The software team is stuck trying to implement a function to satisfy a customer. After speaking with the team, you realize that within the team there are two incompatible views of what the function should do. You arrange for a few team members to visit the customer and watch how they use the product.

So, in daily work, how do you know when to be transactional ("doing the task right") and when to be transformational ("doing the right task")? The first step is go see the problem for yourself. In lean manufacturing, it's called "Go to Gemba"—Gemba is sometimes translated as "the place where the action happens." Problem on the factory floor? Go see the machine. Customer angry? Visit the customer. Engineer not able to finish a key task? Sit with the engineer until you comprehend the issue. If you try to manage a project from behind a desk, you'll be relying on the interpretation of other people. You won't get the best results. We'll talk more about "Going to Gemba" in Section 7.3.2.

When you encounter a problem, the more unexpected the problem, the more likely the solution will be transformational. All the things that make up the "culture" at the company—the standard processes, the technical competencies, and the served markets—have grown up around problems the company has been solving for a while. The more a problem fits into something solved before, the more likely project discipline (transactional leadership) will resolve it. On the other hand, the more a problem is out of the experience of the organization, the less likely the culture will be able to address it. Table 3.1 shows a few examples.

TABLE 3.1 Problems That are More Likely to Require Transaction versus Transformational Leadership

	Probably Transactional	Probably Transformational
We've been stuck at the same place for weeks. Every week John says it will be done the next week.		✓
We're running late, but we make measurable progress each day.	✓	
We suddenly encountered a large delay because a task was misunderstood at the start.		✓
The lead designer was out 4 days ill and now we're a week behind.	✓	
We encountered a failure mode in life test we've never seen before.		✓

In summary, your daily work as a PM starts with seeing the problems the team faces and then applying your knowledge to help untangle the knots. Sometimes you need to bring discipline so the team keeps moving through the process. Other times you need to be asking if the process is right for the problem.

3.1.3 Manage Change

Manage change

Change comes to almost every product development project. As one author puts it, "Projects have a life of their own: they grow, join, change, shrink, and split"[3]. It can come from the customer or the Marketing department in the form of scope creep. It can come from risks that mature into issues, consuming more budget and time than was allocated. It can come from senior management who may reallocate team members to competing projects or withhold approval for expenses. Your job is to manage the change (Figure 3.2). You can leverage the Iron Triangle—everyone in your organization should be aware that any change in one leg of the triangle can force changes in the other two. As PM, try not to get emotionally involved whether proposed change is accepted or not. Focus on providing the information the organization needs to decide upon expanding the scope of the project. To rephrase a famous Stephen Spielberg comment: "the PM doesn't decide 'yes' or 'no'. The PM says 'yes' and how much. Management says 'yes' or 'no'."

Consider *scope creep*, the phenomenon where features or performance requirements are added to the functional specification bit by bit. Scope creep occurs most often when there is little "cost" to modifying the requirements. The

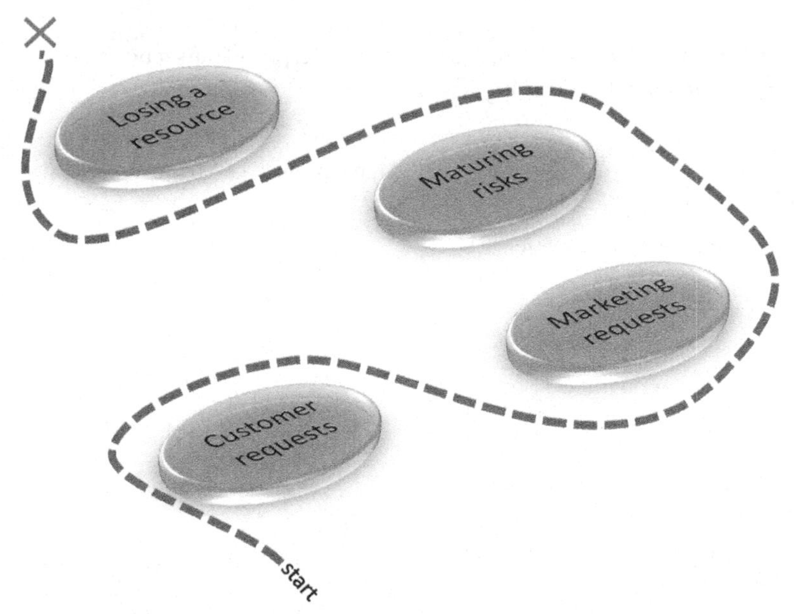

FIGURE 3.2 The PM charts a path through many sources of change.

PM needs to ensure there is an appropriate cost (in terms of time and effort) to those who propose changes. When Olivia in marketing tells you she wants to add a second TCP/IP port to the new Model XJ3, you can make a rough estimate of the budget overrun, extra resources that will be required, and effect on launch date. Then, if it's practical, ask her to get approval from the project sponsor to expand the scope; until then, you probably should stay on the original plan.

Losing resources is another change you will need to cope with from time to time. You can protect the project by first having an adequate plan before the project starts; if you see the project needs more resources, renegotiate quickly. Avoid the temptation to steal resources—they will be hard to hold onto if needed elsewhere and they are not in the project plan. If the project team is robbed of needed people, you have little choice but to revise the project plan to let management know that the project will be delayed. Again, stay neutral—if those resources are needed more elsewhere and the company accepts a delay in your project, accept the decision.

Matured risk—risk that turns into a real problem—is another type of change and usually the most difficult to manage. Typically it will involve explaining the risk and negotiating for more time, budget, and/or people. That's something you would normally deal with at a staff review, so we'll cover that later in this chapter.

3.1.4 Update Project Plans

Update project plans

Once or twice a month, be certain your project documents are up to date—budgets, schedule, and resources should all be realistic. Certainly be sure all documents are up to date before any staff review—a question might be asked that requires a document you hadn't planned to show.

3.1.5 Visit Customers

Visit Customers

From time to time, you need to visit customers and potential customers. It will provide the knowledge needed to make tough calls in the project. A solid understanding of what the customer values enables better tradeoffs. You'll also be more convincing with the team and the senior staff.

Usually you'll want to visit with the sales or marketing leader since they are normally the customer liaison. When possible, take one or two other team members—the more the developers that see applications, the better the team as a whole will understand the products they are working on. This will enable them to make better decisions as they work.

3.1.6 Visit Trade Shows

Visit Trade Shows

Also from time to time, visit a trade show that is related to the product under development. They can hold a wealth of information. You can see competitive alternatives to your new product, which will confirm or challenge the product definition from sales/marketing. You can watch products similar to yours operate and perhaps see opportunities for your product to improve the customer's experience. You can see the trends that are developing during the project lifetime, allowing fast reaction to changes in the market place.

3.2 LEADING TEAM MEETINGS

PM LEADING
Team Meetings

Team meetings should provide the richest dialogue about the project. Regular team meetings are typically weekly. Here, you're usually going over progress since the last meeting and planning activities until the next. There are also team meetings for one-time events, such as design reviews. We'll start with items to cover in the weekly team meeting and then move to special events.

Team meetings have changed considerably over the last 10 or 15 years. Today's development team is more likely to include remote team members who call in for video conferences or telecons. You may have a subteam working from another country. All of this poses new challenges—communication is often called out as the source of project failure [4] and communication is much harder with distributed teams. Different time zones restrict meeting times. Language can be a serious problem, made more difficult when relying wholly on a telephone since so much communication is normally through nonverbal cues. If someone is in a room with you, you know if they are paying attention, if they are angry or happy, or if they understand you, at least in part. When you're limited to a phone, even those things are unclear. And, at least in my experience, when you see someone's face whether in person or by web cam, you become more patient with them. So, if you have a distributed team, get the tools you need. If possible, set up a video conference. If not, at least get a web cam and screen-sharing/whiteboard application like Webex® or GoToMeeting®.

3.2.1 Review Tasks Recently Completed

Review tasks recently completed

The weekly team meeting normally starts with reviewing progress since the last team meeting. You'll probably want to project the project schedule on screen during this discussion.

Let's take the example of the planned project in the previous chapter (Figure 3.3). Let's say we're meeting October 22 (cell C2). The kick-off meeting was held on time, so it gets a "yes" in column K, and that turns the cell in column M green. So far, so good. But, Task 2, "Initial design package," was due Oct 17 and it's still not done. That turns cells for that task (columns M–O) red. Not good. Is this going to delay the whole project? Can we proceed with the proof of concept build without the full design package? How much risk do we have of more delays? Or is this just a bump in the road because a few minor drawings are still in process?

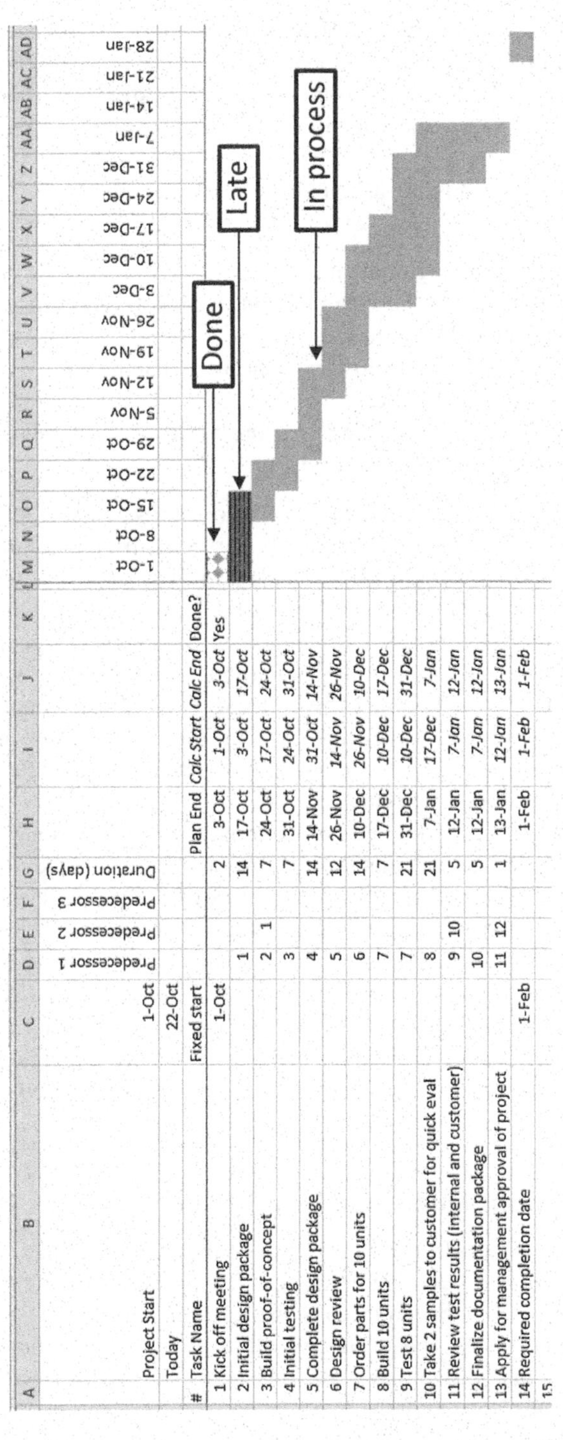

#	Task Name	Fixed start	Predecessor 1	Predecessor 2	Predecessor 3	Duration (days)	Plan End	Calc Start	Calc End	Done?
	Project Start	1-Oct								
	Today	22-Oct								
1	Kick off meeting	1-Oct				2	3-Oct	1-Oct	3-Oct	Yes
2	Initial design package		1			14	17-Oct	3-Oct	17-Oct	
3	Build proof-of-concept		2	1		7	24-Oct	17-Oct	24-Oct	
4	Initial testing		3			7	31-Oct	24-Oct	31-Oct	
5	Complete design package		4			14	14-Nov	31-Oct	14-Nov	
6	Design review		5			12	26-Nov	14-Nov	26-Nov	
7	Order parts for 10 units		6			14	10-Dec	26-Nov	10-Dec	
8	Build 10 units		7			7	17-Dec	10-Dec	17-Dec	
9	Test 8 units		7			21	31-Dec	10-Dec	31-Dec	
10	Take 2 samples to customer for quick eval		8			21	7-Jan	17-Dec	7-Jan	
11	Review test results (internal and customer)		9	10		5	12-Jan	7-Jan	12-Jan	
12	Finalize documentation package		10			5	12-Jan	7-Jan	12-Jan	
13	Apply for management approval of project		11	12		1	13-Jan	12-Jan	13-Jan	
14	Required completion date	1-Feb					1-Feb	1-Feb	1-Feb	
15										

FIGURE 3.3 Reviewing schedule on Oct 22.

When a task that should be complete is not completed, you need to make a choice: is this slip a concern? If not, ask the owner for a new commit date and adjust the schedule. Ensure that no milestones or customer commitments are compromised. Again, have the schedule on-screen so everyone understands the new commitment.

To determine if this is an issue, ask yourself a few questions:

- Is this on the critical path?
- Has this area had ongoing problems?
- Does this look like a risk maturing into an issue (schedule slips are often the first indication of a risk maturing).
- Does the team lack the capability and capacity to handle this?
- Does it seem like we're stuck?
- How many more slips can we tolerate before we delay a customer commitment or a milestone we're being measured on?

The more "yeses" and "maybes", the more likely you've got something to be concerned about. If you're ready to start a team discussion, introduce the topic from a neutral point of view: "We can't get the seal to pass pressure tests. Christian has tried all the original design concepts, but none seem to work." Provide a supportive, blameless environment so people will be open at this meeting and in the future. Perhaps there are one or two people who don't feel Christian is doing the greatest job. You might not be so sure about Christian yourself. Nevertheless, keep the conversation blame free: what's wrong and how can the team address it. Normally, the team will come together and find solutions.

One personal decision you need to make is whether you want to use automatic predecessor-based calculations of start dates as most of the tasks in Figure 3.3 do. The advantage of this is in planning because the schedule is easy to create—you link the tasks and set the durations. The disadvantages come later when new data requires project innovation, which often forces you to change several predecessor links and durations. It's tedious work and it's easy to make mistakes that corrupt the schedule. Over my career, I've chosen to avoid automated calculations, accepting the extra manual work in the beginning in favor of increased flexibility later. However, automatic predecessor-based calculations are popular and can work well too.

3.2.2 Replan Items Due in the Near Future

Replan items due in near future

In the second part of the team meeting, look at work coming in the next few weeks. These items may have been placed in the schedule at the start of the project. If you're half way through the project, you'll know more about those tasks than you did during planning. Do they need to be adjusted? Do they need to be broken into finer granularity now that the time to work on them is nearing? If the task "prepare for focus group" is approaching and was originally a 6-week task, it probably makes sense to break that into four or five smaller tasks over the same time period. Anytime you change tasks, ensure each has an owner and reaffirm their commitments.

3.2.3 Identify Barriers to Progress

Identify barriers to progress

Now find out if there are any barriers—things that might slow the team down but are not related to a particular task. Does everyone have the tools they need? What about training? Are communications working with a remote team member? Are relationships good with outside support groups like the test lab? Are vacations coming up that haven't been accounted for? Find out what your team needs so they can keep working efficiently.

3.2.4 Review Key Deliverables

Review key deliverables

The final step of the team meeting can be to review the key deliverables: customer commitments and milestones (milestones are commitments to the organization). Review those to ensure the team is on track. It's easy to be consumed with daily work and slowly get off course for the longer-term deliverables. Pull up from the detail and look at the project at a high level at least long enough to ensure there are no obvious blockers to producing the key deliverables on time.

The team meetings proceed every week or so through the life of the project. At each meeting, the team moves a short way through the task list. Over time the original tasks are completed, or modified and then completed; new tasks are added. Tasks originally planned with coarse granularity are broken into finer granularity as the team begins working on them. Risks are discovered, some of which mature into issues that, in turn, generate new tasks. The process continues as the team churns through the task list.

The team meetings create the PM's best opportunities to show leadership. He must be a strong transactional leader, week after week keeping the details straight, holding people accountable, and following process. But he must also be a strong transformational leader, inspiring the team to do their best by being positive, innovative, forward thinking, displaying strong character, and connecting with the team on a personal level.

3.2.5 Team Events

Team events

Team events are those meetings called for a special purpose during the project. For example, a kick-off meeting or a design review usually requires the whole team to attend. These meetings are often prescribed by company's development process. If you're new to project management, you probably should not lead one of these events until you've sat through one or two. You can sit in on similar events for other projects or you can ask a more senior PM to lead yours while you listen and learn.

3.2.6 Team Preparation for Staff Reviews

Team preparation for staff reviews

The final activity we'll discuss for team meetings is preparation for a staff review. If an important milestone is approaching, pull the team together to review the schedule, budget, risks, and progress. As PM, you will normally prepare the documents. But ask the team to review them to ensure there is consensus and to maximize quality of the information. This avoids the common problem where old data is presented—for example, a key risk was identified but never made it onto the risk list, or an expense is now double what was estimated at project kick off, but the budget wasn't updated. It also helps the team better understand the project as viewed by the company management.

3.3 REPRESENTING THE TEAM AT STAFF REVIEWS

PM REPRESENTING
Staff Reviews

The PM represents the project to the sponsor(s) or *steering committee* at project reviews. This can be the most difficult interaction for the PM. On the one hand, these meetings can move the project forward. Approvals can be granted as can increased budget for people or expenses. On the other hand, the project can have its priority reduced and have to surrender resources, or it can even be put on hold or canceled. Normally the PM will be the center of the review—often other team members don't even attend. So, the pressure is on.

The first point about staff reviews is to avoid presenting unfavorable surprises. If you know the project has a new issue, meet with the sponsor ahead so it's not a surprise. If that's not practical, use a phone call or an email. Surprising the senior management with bad news should be avoided, but hiding it altogether is worse. If you're found out, your reputation could be tarnished. Also, avoid *ambushing* colleagues—disclosing unfavorable information at meetings to pressure someone for a decision or an action. If you're planning to tell staff "operations won't spare any people for component qualification, so we're 4 weeks delayed," make sure the operations leader knows your plan well ahead.

3.3.1 Show Progress against Schedule

Show progress against schedule

When showing the schedule to staff, stay at a high level. Many PMs jump straight into thick detail. Remember, the purpose of a management review is to evaluate the health of the project and determine if adjustments need to be made.

In most cases, staff members are not there to help solve technical problems. The project schedule should probably fit on one screen shot and show all the major items due for the project; if the full schedule doesn't fit that format, create an overview that does. It's common to keep a high-level schedule for staff and a more detailed schedule for team meetings (just be sure they stay aligned). Typical items the staff schedule includes are:

- Kick off
- Proof of concept complete
- Design review
- Prototype tested
- Patents-related activity (searches, applications)
- Customer feedback
  - What the experience positive? Does the customer accept the price?
- Capital approval
- Capital items purchased
- Suppliers qualified
  - All suppliers for this project have completed the company supplier approval process
- Purchased parts qualified
  - All parts purchased for this project have completed the parts approval process
- Preproduction ready
- Marketing collateral ready
  - Websites, flyers, catalogs, etc.
- Production ready
- Certifications received (UL, FDA, TÜV, CE, etc.)
- Product launch

3.3.2 Review Financial Performance

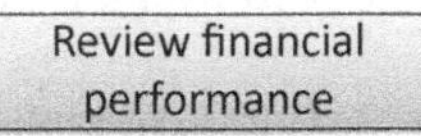

At the financial review, the staff will want to see how profitable the project is likely to be and how reliable the remaining assumptions are. Three major factors need to be reviewed:

1. Are there schedule delays and, if so, are they likely to affect projected revenue and/or total investment for the project?
2. Have product costs increased and, if so, will they affect profitability?
3. Are sales projections well substantiated and is confidence in them increasing as the launch date approaches?

Be prepared for difficult questions. It's common for senior staff to be almost spooky in the way they can pick out the weak parts of a project in a financial review. They often will dig into one of the financial figures. For example, let's say the current estimate for total revenue for next year is $600k. Perhaps a similar product was developed 2 years before and it achieved only $150k in the first year. You might be asked to substantiate the optimistic numbers. At this point, you should be able to turn to sales projections that break down the assumptions into believable portions. If you open a document on sales projections and it's out of date or contradicts the high-level presentation, expect an unhappy response. Know your numbers and make sure they hold together when people dig into the details.

3.3.3 Review Customer and Competitor Activity

Review customer and competitor activity

Be prepared to discuss customer activity (customer visits, results of surveys, focus groups, and interest in betas) and competitors' activity (new products, new large wins, recent patents, and presence at trade shows).

3.3.4 Review Risks

Review risks

Be prepared to review any open risks. Normally the interest will be which risks are resolved, which risks are lingering longer than expected, and which new risks have been identified. If a serious risk has appeared, you may want to present it as an open issue, as discussed immediately below.

3.3.5 Present Open Issues

Present open issues

Be prepared to present any open issues. This can include anything than can have a serious effect on the project: for example, newly identified risks, resource issues, supplier issues, technical problems, customer or competitor concerns, governmental regulations, certification issues, and budget overruns. The focus here is on change, especially unfavorable change. As PM, seek to explain the issue from as neutral a position as possible.

Normally, the best practice is to build a single slide with the information the senior staff is most interested in. Avoid creating a "mystery novel" that starts with "we thought we had a problem…" and 5 min later you finally tell the result. Staff reviews normally go better if you start with a summary. Remember the guidelines for visualization from Section 2.2.13:

- Concise and clear
- Credible
- Drives good action, which starts with you providing strong alternatives

Again, avoid surprises. If you have bad news, let people know ahead, especially your boss and the project sponsor. Bringing detailed back-up data is fine, but start with the summary. Often, people will want to review the data that seems most questionable to them; over time, if your data consistently holds up, you should build confidence and the questions should slow down.

3.3.6 Obtain Approvals

Obtain approvals

Finally, obtain approvals needed to keep the project moving. Some approvals, such as being credited with passing a milestone, can be granted at the meeting. Typically, approvals for large spends take more time. As PM you need to manage the approval process.

If you want to accelerate approval for large spends, make sure you allow enough time for the processes at your company. Approving a $1.2k spend may take 15 min; don't expect the same if you need a check for $65k. Start by determining the day you need the approval to prevent a delay; let's call that "the last minute" (see Figure 3.4). You might be tempted to turn in your request the day or two before and then complain that the project is delayed because you're waiting for the boss. Not a good plan.

Instead, talk to your boss or the head of finance to get an estimate on the approval cycle. Let's say that takes 2 weeks under normal conditions—so allow that plus a little margin. Now, let's assume your first attempt at requesting $65k isn't going to be perfect (that's usually a good assumption for most of us). So, you probably want to add a week or so for corrections. Therefore, your first draft should be ready perhaps 4 weeks before "the last minute."

One other thing you can do to accelerate approvals: take your first draft in person to key stakeholders. You might only be able to get a short meeting; if so, flash the request just in case something obvious is missing. If you can get more time, take it. The better shape your request is going into the process, the faster you'll get approval.

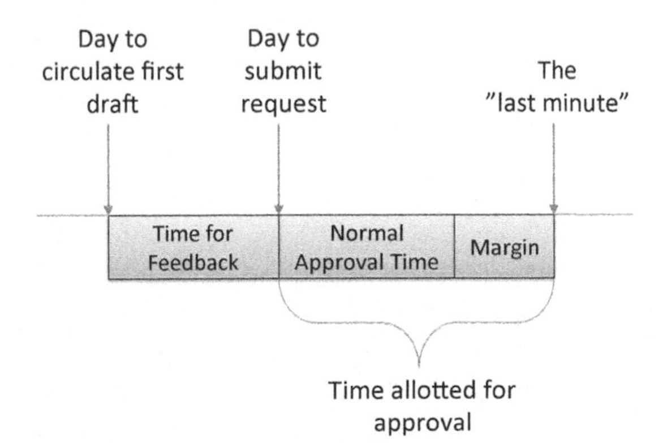

FIGURE 3.4 Circulating a first draft request in time to avoid project delay.

3.4 FINISHING THE PROJECT

As Figure 3.1 shows, all these activities repeat through the life of the project. At times, it may seem the project will never end. Part of being a strong project manager is being able to keep the pace up over a long period—months or even years—weathering the ups and downs while keeping a positive attitude. Almost all projects go through lows and the team will rely on the PM to put the best face on events. But eventually the project will finish. Hopefully, things will go well; if so, the PM's reward will probably be the opportunity to start the next project.

3.5 REFERENCES

[1] Loch CH, De Meyer A, Pich MT. Managing the unknown: a new approach to managing high uncertainty and risk in projects. John Wiley and Sons; 2006. p. 103.

[2] Curlee W, Gordon RL. Complexity theory and project management. John Wiley and Sons; 2011. p. 21.

[3] McManus J. Leadership: project and human capital management. Butterworth-Heinemann; 2006. p. 11.

[4] Curlee W, Gordon RL. Complexity theory and project management. John Wiley and Sons; 2011. p. 25.

Part II

Leadership Skills and Management Methods

Part II will start with a detailed look at leadership in project management. It will then present four of the most popular project management methods used in product development. Each of these methods can be thought of as expansions to the traditional method presented in Chapters 2 and 3.

CHAPTER 4 TOTAL LEADERSHIP FOR PROJECT MANAGERS

Goes beyond project management techniques for an in-depth look at how leadership skills are required for a successful project. Divides leadership two ways: 1) According to transactional leadership (executing existing processes and practices reliably) and transformational leadership (bringing about change using common vision and personal connection to the team). 2) According to skills that improve objectives like meeting a schedule versus focusing on people like displaying strong character or listening well. This view leads to the "Total Leadership Matrix," a view of leadership that allows project managers (PMs) to understand the many ways that their leadership skills can guide and inspire the project team.

CHAPTER 5 PHASE–GATE: EXTENDING THE CRITICAL PATH METHOD

Phase–Gate project management is an extension of the critical path method with many of the requirements of the projects defined as standard work. Phase–Gate also provides defined approval steps at critical points for investment, which fall at similar points in most projects. Phase–Gate is certainly the most popular method in industry today for complex hardware projects.

Phase–Gate is the method against which other project management methods are compared. It can work well, but its largest weakness is the inability to deliver projects on time. The alternative methods of Chapters 6–8 offer improvements, especially in the area of schedule. The positive experiences of companies that use these methods make it clear that improvements can be had, but an understanding of why one method works better than another remains elusive. This will be discussed in detail at the end of this chapter.

CHAPTER 6 CRITICAL CHAIN PROJECT MANAGEMENT

Critical chain project management is an extension of critical path project management that places increased attention to focusing on critical tasks and to rich collaboration among the team. It merges multiple techniques: aggressive schedule assumptions with a large safety buffer at the project end, a unique metric (buffer burn), and minimization of counterproductive human behavior (especially multitasking). In that sense, it can be thought of as a coherent collection of techniques rather than a wholly new method.

CHAPTER 7 LEAN PRODUCT DEVELOPMENT

Lean product development borrows techniques from lean manufacturing and applies them to product development, especially the removal of waste, improving cross-functional collaboration and the use of continuous improvement. Lean is more a mindset than a method, providing many paths from which to choose.

CHAPTER 8 AGILE SCRUM, EXTREME PROGRAMMING, AND SCRUMBAN

Agile Scrum and Scrumban project management are two related techniques that focus on software development. Software development has a "lower cost of iteration" than most hardware development projects and so can more easily be executed iteratively, with less reliance on early planning. Scrumban is sometimes grouped with lean product development but is covered here with Agile because the methods are so similar.

Chapter 4

Total Leadership for Project Managers

The first thing to understand about leadership is you can do it. It doesn't matter whether or not you were the smartest, the best looking, or the most athletic kid in school. It doesn't matter if you're charismatic, outspoken, or reserved. It doesn't matter where you sit on the org chart. If you want to be a leader, you just need to behave like one. Don't believe it? There's a huge body of research on leadership that points to a consistent result: learn a handful of principles, apply them diligently and sincerely, and soon you'll be leading.

In this chapter we're going to review some of the most important research on leadership in the workplace. This will give a view of what's known and show how it applies to project management. The goal is to help you build a broad set of leadership skills that you can use every day.

4.1 WHAT IS LEADERSHIP?

Defining leadership is a vexing task. Our first thoughts of leaders are figures that changed history. Perhaps your favorite leader is Abraham Lincoln, Julius Caesar, Socrates, Andrew Carnegie, or Pope Francis. But what makes someone a great leader? Is it the magnitude of their accomplishments? Is it the methods they devised or is it their charisma? Is it internal compass: their ability to act upon what they thought was right, no matter what others said? Leadership has been argued for millennia and today is still among the most popular topics in nonfiction writing.

To our more modest topic: What is leadership in the context of project management? Most people intrinsically understand the need for leadership in business endeavors. A team of talented, hard-working, and creative people accomplish little if they don't work together. And in most cases, people don't naturally work together well—they need a leader to create common purpose, organize work, make decisions, resolve conflicts, and keep focus on critical tasks. So, most people understand that good leadership is required at work, but still, it's difficult to precisely define what good leadership is.

One way to measure leadership is to evaluate results, success over a long period of time. Peter Drucker said: "Effective leadership is not about making speeches or being liked; leadership is defined by results" [1]. Eisenhower once

Project Management in Product Development. http://dx.doi.org/10.1016/B978-0-12-802322-8.00004-8

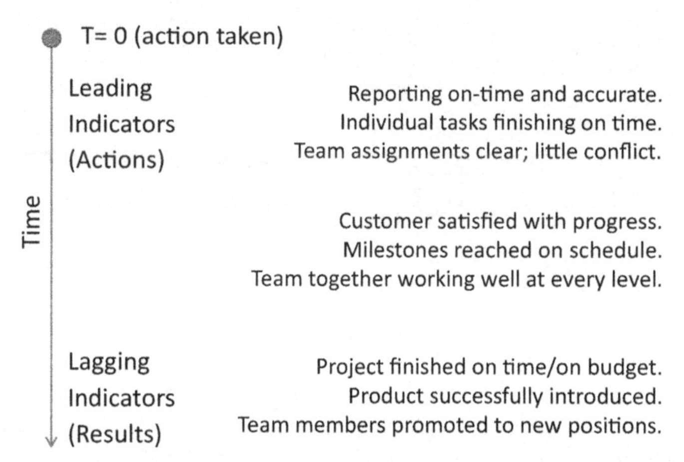

FIGURE 4.1 Example of leading and lagging indicators of project leadership.

said: "Leadership is the art of getting someone else to do something you want done because he wants to do it." Warren Bennis, a man the *Financial Times* called "the professor who established leadership as a respectable academic field" [2] said simply: "Leadership is the capacity to translate vision into reality." Results are the final arbiter of what good leadership is, but results are *lagging indicators*—they can take years to materialize (see Figure 4.1). For those of us who want to improve, relying wholly on results creates a process that takes far too long. We need a *leading indicator*—something that we can learn from quickly. So, let's look directly at the actions leaders take rather than waiting for the results those actions create.

4.1.1 Actions of a Leader

There are a wide range of opinions regarding what makes any given action that of a good leader. Douglas MacArthur said: "A true leader has the confidence to stand alone, the courage to make tough decisions, and the compassion to listen to the needs of others. He does not set out to be a leader, but becomes one by the equality of his actions and the integrity of his intent." Here we see a balance of confidence and character. John Maxwell, a popular writer and speaker on leadership, finds three critical factors: "A leader is one who knows the way, goes the way, and shows the way." These types of quotations are valuable because they provide a window into the way successful people have led. But, you can't learn how to lead a team of developers simply by compiling a set of quotes from great leaders. A more comprehensive approach is called for. Fortunately, over the past several decades, there's been a great deal of research on leadership in the workplace, which provides information that can help each of us lead better.

In the early 1980s Kouzes and Posner started to methodically study leadership actions. They developed "The Five Practices of Exemplary Leadership™" by conducting tens of thousands of surveys and then seeking to identify common themes in the responses. They asked the question: "What do you do as a leader when you're performing at your personal best?" They emerged with [3]:

- Model the Way
  Leaders should demonstrate admirable character. Honesty and integrity are commonly seen as critical for leaders and a large share of famous quotes on leadership confirm this. Stanley McChrystal asked a good question: should you follow someone who cheats on their taxes? He said he wouldn't because someone who will lie to the government will probably lie to you.
- Inspire a Shared Vision
  Vision is probably the single most identified trait of strong leaders. It's the ability to clearly identify worthy goals and then build common purpose with a group to achieve them. As James Lewis says: "A person with no followers is no leader, and people will not become followers until they accept a vision as their own. You cannot command commitment, you can only inspire it" [4].
- Challenge the Process
  Leaders encourage their team to improve the system. This is especially important in project management, a discipline that sometimes fosters a slavish devotion to outdated processes and practices. Strong leaders know that continuous improvement comes from constantly asking "what can we do better?"
- Enable Others
  Leaders build an environment of trust so team members can experiment and take risks. Team members can stretch to reach difficult goals and feel safe: they know their leader will protect them if things don't go as expected.
- Encourage the Heart
  According to Kouzes, this is the least practiced of the five concepts, but it's essential. People must know that their accomplishments and efforts are recognized. Celebrate the successes and reward people for demonstrating the right behavior even if the results are not always what you hoped for.

Kouzes and Posner's five practices are a good example of how modern authors categorize actions of successful leaders. But they didn't so much define leadership as measure it using a vast number of surveys. Other authors have used different techniques and created competing views. But in modern views of leadership, the same themes are repeated again and again: integrity, vision, inspiration, connecting on a personal level, and the ability to get results. One thing is clear: leadership is a multidimensional skill set. The fact that there is no single dominant characteristic helps explain why the term is so difficult to define.

4.2 MOTIVATION AND INSPIRATION

Leadership has a large breadth of context. Leaders in the military, politics, religion, business, and academia have one thing in common: they are able to get results through the efforts of those that follow them. But the things those different leaders do are so varied that it's hard to identify common themes—it's challenging to list a characteristic shared between Mahatma Gandhi and George Patton.

Here we're focused on leadership for project managers (PMs) where the primary goal is to guide a team to finish project requirements on time and within budget. In this context, one of the most important results of good leadership is a highly motivated team. People are motivated by a wide range of factors. A cheering crowd, a proud family, cash rewards, and aversion to punishment all can cause someone to do something they otherwise would not. These are all examples of *extrinsic motivators*—factors from outside a person that dispose her to action. Another set of motivators—*intrinsic motivators*—come from within, for example, the satisfaction of finishing the task well, enjoying the collaboration with others, and the sense of learning and advancing.

The study of intrinsic motivation began in earnest late in the twentieth century. It's been observed in animals that play enthusiastically even when there is no extrinsic reward. Studies in people have demonstrated that intrinsic motivation is critical in cognitive development [5]. For example, students who are intrinsically motivated do better in educational tasks [6]. Intrinsic motivation was increased by:

- Having a sense of self-direction or *autonomy*. Particularly relevant to project management, autonomy can be created when a leader defines what must be done, but not how to do it (sometimes called *autonomy of method*).
- Having an innate desire to master a topic rather than being driven primarily by good grades or some other reward.
- Being confident that they have the skills to complete the task successfully (sometimes called *self-efficacy*).

Self-efficacy is a measure of confidence a person has in their ability to complete tasks and reach goals. There are several sources of self-efficacy including the person's perception of (1) their own technical ability, (2) the perceived strength of the approach, (3) time constraints, and (4) available support throughout the process.

4.2.1 Herzberg's Motivation-Hygiene Theory

Frederick Herzberg studied motivation in the workplace. He developed the *motivation-hygiene* theory (sometimes called the *two-* or *dual-factor theory*), which states that job satisfaction is created by factors that are generally independent of those that cause dissatisfaction [7,8]. He called the factors that increase job satisfaction *motivators*; they are generally intrinsic. He saw motivators as uniquely human because they cause us to experience psychological growth. Motivational factors answer the question "is my work meaningful?"

Those factors that relate to dissatisfaction Herzberg called *hygienic*; they are extrinsic: for example, monetary rewards, supervisor actions, and company policies. In Herzberg's construction, these items do not cause satisfaction, but if they are missing or done poorly they do cause dissatisfaction [9]. Hygiene answers the question "how am I treated at work?" One common misinterpretation of Herzberg's theory is that motivators are more important than hygienic factors. In fact, in Herzberg's view both were necessary. They have different purposes and those in leadership need to understand both [10].

Herzberg's research included studying more than 3500 job-related events that led to either extreme satisfaction or dissatisfaction. Some factors, the *motivators*, appeared predominantly in the events that were reported as leading to satisfaction; for example, the sense of personal achievement was the dominant source of satisfaction, but its lack hardly appeared in events that led to dissatisfaction. On the other hand, company policy/administration and supervision were the dominant hygienic effects. They appeared often in events that caused dissatisfaction but rarely in those causing satisfaction.

To be clear, none of Herzberg's factors related exactly to satisfaction or dissatisfaction, but the pattern that the results were driven by different mechanisms is well established from the sum of his data. A few factors, salary and relationships with peers and subordinates, occurred more in dissatisfaction events, but only by a modest amount. He classified these as hygienic—he used only the two categories—but they might be thought of as being somewhere between motivator and hygiene.

Salary is in interesting example of a hygienic factor. It's commonly used by companies as a motivator. But money doesn't make people passionate about their jobs [11], at least not for long. Few of us show up to work every day excited about increasing that next salary bump. But if someone learns they are underpaid—for example, when another company offers a substantially higher salary for the same job—it can lead to severe dissatisfaction. For most of us, after we perceive we are fairly compensated, money has little motivational effect.

The motivational factors Herzberg identified were: a sense of achievement, recognition, the pleasure of the work itself, a sufficient level of responsibility, personal advancement, and personal growth and learning. There were a larger number of hygienic factors, but they were dominated by company policies and administration, supervision, relationship with supervisor, work conditions, salary, and relationship with peers and subordinates (see Figure 4.2). In addition, Rosenau considers inadequate resource availability as a factor that can cause

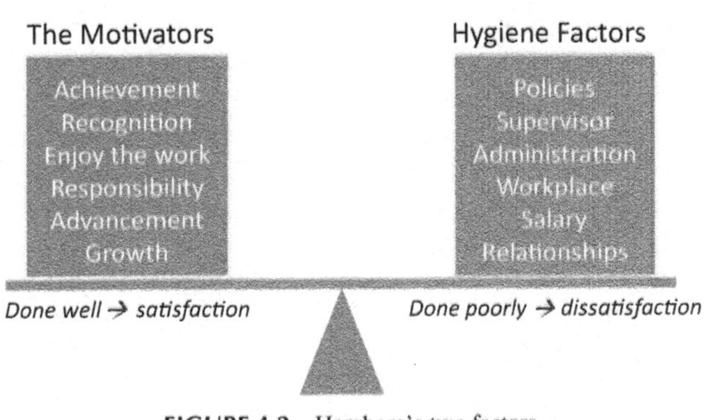

FIGURE 4.2 Herzberg's two factors.

dissatisfaction [12], so it might be added to the hygienic factors. This isn't prominent in Herzberg's research, but my experience in product development is that the lack of adequate resources is a common source of dissatisfaction for PMs and team members.

A similar view of intrinsic motivational factors is offered by The Forum Corporation (www.forum.com), a company formed in the 1970s to help client companies improve their performance with "people-driven" solutions. They focus on employee "engagement," a sense of ownership the employee feels for the organization and the accompanying commitment this brings to their work. They see three factors that determine engagement: a good working environment, trust in leadership, and five intrinsic needs of the employee [13]:

1. A sense of accomplishment
2. Recognition and appreciation for their contribution
3. Taking pleasure in the work itself
4. A sense of belonging to a team
5. Career advancement

Yet another alternative is provided by Daniel Pink, who describes three types of motivation, which he likens to human "operating systems" that he calls M1.0, M2.0, and M3.0 [14]. M1.0 is basic survival, which dominated motivation for most of human history. M2.0 is a system of rewards and punishment. It replaces the survival mentality of M1.0 with the premise: "Rewarding an activity will get you more of it. Punishing an activity will get you less of it" [15]. He portrays M2.0 as outdated for the workplace and points out many companies are yet to recognize that.

M3.0 is based on intrinsic motivators and it is particularly effective for professions relying on creativity. For Pink, there are three main factors: autonomy, mastery, and purpose. He offers many types of autonomy: task autonomy (what you do), time autonomy (when you do it), technique autonomy (how you do it), and team autonomy. Each type defines different ways managers can offer autonomy.

Mastery is the innate desire to be good at something—it's a powerful motivator for many professionals. Technical people are driven to be good at what they do—it's how most engineers get through university where the dropout rate is about 50% [16] (compare to the famously difficult Marine Corp boot camp training with a dropout rate under 15% [17]). Research and intuition tells that mastery is a large motivational factor for many throughout their careers.

Finally, Pink lists purpose—the desire to be part of something larger than yourself. Feeling a part of something larger than yourself has long been known to be an important part of achieving personal happiness (Seligman calls it "meaning" [18]). Its position as a motivator may not be as well established, but it seems intuitive. Perhaps the current focus on corporate vision/mission statements is meant to leverage this; many of these statements portray companies as much more than profit-seeking enterprises. For example, 3M describes their "brand essence" as "Harnessing the chain reaction of new ideas" [19]. GE's vision statement includes: "We have a relentless drive to invent things that matter: innovations that build, power, move and help cure the world" [20].

One example provided by Pink to demonstrate the power of intrinsic motivation is Encarta versus Wikipedia [21]. Encarta was a software encyclopedia undertaken by Microsoft in the 1990s; it was a large effort by the standards of the time. Microsoft had the full range of incentives available to a company offering a for-profit product. On the other hand, Wikipedia has never paid for articles—they rely heavily on the intrinsic motivation of their contributors and editors. Of course, Wikipedia's success has been phenomenal. The English version of Wikipedia had nearly five million articles as of the printing of this book.[1] By comparison, Encarta, which, ironically, you can read about in Wikipedia, ended in 2009 with 62,000 articles.[2]

4.2.2 Creativity and Motivation

Herzberg's research was on a broad swath of professions: maintenance personnel, managers, teachers, food handlers, among others [8]. Many of the respondents were in careers with little emphasis on creativity. Fostering creativity is an important aspect of project management and so deserves special attention here.

Teresa Amabile begins an article on creativity with this chilling observation about modern organizations: "creativity gets killed much more often than it gets supported" [22]. She believes creativity requires three things: technical expertise, motivation, and the ability to think flexibly and with imagination. She lists five "levers" managers can pull to improve the creative environment: magnitude of challenge, degree of freedom, work group structure, amount of encouragement, and organizational support.

Regarding challenge, good managers need to match employees with tasks that stretch them enough to make the job interesting but not so much they lose "self-efficacy" as discussed earlier [23]. This is well aligned with Herzberg's belief that every job should be a "learning experience." Amabile's experience is organizations often kill the creativity of their associates with fake or impossibly short schedule demands. My own experience lines up with that: I once managed a fellow who was an outstanding engineer, but when he was overwhelmed, he stopped working and read technical journals. That experience was an early lesson in how counterproductive it can be to overload team members.

Amabile emphasizes team interaction. She believes the team brings many benefits including diverse ways of thinking, having a shared excitement over the team's challenges, being able to get help more easily from peers who understand your tasks, and sharing a sense of respect for each other [24]. She also

1. http://en.wikipedia.org/wiki/Wikipedia.
2. http://en.wikipedia.org/wiki/Encarta.

emphasizes the supervisor's role in recognizing contributions from team members, by greeting new ideas with an open mind, and by valuing ideas for the process that generated them rather than simple-minded evaluation of the results. This last point is crucial for creating a safe environment—leaders should reward people for doing the right thing based on the information available to them at the time. If you reward only positive outcomes, you'll risk making your team overly conservative. Wise risk taking ought to be rewarded, so a strong leader will understand that even when people do the right thing, the outcome cannot always be predicted.

Pink recalls a few experiments that make it clear that extrinsic motivation (specifically monetary reward) not only fails to improve performance in creative activities, but can significantly degrade it. He quotes a study done by the Federal Reserve Bank of Boston: "In eight of nine tasks we examined…higher incentives led to worse performance" [25]. He also quotes data from the famous "candle problem" where a subject is given a box of tacks, a candle, and a set of matches on a table. The subject is told to use those objects to hold the candle to a wall above a table in a way that wax does not drip onto the table as the candle burns. I won't spoil the puzzle (see Ref. [26] for the solution), but most people figure it out in a few minutes absent any reward. When people are given small monetary rewards to complete the task quickly, no improvement is observed. Most interesting, when a large reward is offered to speed completion, the time to solve the problem becomes 3.5 times *longer*! It's widely believed that the reward focuses the mind so narrowly that a creative solution is more difficult to imagine. In other words, monetary reward thwarted the creative process.

We have discussed a few approaches to understanding motivation in the workplace. It is a sample of the large body of thought on the topic. There are differences in the four trains of thought presented here (see Table 4.1), but they are small, especially when compared to the prevalent thinking in many companies that focus on rewards and punishment to motivate their teams. Reviewing Table 4.1, the parallels are clear and the differences seem more a matter of weighting than any deep-seated disagreement.

TABLE 4.1 Different Views of Intrinsic Motivators

Herzberg	Forum.com	Pink	Amabile
Sense of achievement, recognition, pleasure of the work, level of responsibility, personal advancement, and growth.	Five engagement needs: a sense of achievement, recognition, pleasure of the work itself, sense of belonging to a team, career advancement.	Autonomy, mastery, purpose.	Challenge, freedom (autonomy), properly designed work groups (teams), encouragement, organizational support.

All of this is good news for PMs. Many PMs believe their ability to motivate their teams is limited because, in most cases, team members don't report directly to PMs. As a result, their ability to create traditional rewards (promotions and raises) and punishment is limited. But from the research it should be clear that PMs have many levers to channel intrinsic motivation. They can have a direct effect on recognition, growth (through proper assignment of tasks), a sense of belonging, and most of the other items in Table 4.1.

4.2.3 Inspiration and Project Management

The term *inspiration* can be defined as "being mentally stimulated to take action, especially toward something creative." For many people, the term seems more appropriate for a religious meeting or a political campaign. But what does inspiration mean in the context of a project?

One measure of inspiration from a PM is the performance improvement in each team member due to the PM's leadership. This improvement comes through the ability to capture an employee's *discretionary energy*, the effect of when people willingly give "personal effort, time, and mindshare to the organization, above and beyond what is expected" [27]. Project managers can inspire their team members by presenting a clear vision of the value of the project, having a personal connection with each team member, and be being able to organize and manage the daily work, thus giving the team confidence they will be successful. In these cases, the team will be operating near the peak of their ability because of the inspiration provided by the PM. Alternatively, if the PM expresses no vision of the project, lacks behavior that people expect of a leader, or cannot manage the project workload, team performance can tumble. Of course, the PM's leadership skills are just one factor. Factors outside the PMs direct influence must also be considered. Here are three categories, which are diagrammed in Figure 4.3:

1. Rewards and coercion: Each team member will experience a certain amount of motivation due to extrinsic factors: salary, working conditions, and so on. This creates a "floor" in the sense that less performance is probably not accepted at the company.
2. Company culture: The company culture and management will channel some intrinsic motivational factors. The team member may believe strongly in the company's goals or may value the autonomy the company normally grants to people at his level. These things will result in higher performance than the extrinsic motivators alone.
3. Personal leadership: The leadership skills of the PM create the last factor.

Our focus here is on the third factor. A strong PM will increase the performance of each team member to reach or nearly reach the "ceiling," which is the limit imposed by factors outside the PM's influence: the company, the team member's personal life, and, most important, the strength of the individual's intrinsic

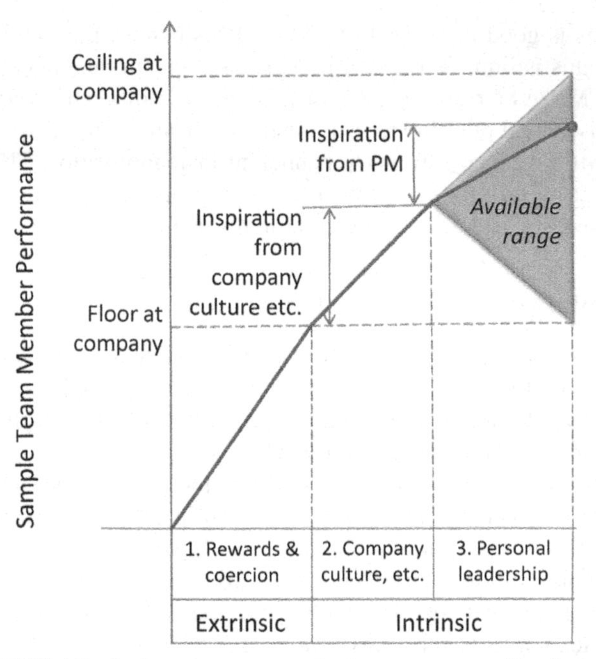

FIGURE 4.3 Performance versus motivation as a measure of inspiration.

motivators. Weak project managers can be so demotivating that they push performance down to near the floor. The difference between the two extremes is the "available range" of results.

To inspire people, their intrinsic motivators must be fed. But every person has different intrinsic motivators. One may value the freedom to solve challenging technical problems in unconventional ways. Another may value high-quality team interactions. Still another will want the tasks that best prepare her for a promotion. The PM must have a relationship with each team member to know which intrinsic motivators to channel. By recognizing what is valued by each team member and being skilled enough to provide it, the PM can inspire the entire team.

4.3 THE TOTAL LEADERSHIP MATRIX

Your leadership skills will allow you to respond to wide range of issues such as difficult customers, pressure from senior management, team members not able to complete a key task, conflict between team members, and the need to recognize a contribution. To do all this, you'll need a big tool box. In this section, we'll look at leadership skills as tools to solve problems. We'll start by looking at what people express about their jobs, especially what they're unhappy with (the "problem"), which is similar to Herzberg's approach. Then we'll focus on the leadership skill that can help resolve that problem.

4.3.1 Why People Don't Like Their Jobs

Let's begin by listing the complaints people frequently have about their jobs. Such lists are commonplace; below is a list of 30 reasons merged from three places: DeCarlo [28], the Huffington Post [29], and Herzberg's results:

- The work is boring/I'm not challenged
- I don't understand the business value of what I'm doing
- My function here is unclear
- I have no say in how we do things here
- I'm not confident my career is going where I want
- I don't like my teammates
- There is too much bureaucracy here
- I get in trouble when I deliver bad news
- I'm not capable of doing this job
- My team is not capable of doing this job
- We are underresourced
- Our project manager has not been successful
- The goals of this project are not realistic
- This project is too complex
- The politics around here are against this project
- Our company isn't good at this type of project
- Our sponsor is weak
- The customer is not engaged
- Our suppliers are not capable
- Our meetings are ineffective
- We don't resolve issues—they just linger on
- We can't make a decision
- The support system around here is poor
- The workplace is bad
- My boss is bad
- My salary is too low
- My job is not secure
- I'm not valued here
- My values are not the same as the company's
- Problems in my personal life

4.3.2 Understanding the Categories

We can sort this list out two ways: according to a lack of transactional leadership versus transformational leadership, and according to whether it's related to company objectives or to people (Table 4.2). As discussed in Chapter 1, leadership skills are commonly divided between transformation and transaction. *Transactional leadership*, commonly called *management*, is for cases that repeat in a recognized pattern. For example, each task in a project has a description,

TABLE 4.2 Transactional versus Transformative Leadership

	Transactional Leadership	Transformational Leadership
How it works	Works within a system.	Works to change a system.
	Starts solving by fitting experiences to a known pattern.	Starts solving by finding experiences that show the old pattern doesn't fit.
	Asks "where's the step-by-step?"	Asks "what do we need to change?"
What it does	Minimizes variation of the organization.	Maximizes the capability of each person.
	Expects the entire team meet a standard.	Inspires each person to give their best.
	Can be duplicated and sustained.	Requires minimal structure.
	Best at delivering defined results.	Best at delivering innovation.

an owner, an end date, and so on. This repeated pattern allows development of processes and practices so results can be measured and variation reduced. Great transactional leadership makes the team better and better at what they do often. It delivers defined results on time and within budget.

Transformational leadership is for cases where change is needed. For example, you notice that several times this year, a team member is finishing a customer deliverable on time, but the customers still seem unhappy. Recent experiences tell you he doesn't fully understand customer needs, probably because he doesn't spend enough time listening to customers. You start sending him to customer sites and it gets results. People get excited when they see transformation. Great transformational leadership enables the team to do things they didn't think they could.

Another dimension is organizational objectives versus people. *Objectives* include things like creating profit, releasing a new product, winning market share, and reducing quality defects. Some objectives are easy to measure: sales in Q1 of this year. Others are difficult: how well does our 3-year product roadmap meet our customer's needs or what will our customers demand for delivery time and how much will we have to invest to meet that demand?

People in this context includes all the factors that relate directly to team members. Some activities around people are easy to measure: annual reviews and company surveys of satisfaction, salary actions, and ethical guidelines. Others are challenging to measure: Do team members feel empowered? Do they feel their careers are moving the right way? And, are they being given the opportunity to master their field of interest?

4.3.3 Sorting out the List

Now let's sort the list of Section 4.3.1. For example, the first item was "The work is boring." This is a symptom that could have a few root causes. A likely cause is: this is below my capability; it should be done by a more junior person. This root cause can be categorized as:

- *Transformational*, because the proper assignment of tasks demands an evaluation of the task requirements against the individual's capabilities and desires. It requires a relationship with the team members and good judgment; usually, it cannot be left to a process.
- *People*, since an individual's needs for challenge and enjoyment of work are the primary issues.

You can imagine other root causes and, in fact, many items in the list have multiple possible causes. The entire set of 30 reasons is sorted by likely root causes in Appendix B. As with Herzberg's motivation and hygiene (or any other study in this area), there's not always a sharp line defining the boundaries, but the categorization is nevertheless illuminating. The results from Appendix B are:

- More than 50 possible root causes were pulled from the 30 reasons.
- About two-thirds of the causes were related to people rather than objectives. This is expected—most of Herzberg's hygienic factors relate to people issues.
- About two-thirds of the issues were related to transformation rather than transaction. This makes sense for two reasons: first, transformation is often lacking because it is more difficult to build in an organization—it's hard to replicate because it doesn't follow a step-by-step process. A second reason is problems of transaction are not always apparent to team members and so may be underreported. For example, if you are unaware there exist strong processes for team meetings (consistent agendas, software tools for reliable follow-up) you're likely to blame the meeting leader for ineffective meetings.

So skills to deal with employee problems revolve around to these two criteria: transaction versus transformation and objectives versus people. Combining the two, there are four categories:

- Transformation for objectives, which creates *vision*: the ability to define a goal and create common purpose in the team to achieve that goal.
- Transformation for people, which creates *connection*: the ability to relate to each individual, to understand their needs and abilities so they can contribute at the highest level.

TOTAL LEADERSHIP	OBJECTIVES	PEOPLE
TRANSFORMATION	VISION	CONNECTION
TRANSACTION	PROCESS	POLICY

FIGURE 4.4 The Total Leadership Matrix.

- Transaction for objectives, which creates *process*: standard work around how you develop products. This can be the total development process, the subprocesses for individual events like design reviews, and the processes for repeated events like daily meetings. The more activities you can move to standard work, the more opportunity to drive continuous improvement and to increase efficiency.
- Transaction for people, which creates *policy*: the standard work around people such as annual reviews, meaningful personal objectives, specific development plans, reliable career planning activities, ethics training, and respect for diversity.

These four categories can be displayed as the Total Leadership Matrix, which is shown in Figure 4.4.

4.4 LEADERSHIP IN PROJECT MANAGEMENT

One of the most important functions of leadership in project management is to channel your team's intrinsic motivators. Unfortunately, this function is often neglected. As pointed out in Chapter 1, the PMBOK® Guide gives little weight to leadership [30] in general and to inspiring team members. And it's not just the PMBOK® Guide. A survey of the top 10 concerns of IT directors placed managing budgets as the top priority and project management as their lowest priority. "Motivating subordinates and team members did not make the list" [31].

In fact, having a motivated team can be the difference between the success and failure of a project. As we've discussed already, the PM has considerable influence in this area. In this section, we'll look in more detail at different leadership skills that can help the PM inspire the team, dividing these skills among the four categories in the Total Leadership Matrix. The approach is to treat the many different leadership skills as tools and this section will discuss how to choose the right tool for a given situation.

4.4.1 Transformational Leadership Skills—Vision

Vision is the ability to see the larger picture—where the team is going and why. Great companies are often led by visionaries like Steve Jobs. The goal of a PM is certainly modest in comparison; your vision will be focused on the project, the team, and the product under development. But even if it is modest, the team will value your ability to imagine how the project will change things for the better.

The most obvious goal of most product-development projects is to generate profit. But think past the numbers—more profit can help people at your company in many ways—stabilizing jobs, creating more funds for training and equipment, and reducing pressure on the immediate thus allowing time to work on longer-term goals. And profit is a result—the product must create value to be able to generate profit. So, include the value directly in your vision. Perhaps it makes something safer, easier to use, or provides more enjoyment for people's personal time. The project also can provide opportunities for your team to grow in their chosen field of expertise through the challenging tasks it requires. As you explain the vision you have for the project, include the many ways people's lives can be improved by it.

Here are a few more guidelines you can use as you improve your ability to create and articulate your vision:

- Connect each team member's contribution back to the product's vision so they understand why their work is important.
- Stay aligned with your company's vision so your vision is credible. Ideally the vision for your project/product should derive directly from your company's wider purpose: "Our new product XLX is not only going to beat the product from our competitor but it's also going to better position us with new technologies for future success in this strategic market."
- Listen to customers. If you can see how your product will improve your customer's life, you can build it into your vision. Let's say you're working on a software project to halve the time a person has to spend entering data into a terminal. That brings value to the person's employer by reducing time spent, but it might also improve the person's life (not many people enjoy keying in data).
- Know the market and the competition so you can present a credible vision. If you say the new product will bring more value than what is on the market today, you have to know what's on the market before you'll be believed.
- Understand what your team is capable of and be able to articulate how your team will achieve success. This will help give them the confidence they need to stay motivated through the project.

4.4.2 Transformational Leadership Skills—Connection

Connection describes all the ways you build relationships with your team. Character is the most important component of building a connection. Jim Kouzes said: "Clearly, credibility makes a difference, and leaders must take this personally. Loyalty, commitment, energy and productivity depend on it." Dwight Eisenhower said simply: "The supreme quality of leadership is integrity." If you don't have the highest level of integrity, your team will eventually discover it and your connection to them will be weakened. There are other aspects of character that are required of a leader: consistency, courage, resilience, humility, and empathy come to mind. So does followership—Aristotle is credited with saying: "He who has not learned to obey cannot be a good commander."

Let team members operate at their maximum level of responsibility. As Kouzes and Posner put it, team members "…neither perform their best nor stick around for very long if you make them feel weak, dependent, or alienated" [32]. More junior associates need guidance on how to do things as well as on what to do. More senior associates usually do better with guidance on the goals; let them determine the "how." Telling someone the "how" who needs only the "what" is a quick way to become a micromanager, one of the more scornful terms people apply to project managers. A few quotes from three well-known leaders help discourage micromanagement:

- The best executive is the one who has sense enough to pick good men to do what he wants done, and self-restraint enough to keep from meddling with them while they do it.—Theodore Roosevelt
- Never tell people how to do things. Tell them what to do and they will surprise you with their ingenuity.—General George Patton
- Outstanding leaders go out of their way to boost the self-esteem of their personnel. If people believe in themselves, it's amazing what they can accomplish.—Sam Walton

Here are a few more guidelines you can use to improve your connection with your team:

- Stay positive. Be reliable through the normal ups and downs of the project. Celebrate the team's small successes throughout the project.
- Provide freedom where you can—work hours, methods, collaboration, and reporting. As we discussed above, autonomy is one of the most important factors in team member job satisfaction.
- Provide a safe, blame-free environment so members can take risks and be open with one other. Protect the team from unfair criticism and unreasonable expectations.
- Help resolve interpersonal conflicts fairly. Take appropriate action when a team member acts out to maintain a healthy team atmosphere; of course, always conduct sensitive conversations in private.
- Have rich communication in group meetings and in one-on-one conversations. Recognize people when they've done well and give open feedback when things are not going well. Keep feedback balanced and avoid management by exception where the only time the team hears from you is when they've made a mistake.
- Help your team members build their careers. Learn where each wants to go professionally and seek assignments for them in the project that will help them move that direction. Help them get the training and tools they need.
- Seek feedback from your team about your own performance. I often ask my team at the end of a meeting, "on a scale of 1–10, how well do you think we did"—I always learn something. A few times a year I ask a few of my reports to tell me how I'm doing in general. It's some of the most valuable feedback I get.
- Challenge team members—give them the most difficult tasks they can handle. Challenge them, but help them avoid being overwhelmed by being thoughtful about how much you assign them and encouraging them to circle back to you when they discover their schedule is overloaded.

4.4.3 Transactional Leadership Skills—Process

Process is the backbone of any development group. Process moves most of what the team does into standard work. Without process, the team will be forever swimming in the details of the project. Each time process directs the team members in ordinary activities, the efficiency of the team member and the PM improve.

The PM's first responsibility here is to ensure that the team diligently follows the processes the company has specified. Review your team's work regularly to be sure they each meet the intent of the processes in your company. It's too easy for team members to fall into a "check the box" mentality if no one ever looks over their work.

Here are a few more ideas that may help:

- Be organized in every aspect of your work. Keep issue lists from meetings and consistently drive each issue to resolution. Be prepared for meetings and be on time. Be consistent about where information is stored and keep it up to date.
- Manage change within the process. Change management is one of most challenging transactional skills because change comes from so many quarters: customers asking for more features, senior leaders trimming budgets or pulling in schedules, and unexpected technical barriers are a few. Unmanaged change brings chaos.
- Keep process alive—process should improve your project. When parts of process are outdated or inadequate, work with team members to change them. The best processes are constantly adapting to meet the changing demands of the organization.

4.4.4 Transactional Leadership Skills—Policy

Policy describes the procedures organizations use for their people. They include ethics and diversity requirements, annual performance reviews, and company education programs. Here are a few suggestions.

- Ensure team member personal objectives are aligned with their role in the project.
- Ensure team members are able to take advantage of company training programs.
- Take part in the annual reviews of the team members to the extent their supervisors allow. For example, you could provide feedback to their supervisor on performance and behavior ahead of the review.
- Ensure your team is using the standard tools of your company. There are often two issues here: make sure they get the benefits of standard tools like up-to-date licenses and supplier training. On the other hand, once they are enabled, follow up to encourage them to stay in standard tools. Engineers are famous for holding onto outdated tools because they are familiar.

4.4.5 Total Leadership Skills Matrix for Project Management

Each of the leadership skills in this section can be placed inside the Total Leadership Matrix as shown in Figure 4.5. In this figure, skills are divided among the four categories, with those skills thought to be more important for a project manager shown in larger fonts. This creates a "Total Leadership Skills Matrix" for project managers. This is a tool you can use in your daily work. For example, make a copy of Figure 4.5 and mark off which of the skills you're strongest at in green, functioning well at in yellow, and need to improve in red. Or ask a colleague or project team member to name a few you're doing well with (green) and a few that have the most room for improvement (red).

Figure 4.5 is a sampling of the skills for PMs. Reviewing the literature and seeing how different authors have come to different (though similar) conclusions, it seems unlikely that a definitive list exists. Leadership needs change depending on the situation—a PM for a highly skilled team will need different skills compared to a PM for a weaker team. And, of course, PMs with different personalities will rely more heavily on different skills. Each of us will have a unique leadership journey.

There are many leadership styles that work—Eisenhower and Patton certainly lead in very different ways, but they both got results. The Total Leadership Skills Matrix relies on many well-established principles; but don't try to follow them in lock step. If your situation demands a different emphasis, hold to well-established principles, but change the specific skill set list you want to match your situation and your leadership style.

This discussion began with the goal of presenting many leadership skills as a tool set you could use to solve different problems. Then, the focus was placed on inspiring your team by channeling the most common intrinsic motivators.

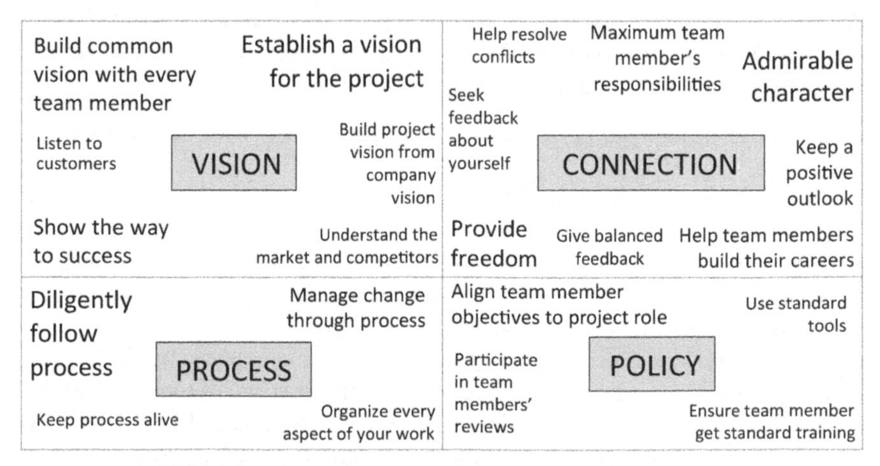

FIGURE 4.5 The Total Leadership Skills Matrix for project managers.

Reviewing Section 4.2.1, we can pull 10 intrinsic motivators that have broad support in the literature:

1. Sense of achievement
2. Recognition and encouragement
3. Pleasure of the work itself
4. Level of responsibility and challenge
5. Personal advancement
6. Belonging to a team
7. Personal growth/mastery of a subject
8. Autonomy
9. Confidence in organizational support
10. Self-efficacy (roughly, confidence in yourself)

So, let's devise a simple test. If the skill set of the Total Leadership Skills Matrix is reflective of the skills the PM needs, we ought to be able to build a table that shows how each skill can be used to help channel each of the intrinsic motivators. Further, that table should show broad coverage of intrinsic motivators, with different skills helping in different ways. Just such a table is shown in Figure 4.6.

Figure 4.6 shows some interesting relationships. For example, Connection helps every intrinsic motivator. That aligns well with so much of modern writing, which focuses on the personal connection. At the other end of the spectrum, Process helps mostly with self-efficacy. Process isn't "inspiring" in the normal sense of the word—people don't get excited about following a list of instructions. But knowing that you work in an organization that can get things done does give confidence that your hard work will be translated into results. Also, displaying admirable character is necessary to help the broadest set of intrinsic motivators—if people don't believe or trust you, there's very little you can do to inspire them. There's a lot of room for interpretating which boxes to check in Figure 4.6 (as there seems to be in every area of leadership study), but taken together you can see a broad relationship between the many ways people are inspired through internal motivators and the many skills available to a PM.

4.5 THE INTERSECTION OF TRANSACTION AND TRANSFORMATION

Unfortunately, it seems two opposing schools of thought have emerged in project management. One favors transaction almost to the exclusion of transformation; the other, perhaps in reaction, espouses transformation and eschews transaction. In many traditional project management books there is an overemphasis on transactional leadership. For example, consider Curlee's comments on the PMBOK® Guide:

- "Understanding the limitations of a project manager's communication skills is often a good first step to becoming…more effective. This…is all but ignored in the PMBOK® guide" [33].

Total Leadership Skills	Key Intrinsic Motivators: Sense of achievement	Recognition/encouragement	Pleasure of the work	Level of responsibility/challenge	Personal advancement	Belonging to a team	Personal growth/mastery	Autonomy	Organizational support	Self efficacy
Vision	☑								☑	☑
Establish a vision for the project	✓									
Tie each person to the vision	✓									
Show the way to success										✓
Understand the market and competitors										✓
Listen to customers										✓
Build project vision from company vision									✓	
Connection	☑	☑	☑	☑	☑	☑	☑	☑	☑	☑
Display admirable character	✓	✓			✓	✓			✓	✓
Maximize team member's responsibilities				✓			✓			
Provide freedom everywhere possible	✓		✓					✓		
Keep a positive outlook										✓
Give balanced feedback		✓		✓						
Help team members build their careers				✓	✓		✓			
Seek feedback about yourself						✓				
Help resolve conflicts						✓				
Provide a safe environment						✓				
Avoid micromanaging	✓		✓					✓		
Process										☑
Diligently follow process										✓
Manage change through process										✓
Organize every aspect of your work										✓
Keep process alive										✓
Policy		☑			☑				☑	
Participate in team members' reviews		✓			✓					
Use standard tools									✓	
Ensure team members get standard training									✓	

FIGURE 4.6 Relating Total Leadership Skills to key intrinsic motivators.

- "Since several studies have supported the idea that transformational leaders are more successful project managers…the guide should offer transformational leadership as an alternative…" [34].

It seems clear that some sources overemphasize transactional leadership. In reaction, many authors have put more emphasis on transformational leadership.

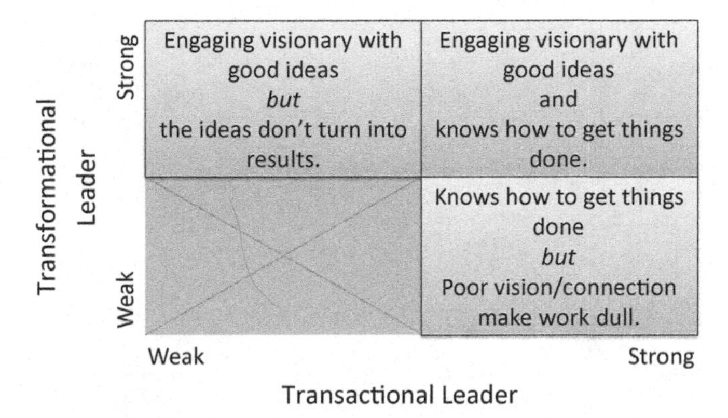

FIGURE 4.7 The intersection of transaction and transformation.

However, in some cases, the pendulum may have swung too far. Rear Admiral Hopper's famous quote "You manage things; you lead people" is perhaps the most well-known quote demonstrating this. Another example comes from an author who suggests adding rigor to the process is backward, that we should relax controls and accept that uncertainty reigns [35]. And "Adaptability is more important than predictability" [36]. And finally, "Traditional project management is past oriented. eXtreme PM is future oriented" [37] (eXtreme PM is a less transactional-oriented management style).

However, as Figure 4.7 shows, being strong in both transaction and transformation make the complete leader. Leaders weak in transaction are unable to guide a team to make a vision into reality. On the other hand, people rarely give their full commitment to leaders who are weak in transformation.

4.5.1 Transaction and Transformation Working Together

Transaction and transformation can actually work together on a single problem. For example, consider the process flow chart in Figure 4.8 where transactional and transformational leadership feed each other to create and sustain improvement.

The first step is identifying an opportunity to improve. This can come from transactional leadership such as measurements in the factory that show an increasing quality problem (1a). In this case, a process in the organization is constructed to detect a problem. Working within the system normally identifies the issue; discipline, process, and attention to detail will usually bring success.

But, you can also identify an opportunity through transformational leadership (1b). These are the places you find by looking outside the process. Are design reviews accomplishing what projects really need them to? Are people confused about what the goals of the projects are? Your intuition fed by your connection to your team guides you in finding these types of opportunities.

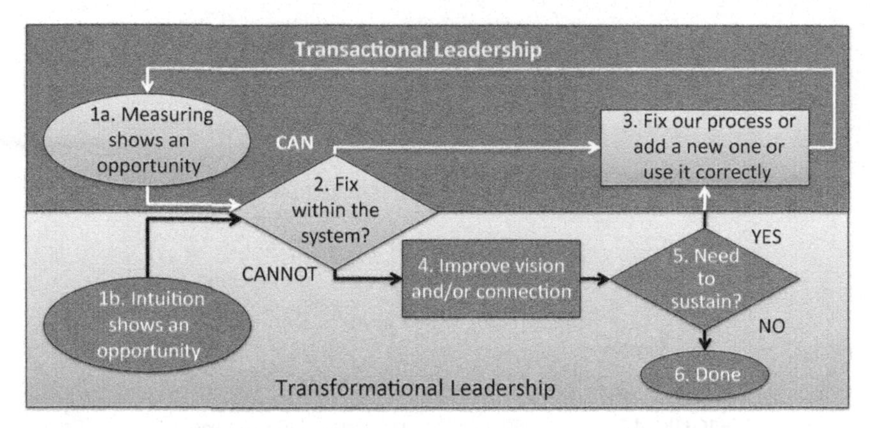

FIGURE 4.8 Improvement using transaction and transformation.

In the next step (2), you must decide whether this problem can be fixed within the system or not. Do you need to better follow the process of how you qualify suppliers to prevent a quality problem from returning? That's transactional leadership (3). Or do you need to sit down with a team member to better understand what personal growth they are looking for in the project so you can help them find tasks that will be more satisfying. That's transformational leadership (4). For those solved with transformation, you need to decide if this type of improvement must be sustained (5)—if so, you'll need to improve transactional leadership too. For the example of the team member above, perhaps you need to create a career growth plan and then schedule a meeting every 2 or 3 months to review progress (3); that's using transaction to sustain transformation. Or you may decide that one conversation and a few follow-up actions are sufficient to resolve this problem (6). Transaction usually adds an enduring burden of measurement and review, so apply it only where it's needed.

So, transactional and transformational leadership work together. Each set of skills helps identify different areas to improve; each gives different ways to bring the improvement about. Yesterday's transformation can be sustained with today's transaction. Today's transaction makes the team more efficient, freeing up time to work in new areas with future transformation.

4.6 COMMUNICATION TOOLS

Communication is a critical skill for PMs. When team members are informed late or misinformed, their performance suffers as does their morale. When stakeholders feel they are out of the loop, dissatisfaction with the PM will increase. At the other end of the spectrum, when the PM creates excessive communications—too many emails and meetings—it wastes time and can make the project seem chaotic. The PM should then strike a balance using the available tools.

In this section, we'll discuss the four basic ways to communicate information about the project to the stakeholders: broadcast, conversation, conference, and recorded collaboration. Each has its advantages and disadvantages, which are

TABLE 4.3 Review of the Four Ways to Communicate

Type	Broadcast	Recorded Collaboration	Conversation	Conference
Common examples	Email, internal blog.	Project schedule, functional specification, marketing plan.	One-on-one discussions, small group meetings.	Team meetings, design reviews, kickoff meetings.
Strengths	Fast and easy way to inform a wide group.	Allows the work of many to be woven together.	Helps PM and team members understand each other.	The most accepted forum to set project direction.
	Easy to maintain a record.	Supports consensus by having a single place of record.	Builds relationships, helps resolve conflicts.	Pulls in everyone needed for the conversation.
	Supports a rich set of media.	Can be tracked over long periods of time.	Works well for most personality types.	Accelerates decision process.
Weaknesses	Poor for presenting complex topics.	Substantial amount of work to build.	Creates different understandings within the team.	Expensive time consuming for many people.
	Unreliable. Don't know if information was received, read, and understood.	Must be sustained over time.	Can be inefficient because consensus has to be rebuilt many times.	Generally dominated by strong personalities.
	Tends to extend conflict, especially argue-with-an-audience emails.	Are not read and understood by all.	Can cause conflict if a team member is consistently left out.	Scheduling conflicts make it difficult to create a quorum quickly.
Most valuable for…	Informing stakeholders about routine events of wide interest.	Improving a process; creating an action plan; defining products.	Issues where the small group is clearly empowered to take on a topic.	Solving tough problems; maintaining team consensus; accelerating decisions.

summarized in Table 4.3. These methods will all be familiar to you; the goal here is to provide analysis for familiar techniques to help provide insights into their use.

4.6.1 Broadcast (Emails and Internal Blogs)

Broadcasting is the fastest way to convey a single message to many people. Modern broadcasting for PMs—mostly emails and blog entries—is easy to generate and can contain a rich set of media: text documents, slide presentations, photographs, video, and links to websites. Both emails and blogs come with a simple record system—it's easy to know when you sent the information and who should have received it.

Broadcasting also has a number of well-known weaknesses. It's poor at presenting complex topics. Most people will invest only so much time in understanding a complicated email or blog entry; if more than the expected mindshare is required, the message will be only partly understood. Broadcasting is unreliable: you don't know if the recipients read and understood the information. The reply threads to broadcast messages can weaken relationships, especially the "argue with an audience" emails, the unfortunate tendency for a large-copy-list email to contain an argument between two people who keep clicking "reply all" when one of them should just call the other.

Broadcasting shows its greatest strength when telling many people about routine events and decisions. Avoid using them when those on the copy list are unlikely to understand after a few minutes of reading; always avoid sensitive subjects—emails are poor at building consensus in the face of discord.

4.6.2 Recorded Collaboration (Project and Product Documents)

Recorded collaboration is the technique of using a document to combine the effort of many people into a single work product. Project schedules and product specifications are examples. In a sense, this is the complement to broadcasting. A single email may generate half a dozen responses that will likely have some contradictions. Only through the hard work of collaboration can the email thread be knit into a single path. This method records that collaboration, typically in a project or product document provided to all stakeholders through a shared storage site.

Recorded collaboration has weaknesses, especially a requirement for significant effort to create and sustain documents. Helping the team adopt new documents is time-consuming as well—perhaps only a fraction of the team will fully comprehend the content at the outset; the PM may have to invest considerable effort to get all team members to understand and use the document. So, apply the technique to those areas that most need it. Otherwise you risk spending energy on a document that sits unused. Examples abound from a SharePoint vacation schedule only half the team uses, to a Design Review document that was never updated after the first meeting, to a competitive analysis document that isn't corrected after it was discovered that a strong alternative technology was omitted.

Minimize the number of documents where possible to make it simple for the team to understand which facets of the project are controlled by formal documents. If you want to add a formal action list for a marketing plan, think about putting the list in the plan rather than creating yet another document. Merging documents works well when multiple documents have the same owner; otherwise, you risk confusion about who will maintain the document over time. Recorded collaboration works best when applied to a limited number of areas where the team must be fully synchronized. Before investing the effort to build the document and train the team, be certain you are prepared to manage the document over its intended lifetime.

4.6.3 Conversation (One-on-One and Small Group Meetings)

Conversation refers to the communication of a handful of people talking about a topic—the group is small so there is no need for a defined leader or an agenda. The goal is to create alignment through shared understanding of an issue among the key people involved. We do it every day—the two electronics experts on the project may get together to work out how to speed up memory access for that new processor. The manufacturing team may get together to walk through a few supplier proposals for tooling. Conversation is the core of teams working together on a problem. It also builds relationships; when colleagues work together to solve a problem, they will usually walk away with more trust in and respect for the others involved.

Conversation has its weaknesses as well. It can create conflicting understandings of a problem, especially when the small group is not in alignment with the rest of the team. The small group may even be seen as illegitimately redirecting the project by the larger team if they are taking actions beyond their responsibility. This can happen when the sponsor pulls two people off to quietly look at a new alternative that might obsolete other team member's work, or when those two people take on that task without anyone's bidding. Other weaknesses include when a small group consistently leaves out a team member who feels they should be included, which will create discord. And it can also be inefficient if consensus must be rebuilt as people are introduced to the results of the small group meetings one by one. Conversation is probably the most difficult communication method to record for future reference.

Conversation is a powerful tool for problem solving when the right people are talking. But problems abound when conversation is used for communication that should take place with the larger team. When a small group that is not seen as legitimately empowered to take on a task, this will cause discord in the team.

4.6.4 Conference (Large-Group Meetings)

Conference is the communication type typically used for larger meetings—say more than three or four people. In conference, there is usually a meeting leader and an agenda with a fixed time. In a project, this communication type is most commonly seen in regular team meetings and team events like design reviews. Team meetings are strong tools for setting project direction—everyone is invited and each person can voice concerns. There is an intuitive legitimacy given to decisions reached in a conference setting. It can produce rapid and solid decisions when all functions of the team are represented.

Conference has its weaknesses, most famously because it is being expensive. A team of 12 people meeting for 90 min is the equivalent of more than 2 people-days. Also, unless the PM is a strong facilitator, large meetings are generally dominated by a few strong personalities. Many people are uncomfortable expressing their opinions in a large group. And large meetings are difficult to schedule and so are often a poor choice when trying to react quickly to a problem.

Conference is another powerful communication tool. Its high costs demand the PM employ it sparingly. Use it to keep the team aligned (regular meetings) and when a tough problem demands the whole team can have input (team events).

4.6.5 Practical Tips for PMs on Communication

The discussion above presented an overview of four methods of communication. This section will use those principles to provide seven practical tips for the PM:

- Watch for email threads that need to be converted to project documents (*broadcast* that should be *recorded collaboration*). If you see long email threads where people are adding suggestions on a direction important for the team, decide if a document is needed. For example, if you see a long thread on how details of a test will be executed, it may be time for a formal test plan stored on a shared folder.
- Ensure your team has the right tools for *conversation* and *conference*. With today's distributed teams, there's no reason to rely wholly on the telephone when talking with remote team members. If your company has video conferencing, use it. If not, use internal chat rooms, conferencing software and web cams. Ensure your team has high-quality audio equipment or at least avoid inexpensive speaker phones—this can help immensely with team members who speak English as a second language.
- Get people talking when you sense tension. Stop "argue-with-an-audience emails"; set up a meeting or conference call to get team members talking.
- Create shared folders for team documents. Ensure everyone has access. Pull important documents up frequently in team meetings to speed the adoption process within the team.
- When facilitating team meetings, help everyone participate. Many people feel more comfortable giving their opinion when asked, so ask for it. And be prepared to reign in those to contribute so often that others cannot express their thoughts. React decisively against ridicule expressed in your meetings.

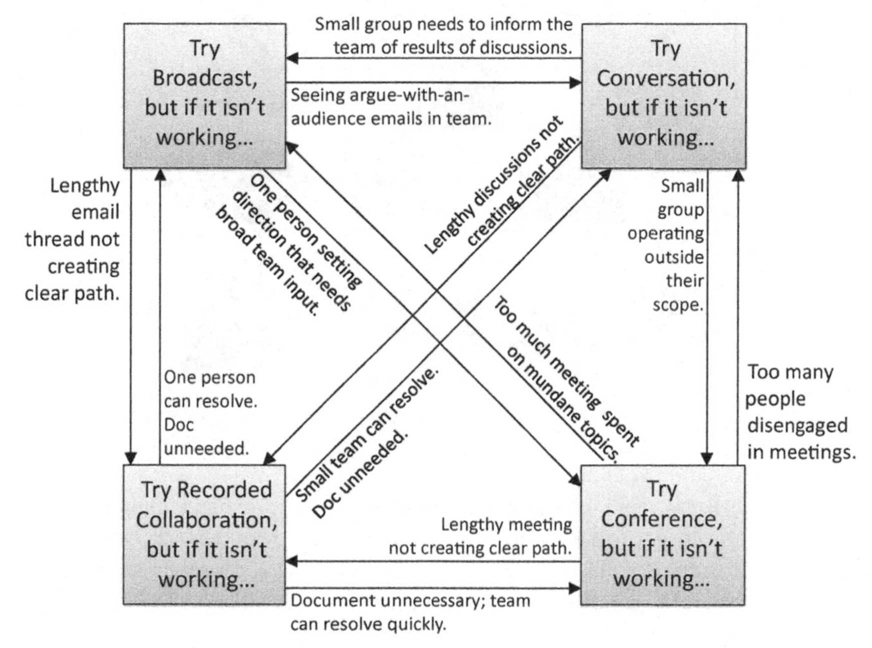

FIGURE 4.9 Examples of when to change communication types.

- Poll your team to get their thoughts on whether there are too many or not enough team meetings. Don't be surprised if there's a mild bias toward "too many" (meetings are commonly undervalued by technical people). However, if a large number of team members feel significant time is being wasted in too frequent or too long meetings, make adjustments.
- Finally, when you are using one communication method and it doesn't seem to be working, consider which of the other three might be a better fit. Figure 4.9 gives 12 examples that may help make the process clearer.

4.7 TEAM DISPERSION

In the final section of this chapter we discuss some of the challenges for project managers depending on the degree to which the team is dispersed. We'll look at four ways teams are commonly dispersed; each brings its own set of challenges.

4.7.1 The Traditional Team

The traditional project team is a cross-functional group with most people located at one site. Typically the team will be seated in their functional groups, spread across the building or building complex (Figure 4.10). In many cases, a few people will be remote, perhaps working from home or at a different site within the company.

FIGURE 4.10 The traditional team.

The traditional team has many advantages for the project manager. It's relatively easy to get most of the team in a meeting room, so the team gets a lot of face time. People who meet more understand each other's problems better and generally get along better. But there are a few things the PM has to watch out for:

- Are team members focused on the project? Since people are seated in their functional groups, they can get pulled into issues that compete with the project.
- Does the team communicate well? Keep an eye on email traffic. When people are walled off, it's easy to send an email when face-to-face would be more efficient. PMs often need to step in when there are long email trails with issues remaining unresolved.
- Are the views in the main site overrepresented? For example, if you're working on a product for distribution across the Americas, but the only site working on the project is in Colorado, how much will the team "project" the Colorado views onto the entire market? Customers in New York, Sao Paulo, and Vancouver will likely value the same product in different ways. Customer face time helps, so get your team out of the building when possible.

4.7.2 The Global Team

The global team is the natural extension of the traditional team for global organizations (Figure 4.11). Here there may be two or three groups working out of different sites spread across the country or around the world. Again one or two people are likely to be remote from any site. Within the sites, people normally sit in their functional areas. One of the largest advantages of the global team is the diversity of thought many sites bring—if the project has one team in Shanghai, one in Pittsburgh, and another in London, you increase the likelihood that your product will be appropriate for worldwide distribution.

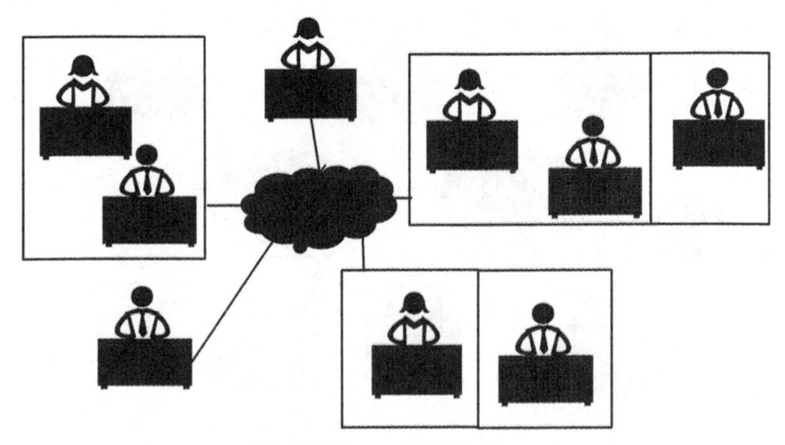

FIGURE 4.11 The global team.

The global team brings more complexity for the PM. Here are a few things to look for:

- Is the team focused? As with the traditional team, people are still likely to get distracted, but the global PM has much less visibility into what's happening at the other sites. He may find out part of the team is not working on the project only after a milestone is missed. The PM should visit the sites working on the project on a regular basis.
- Communication is a larger issue with the global team as well. In addition to the issues the traditional team dealt with, there may be significant cultural, language, and time-zone differences. A team in Seattle doesn't share a single working hour with a team in Rome, so people rely more on email and less on face time. Here the team likely won't understand each other's problems as well and this can lead to more conflict within the team. Using conferencing software and web cams during calls gives everyone a more human feel—it's an inexpensive way to improve the conversation. Finally, pull the team together when possible, especially at critical points of the project—kick-off meetings, software integration, and production pilot runs, for example.
- Does the PM favor the site she works out of? It's so easy to favor your home site, you might not be aware you're doing it. However, people in other sites will see it. Again, travel to all the sites often to reduce the fact and the perception of favoritism. And share the "time-zone pain" so the same team isn't always having to join at 6:00AM or stay at work until 8:00PM.
- Be culturally sensitive. If you treat everyone as you would treat people in your region of the world, you're likely to get poor results.

4.7.3 The Virtual Team

The virtual team is a relatively new concept for product development (Figure 4.12). Here, the team can be distributed anywhere—in the most extreme case, no two

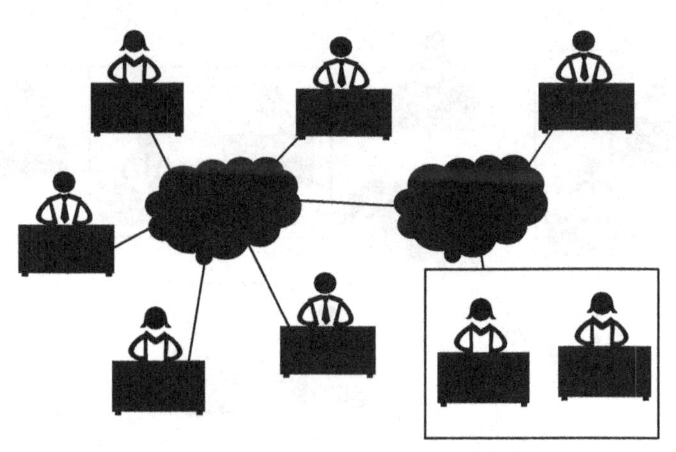

FIGURE 4.12 The virtual team.

team members sit in the same building. The virtual team brings many advantages especially in its ability to allow you to choose employees from anywhere in the world. It also allows you to attract people that don't want to work in an office building. For communication, you'll rely heavily on internal chat rooms, video, web-based screen sharing, and telecons; fortunately these tools are becoming better and less expensive over time. Some studies show an improvement in team efficiency for virtual teams [38].

For all its advantages, the virtual team probably puts the most pressure on the PM. The PM likely cannot visit each person often; the team may be able to meet only rarely. The virtual team can have more issues with communication because there is no face time, perhaps for months at a time. Here are a few other items to consider:

- Concerns about team member focus may be increased because the PM is not able to work closely with team members on a regular basis. On the other hand, this effect can be offset since virtual team members are not sitting in functional groups where they are easily distracted by competing daily work.
- Virtual teams rely on technology for communication more than any other structure. So be sure you've invested in the right tools. Get people headsets or high-quality speaker phones, and at least use web cam video. Ensure the whole team has high-speed internet and invest in internal chat rooms. Get the team together in one place when you can—there's no replacement for sharing a meal or working side by side to solve something together.
- How will team members grow technically? Working in virtual teams means people will spend most of their day alone. Are the members technically competent? If not, they may get stuck on problems or stay hung up with poor work habits. Mentoring and coaching are much harder on the end of telephone line. And it can take a lot longer for the PM to recognize a problem in a virtual team structure, perhaps after plenty of damage has already been done.

FIGURE 4.13 The big (Obeya) room.

- Are your meetings effective? People can easily tune out during long telecons, so the PM must work harder to be organized—have a strong agenda [39] and stay focused. Don't let one or two outspoken people dominate the conversations. And use web cams to find out who's listening carefully and who's working on email.
- Are you accessible? Technology will have to replace the impromptu meeting. So "Skype, tweet, email, and meet in person when necessary" [40]. PMs must work hard in virtual teams to be available when needed.
- As with the global team, share the time-zone pain and be culturally sensitive.

4.7.4 The Big (Obeya) Room

One of the innovations in modern project management feels like something from the past: put the whole team—design, manufacturing, sourcing, and marketing—in a single room (Figure 4.13). Pull the information out of the computers and tape it on the wall. Felt markers and sticky notes take over much of what the Excel® sheet does in a traditional project.

The Obeya (sometimes Oobeya or Obeye) room solves communication problems by putting people close together and displaying what's important for all to see. The method was made famous by the Toyota Prius development and is gaining popularity in the West. A number of well-known companies have published substantial improvements when using the Obeya room. The Obeya room is new for general product development. We'll talk more about it more in Chapter 7, Lean Project Development.

4.8 RECOMMENDED READING

Cloud H. Boundaries for leaders: results, relationships, and being ridiculously in charge. Harper–Collins Publishers; 2013.

Herzberg F. One more time: how do you motivate employees? Harvard Business Review; September/October 1987 (originally published in 1968).

Kouzes, Posner. The leadership challenge: how to make extraordinary things happen in organizations. Jossey Bass; 2012.

Pink DH, Drive: the surprising truth about what motivates us. Riverhead Books; 2009. p. 18–21.

4.9 REFERENCES

[1] Kruse K. 100 best quotes on leadership. Forbes.com; October 16, 2012. http://www.forbes.com/sites/kevinkruse/2012/10/16/quotes-on-leadership/.

[2] Bennis W, Biederman PW. The essential Bennis. Josey Bass; 2009. p. 443.

[3] Kouzes, Posner. The leadership challenge: how to make extraordinary things happen in organizations; 2012.

[4] Lewis JP. Project planning, scheduling, and control. 3rd ed. McGraw-Hill; 2001. p. 363.

[5] Ryan R, Deci EL. Intrinsic and extrinsic motivations: classic definitions and new directions. Contemp Educ Psychol 2000;25(1):54–67. http://dx.doi.org/10.1006/ceps.1999.1020.

[6] Wigfield A, Guthrie JT, Tonks S, Perencevich KC. Children's motivation for reading: domain specificity and instructional influences. J Educ Res 2004;97:299–309.

[7] Herzberg F, Mausner B, Snyderman BB. The motivation to work. 2nd ed. New York: John Wiley; 1959. ISBN: 0471373893.

[8] Herzberg F. One more time: how do you motivate employees? Harvard Business Review; September/October 1987 (originally published in 1968).

[9] Hackman JR, Oldham GR. Motivation through the design of work: test of a theory. Organ Behav Hum Perform August 1976;16(2):250–79. http://dx.doi.org/10.1016/0030-5073(76)90016-7.

[10] Herzberg F. Jumping for jelly beans. The Business Collection, BBC; 1973. Currently available at: https://www.youtube.com/watch?v=o87s-2YtG4Y.

[11] Amabile TM. How to kill creativity. Harvard Business Review; September–October 1998.

[12] Rosenau MD, Githens GD. Successful project management: a step-by-step approach with practical examples. John Wiley and Sons, Inc.; 2005. p. 147f.

[13] How great leaders drive results through employee engagement. The Forum Corporation. Currently available at: http://www.forum.com/_assets/download/6cfc6745-ae45-4f46-962c-435894958ac5.pdf.

[14] Pink DH. Drive: the surprising truth about what motivates us. Riverhead Books; 2009. p. 18–21.

[15] Pink DH. Drive: the surprising truth about what motivates us. Riverhead Books; 2009. p. 34.

[16] http://dondodge.typepad.com/the_next_big_thing/2008/11/50-of-us-engineering-students-dropout--why.html.

[17] http://usmilitary.about.com/od/joiningthemilitary/l/blbasicattrit.htm.

[18] Seligman M. The new era of positive psychology. Ted Talks; February 2004. http://www.ted.com/talks/martin_seligman_on_the_state_of_psychology.

[19] 3M vision, objectives & strategies, http://solutions.3m.com/wps/portal/3M/en_WW/Corp/Identity/Strategies-Policies/Vision/.

[20] GE mission statement. 2013. Currently available at: http://www.strategicmanagementinsight.com/mission-statements/general-electric-mission-statement.html.

[21] Pink DH. Drive: the surprising truth about what motivates us. Riverhead Books; 2009. p. 21.

[22] Amabile TM. How to kill creativity. Harvard Business Review; September–October 1998. p. 110.

[23] Amabile TM. How to kill creativity. Harvard Business Review; September–October 1998. p. 117.
[24] Amabile TM. How to kill creativity. Harvard Business Review; September–October 1998. p. 121.
[25] Pink DH. Drive: the surprising truth about what motivates us. Riverhead Books; 2009. p. 41.
[26] http://en.wikipedia.org/wiki/Candle_problem.
[27] The Forum Corporation.
[28] DeCarlo D. eXtreme project management. Jossey-Bass; 2004. p. 94.
[29] Pozin I. Top 10 reasons people hate their jobs. Small Business, Huffington Post; July 11, 2013. Currently available at: http://www.huffingtonpost.com/2013/07/11/why-people-hate-jobs_n_3579873.html.
[30] Curlee W, Gordon RL. Complexity theory and project management. John Wiley and Sons; 2011. p. 38.
[31] McManus J. Leadership: project and human capital management. Elsevier; 2006. p. 5.
[32] Kouzes, Posner. The leadership challenge: how to make extraordinary things happen in organizations; 2012. p. 22.
[33] Curlee W, Gordon RL. Complexity theory and project management. John Wiley and Sons; 2011. p. 26.
[34] Curlee W, Gordon RL. Complexity theory and project management. John Wiley and Sons; 2011. p. 37.
[35] DeCarlo D. eXtreme project management. Jossey-Bass; 2004. p. 21.
[36] DeCarlo D. eXtreme project management. Jossey-Bass; 2004. p. 16.
[37] DeCarlo D. eXtreme project management. Jossey-Bass; 2004. p. 7.
[38] Goncalves M. Managing virtual projects. McGraw-Hill; 2005. p. 23.
[39] Rosenau MD, Githens GD. Successful project management: a step-by-step approach with practical examples. John Wiley and Sons, Inc.; 2005. p. 199.
[40] Shapiro A. How do Deal when employees are scattered across the world. The Fast Company. Currently available at: http://www.fastcompany.com/3030308/work-smart/5-ways-to-bridge-the-physical-gaps-between-your-international-team.

Chapter 5

Phase–Gate: Extending the Critical Path Method

In this chapter, we'll discuss Phase–Gate project management as an extension of the critical path method (CPM). Together they form what is almost certainly the most popular method for developing hardware products in mature organizations. While Phase–Gate is most commonly used to improve CPM, it can be used with other methods such as critical chain project management (Chapter 6) and lean product development (Chapter 7).

The chapter begins with a detailed description of the method. Then an example phase is created to demonstrate key principles. There is a strong focus on schedule delay in this chapter in large part because projects using this method often struggle with the issue. Three topics related to schedule will be explored: monetizing the need to avoid delay, the factors that commonly cause delay, and various ways to mitigate those factors. The chapter concludes with an analysis of the strengths and weaknesses of the method and where it best fits.

5.1 OVERVIEW OF PHASE–GATE PROJECT MANAGEMENT

The Phase–Gate method deals with the high complexity common in product development that can overwhelm CPM, which was the topic of Chapters 2 and 3. CPM works well on projects of low complexity. But most product development projects require so many tasks that the work breakdown structure (WBS) can become hundreds or thousands of lines long; a 100-line WBS may work well for daily management, but how do you sift through such a list to help decide larger issues? Is the project running on time? Are more resources needed? Is there sufficient customer feedback for current state? Phase–Gate project management augments traditional project management by providing standard work for a higher level view.

For daily management, Phase–Gate projects are similar to CPM: a WBS of cascaded tasks is created with the familiar features: a task owner, predecessors, a due date, and so on. But at a higher level, Phase–Gate divides each project into large subprojects called *stages* or *phases*. These phases are separated by approval steps called *milestones*, *tollgates*, or simply *gates*. The team must complete predefined requirements of a phase to pass a management review at that phase's milestone; only then should the team proceed to the next phase.

Project Management in Product Development. http://dx.doi.org/10.1016/B978-0-12-802322-8.00005-X

The phases are typically defined in the organization's processes to be the same for every project. Further, the main activities and work products of each phase can also be defined and so can be common for all projects as well.
There are four main advantages of Phase–Gate project management:

- Common activities for all projects.
  Every project must traverse the same phases and, within the phases, perform the same major tasks. There is a common understanding of a "product test plan" that might be created in Phase 2 and executed in Phase 3. Over time, a common lexicon is created. There is a sense of order brought about by all stakeholders having a shared understanding of the intent and details of each phase.
- Standardizing approval processes.
  The approval processes for projects can be chaotic when managers pick out details to inspect without a clear method. The results vary greatly depending on who is reviewing so PMs don't know how to prepare. Phase–Gate brings standard work to the approval process—everyone understands most of the high-level questions that will be asked. This allows PMs to prepare better and the approval process to provide more consistent results.
- Opportunities for continuous improvement.
  Phase–Gate project management provides many opportunities for continuous improvement. Deliverables from a phase can be added or improved. A senior manager might say: "The sourcing team says they are always surprised by needs of projects because they get involved too late—let's add 'create sourcing plan' in Phase 2." Or: "Our Phase 3 design reviews don't seem to pick up manufacturability issues very well—let's add a requirement to explicitly review these issues in the design review."
- Provide a common framework for portfolio management.
  Phase–Gate project management creates a common structure that all projects can execute. This framework can be used to understand and compare all projects in the portfolio. Metrics can be derived for each project to support analysis—days late to Milestone 2, all action items complete from Phase 3 design review, and the like. Phase–Gate is the beginning of a portfolio management system for many companies.

Phase–Gate addresses what Kendall and Austin refer to as the universal problems in creating a work breakdown structure: the lack of formal process and the lack of review [1]. Defining the common activities for all projects provides a template for the project manager (PM) to lead the team in planning the next phase of a project. Phase–Gate specifies reviews at many levels, from steering committee reviews to design reviews and *failure modes and effects analysis* (FMEA).

Phase–Gate often removes the need for a project team to build a WBS at all. A phase definition provided by the organization defines the overall direction of the project while a simple task list can be used to manage daily work. This can be positive because complex Gantt charts created for a product development project often prove ineffective—they may be built early in the project and never revisited. Most sponsors lack the time and patience to understand the implications of individual tasks; even the team members may not understand the WBS. Beyond this, the ordinary variation in project execution can change the WBS, perhaps as often as every few weeks, and PMs usually fail to keep the document up to date.

5.1.1 Phase–Gate Issues

There are known issues with the Phase–Gate method. One of the largest reported problems is that projects using it often overrun their schedules and budgets. The delay experienced in Phase–Gate is normally one of the top two or three reasons people advocate alternative methods such as critical chain project management (Chapter 6), lean product development (Chapter 7), and Agile (Chapter 8). We'll discuss schedule delay in detail in Section 5.3.

Phase–Gate is often applied to projects where it doesn't fit. It's typically a heavy process and can drag down simpler projects. Moreover, like an income tax code, it can become heavier over time. It's the nature of organizations that requirements are easier to add than to remove. If the organization is suffering from a new problem (say, pirated parts found in development testing), a task can be added to the standard process (document the sources of key test components); going forward, every project will have to meet that new requirement. When a new problem comes into view, it's easy to motivate action. But 2 years later, when the sourcing team has greatly improved their ability to detect piracy, that step can become unnecessary. Unfortunately, at that point there is likely to be little enthusiasm for removing the requirement. Partly it brings personal risk—no one wants to be the champion for removing a process that prevented a problem if the problem returns later. Partly it's just a lot of work—taking a step out of a company-wide Phase–Gate process requires consensus from many people. So, over time, many requirements are added but few are removed; step by step, the process brings more overhead. The Institute of Electrical and Electronics Engineers (IEEE) Benchmark team lists numerous issues with processes used in industry related to their heavy weight [2]. They refer to development processes that have "gone too far," adding to checklists that are already too long and using overly complex tools. As they put it, "'Time to Market' aspects and project and process efficiency are often not in focus."

Another issue is that Phase–Gate processes are designed for products that have a "high cost of iteration." For example, when designing a product that will require $500k of capital equipment, it's necessary to validate the specifications of that equipment before placing the order. Also, if a weak test plan misses an issue that can only be resolved by changing the product design and that change affected the capital equipment, it could bring expensive, time-consuming rework. Hardware projects typically have a high cost of iteration: supplier tooling, capital equipment, and regulatory certification can all be expensive to change late in the project. By contrast, software projects often have a low cost of iteration. New versions of software can be generated in days or weeks. Thus, Agile methods are often better choices for software projects than Phase–Gate.

5.1.2 Stages or Phases

A stage or phase is a defined by a collection of deliverables required to bring a project to the next level of approval. Deliverables are measurable results of progress; for example, a phase would likely have "complete design review" (a specific event) rather than "complete design" (a vague result). Organizations customize Phase–Gate processes to meet their needs. So, there is variation, but because most Phase–Gate product development processes are driven by similar concerns, there is a lot of commonality.

There are typically about six phases beginning with an initiation and ending with product launch (see Figure 5.1). Phases are usually divided up to maximize the value of the approval that occurs at the end of the phase. For example, one of the most important approvals comes at "Design readiness" because this triggers the process to start large investments: commitments to supplier tooling, purchasing capital equipment, and designing assembly lines. The expenses here are usually high. The project team must show diligence in design reviews, testing, and risk analysis. If major defects in the design are discovered later, it will probably generate expensive rework. Each phase is designed to end in a milestone that minimizes risk before committing to the next level of investment.

5.1.3 Sample Phase–Gate Process

The following section builds a sample six-phase, Phase–Gate process. Each phase ends with an approval step called a milestone (MS). So, Phase 1 ends with MS1, Phase 2 with MS2, and so on.

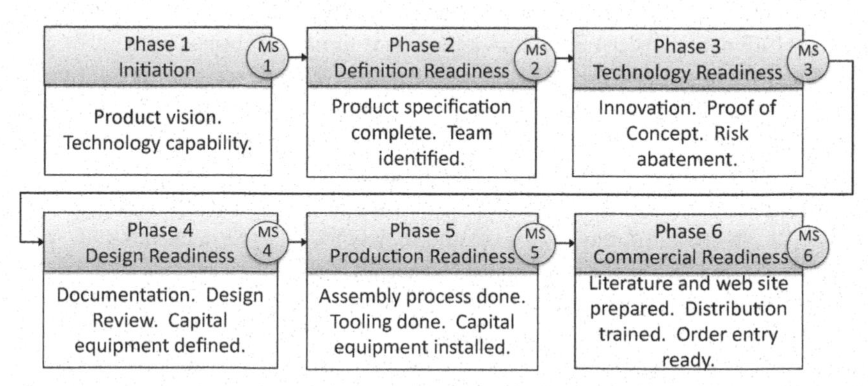

FIGURE 5.1 Typical Phase–Gate process.

1. Initiation
   A small team—often just two or three people—present a vision of a new product. Initiation requires little investment and there is often no formal approval process to start it. A handful of knowledgeable people with passion may be all that's needed.
   For a general market product, the team will address issues such as:
   a. How does the product fulfill needs of our existing markets or the new markets we want to serve?
   b. How are these needs being filled currently? Are there direct competitors or are these needs being met with an inferior technology?
   c. Do we have access to the markets this product is designed for? If not, can we get it?
   d. Do we have the core competences to design, produce, and sell this product?
   e. What advantages will we have over the competition? How long can they be sustained?
   f. What are the major technical and commercial risks?
   ✓ Milestone 1 (MS1) is granted when the leadership believes there is enough merit to fully define the project. Because the next phase also requires a modest investment, the bar for MS1 may be low—if the product is within the company's ability to produce and it lines up with strategic direction, approval is likely.
2. Definition Readiness
   The Definition Readiness phase usually requires a small team: two to four people working part time with most of the focus on product function and market needs. Issues addressed might include:
   a. Completing the functional specification defining product form, fit, and function
   b. Creating price and cost targets
   c. Defining intended markets and projected sales
   d. Specifying targets for project schedule, budget, and required skill set
   ✓ MS2 is granted when:
      - The product is well enough defined to move to Phase 3.
      - The product as defined is attractive: the target markets are right, the product seems to bring value to the customer, and we think we have the ability to be successful.

   While MS2 doesn't gate the largest investment (usually that's MS4), it does approve the first significant investment in the project. So, it's the point the company decides to give the project mindshare and, as a result, MS2 can be a difficult milestone to pass.
3. Technology Readiness
   The Technology Readiness phase brings a full project team together. The major steps are:
   a. Create a design specification (the "how") to accompany the functional specification (the "what").

b. Validate the technology. Build alpha units to prove critical technology. Test early units with customers.
c. Develop a prototype/test plan.
d. Develop an IP strategy.
e. Develop initial manufacturing, sourcing, and regulatory certification plans.
f. Create plans to coordinate with related projects (say, the firmware and hardware development for one product, or compatibility of a printer and its ink cartridges).
g. Validate the product value proposition with a sample of customers.
h. Develop a list of potential customers and their likely purchases.
i. Provide a detailed product–cost analysis.
j. Provide a detailed project schedule, budget, and skill set requirements.
✓ MS3 is granted when:
- The functional specification is complete and appears achievable. The design specification fulfills the requirements of the functional specification.
- The technology development is complete.
- The IP plan is complete.
- The market is clearly defined.
- A sample of customers has given sufficiently positive feedback.
- The financial analysis shows an acceptable profitability.
- The project plan is complete for the remainder of the project.
- The risk identification process has been diligent and all known risks are at an acceptable level.

4. Design Readiness
The Design Readiness phase completes the documentation for the design and the development of manufacturing processes. The major steps include:
a. Complete the product design, drawings, and bills of material.
b. Complete reduced-functionality software and firmware to allow rigorous testing of the hardware.
c. Pass a formal design review.
d. Perform FMEA [3] on the design and on the assembly processes.
e. Build beta units and test against the functional specification.
f. Design the manufacturing processes.
g. Design the production equipment and tooling.
h. Qualify all new suppliers.
i. Obtain potential customer feedback using beta units.
j. Update functional specification, risk analysis, and financial analysis according to information learned during the phase.
✓ MS4 is typically granted when:
- The design has been sufficiently validated through design reviews, FMEAs, testing, and customer feedback.
- The needed capital equipment is fully defined and purchase requisitions are ready for approval and within budget.

- The team has been diligent in identifying risks and all known risks at an acceptable level.

It's common for MS4 to approve the largest single investment because of the need for capital equipment and extensive testing and certification tasks that will take place in the next phase.

5. Production Readiness

The Production Readiness phase completes all steps necessary to show the product can be produced with acceptable quality levels and at acceptable financial margins. The major steps carried out during this phase typically include:

a. All bills of material released into the materials requirement planning system. Release of all component drawings, material/chemical specifications, and software.
b. Create and validate the assembly line(s).
c. Test a wide sample of preproduction units.
d. Qualify all supplier parts.
e. Complete production planning to support forecasted production ramp.
f. Train the factory staff including direct labor, line supervision, maintenance, and procurement.
g. Preproduction units shipped to customers for feedback.
h. Update functional specification, risk analysis, and financial analysis.

✓ MS5 is typically granted when:
- A sufficiently large sample of products have been produced and tested successfully against the functional requirements.
- Manufacturing, quality, and sourcing have accepted responsibility for producing the product to the forecasted schedule and met target costs.
- Customer satisfaction is at a high level of acceptance.
- All known risks are at an acceptable level.

6. Commercial Readiness

The Commercial Readiness phase contains tasks to make the product ready to go to market. It includes the ability to accept orders, the completion of all marketing literature, and creation of a sales plan. Major steps carried out during this phase include:

a. Website/customer service ready to accept orders.
b. All customer documentation complete: manuals, application notes, selection guides.
c. All regulatory certification is complete.
d. Advertisements are scheduled.
e. Sales information such as price lists and competitive comparisons are available.
f. Sales team is trained and has been provided with necessary demonstration equipment.
g. Technical support team is trained and has been provided with necessary equipment.

h. Customer training program ready.
i. Updated risk and financial analysis.
✓ MS6 is typically granted when:
  - The product is ready for sale.
  - All known risks are at an acceptable level.

The Phase–Gate process presented here is just a sample. Companies normally create a process fine-tuned for their products, markets, and organizational culture. The process here has close to 100 separate requirements. Bear in mind, each of these steps should be further defined. What does "Qualify all supplier parts" mean? The process to qualify parts must be documented to provide meaning. And the process shown here is simple by modern standards—it's easy to imagine double or triple the number of steps. All this is simply too much to manage in a single WBS. Phase–Gate project management breaks down the requirements into understandable subprocesses each with a clear set of approval criteria. It makes it possible for every team member to understand the goals of the current phase and at the same time it brings uniformity necessary to build project management expertise through the organization.

Phase–Gate project management also works at the management level. Managers almost never have time to understand the technical details of each project. It is usually not their area of expertise and even if it were, there isn't enough time in the day for senior managers to be intimately familiar with the technical details of every project in their range of responsibility. Phase–Gate drives the review to a higher level—a manager will want to know that a thorough design review was carried out, that the right people were present, that minutes were taken, and that all identified issues are dealt with. However, she is unlikely to want to know many of the specific details that were covered. Having strong process allows the management team to evaluate the project without being technical experts.

5.1.4 Managing the Investment

A Phase–Gate investment approval process is shown in Figure 5.2. This illustrates the actual use of resources by phase in a solid line. It also shows the amount approved in each milestone with a dotted line. You can see that each milestone approves the next increment of investment with the large increments coming after MS2 (for R&D resources), MS3 (for design, test, and manufacturing engineering), and MS4 (for factory equipment and supplier investment). Although the financial investment in Phase 6 is often small, MS5 probably brings the largest commitment from the company because at this point, the product is released to the market. Unlike earlier phases, it's difficult to stop after granting MS6. It's not possible to "unrelease" a product without causing damage—a loss of confidence of the customer base, the distribution channel, and the sales team.

Look over Figure 5.2. Notice that in the first two phases, the investment is small. Changing direction is inexpensive. After MS2, the investment is growing and the cost of change is steadily increasing. It's important that each phase is completed

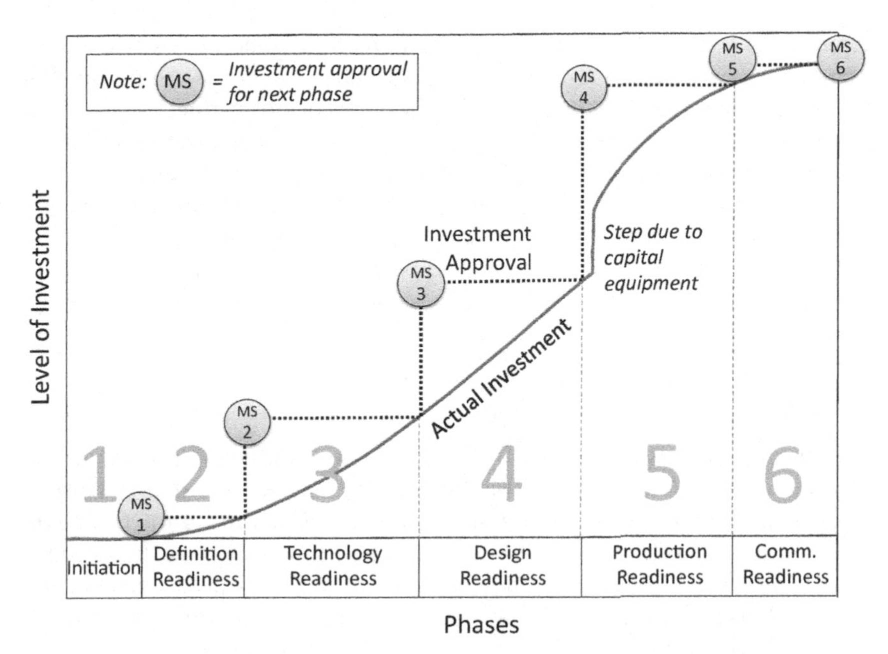

FIGURE 5.2 Investment versus project life: actual use versus approval.

diligently because each becomes the foundation for the next. Notice also that the milestones are the points where the organization commits to the next phase. MS2 represents a general approval for the spend for Phase 3. The diligence of the review presentation should align with the size of the investment being requested.

One question that comes up often is: "While finishing up requirements for a milestone review, is it desirable for the team to get started on the next phase?" After all, isn't it good to keep making progress while waiting for approval? The answer is generally "no" because activities in the next phase often inadvertently create commitments on behalf of the company. For example, let's say in Phase 4 that units are sold to an industrial customer when the process declares units will not be sold until MS5. The customer then starts using the prototypes to design their products. Months later, the review of MS4 reveals the product cannot be produced reliably as designed; so several significant changes must be made. After the changes, the product doesn't work as a component in the customer's products. In such cases the customer is likely to be highly dissatisfied because they have invested in a product that cannot be purchased and so they have to rework their design. This is just one example—there are many commitments that are made as a matter of course: commitments to and from operations, to the sales force, to the distribution channel, to suppliers, and to customers.

None of this means the team cannot start on the earliest steps a few weeks ahead of an approval meeting. However, doing a significant portion of work beyond an unapproved milestone can bring serious problems. I can recall cases where the project team worked for months on later phases without obtaining approval for earlier milestones. By the time the milestone review was done, many of the commitments reserved for management review were essentially made and some turned out to be unwise. This practice of getting far ahead of unapproved milestones robs the value of the approval process, which is primarily to manage risk. It also can break trust with the management team.

Phase–Gate is a powerful technique for managing risk and investment. According to Tom Agan, it is one of three key best practices in product development: "There is an optimal number and proper usage of stage-gates[1]... If implemented correctly, these stage-gates can increase revenue from new products by more than 130%" [4].

5.1.5 Concurrent Engineering

Some years ago it was common for organizations to structure projects to be executed one discipline at a time in a process now called *serial engineering*. Product marketing would create a specification and then hand off to design engineering, who would develop and design a new product to meet the specification. Design would then hand off to operations to build the production system that could make the product. The approach was fraught with problems, two of which were dominant:

1. The various disciplines required to develop a project must start early in a project, one example being that operations planning for new capabilities must start near the beginning of the project to avoid delaying the launch. With serial engineering, departments often saw the project only after they received the handoff.
2. The functions of the disciplines, which should be highly interrelated, are isolated. Marketing must work hand-in-glove with design and operations to ensure that the organization can deliver the product marketing defines. Design must work just as closely with operations to ensure the new product falls within the capability of the organization or that new capabilities can be created within the duration of the project.

The widely accepted alternative to serial engineering is *concurrent engineering* (CE): cross-functional teams working together for the full length of the project. In CE, the concept of handoff becomes irrelevant—each discipline works from start to finish on areas critical to the success of the product. CE, supported by the cross-functional project team, has largely replaced serial engineering (see Figure 1.2 in Chapter 1) and this text has assumed that model. The Phase–Gate process that is the topic of this chapter is usually structured to prescribe CE by defining deliverables from the various disciplines throughout the project. This ensures early involvement from all disciplines.

---

1. Stage-Gate® is a registered trademark of Stage Gate, Inc.

Set-Based Concurrent Engineering

Set-based concurrent engineering (SBCE) is an extension of CE. Some authors describe SBCE as a lean product development (LPD, the topic of Chapter 7) perhaps because it is most famously used at Toyota, the company that defined lean production processes. It can also be described as an alternative to lean because it does not borrow directly from lean production processes the way the techniques in Section 7.3 do.

SBCE seeks to delay critical design decisions as long as possible by developing multiple solutions more fully than traditional project management techniques. Virtually all formal product development processes recommend developing multiple solutions until there is enough data to make a final solution. However, in practice, many teams converge to the solution very early in the project. Other solutions may be considered, but if they are, it is usually with less rigor than the evaluation of the favored solution. Perhaps a proof-of-concept of a secondary solution might be built, but often the competing solutions are ruled out with paper studies. In many cases, it is the organization's process system itself that drives this behavior by creating critical milestones that portray the selection of the solution as progress without ensuring many solutions were evaluated thoroughly. In other words, the process often rewards the selection of a solution with little emphasis placed on diligent evaluation of competing solutions.

SBCE uses process to require thorough evaluation of a *solution set*. The team cannot simply provide the favored solution but must execute the process that first creates a broad set of solutions and narrows the set one by one until the strongest design emerges [5–7]. In fact, communication in the development phase is limited to discussion about the solution space and not any single solution. SBCE teaches that delaying the final solution and expending greater-than-normal effort on many solutions results in acceleration later in the project.

The process for SBCE is:

1. Use formal design guidelines to map the design space. A rich solution set contains solutions with a wide range of competing techniques. After the process starts, this set is practically fixed—the goal is that no new solutions appear part way through the process.
2. Gradually narrow the solution space eliminating weak solutions using objective data and imposing the minimum constraints for the selection.
3. Proceed until the set is narrowed to a single solution. This can happen quite late in the process, well after most project management methods require a design freeze.

CE seems to be dominant in industry today; it's unclear how widely SBCE has been adopted. There are case studies that attest to its efficacy [8] but it's difficult to find the breadth of glowing testimonials generated by other alternative project management methods such as critical chain project management (CCPM), LPD, and Agile methods. Many of the papers recommending the topic focus on its use at Toyota leaving questions about the breadth of products and company cultures in which the method can be effective. Still, the approach is so logical it's compelling. Perhaps enough experience in industry will accumulate so the strengths and weaknesses of the approach will be clearer in the coming years.

5.2 CREATING A WBS FOR A PHASE

In a Phase–Gate process, the major steps in each phase are defined by standard process. For example, in Technology Readiness, typical steps include:

1. MS2
2. Evaluate market need
3. Analyze competitors
4. Complete functional specification
5. Estimate targets for product cost and price
6. Perform initial financial analysis
7. Perform initial risk analysis
8. Build and evaluate alpha prototype units
9. Complete design concept
10. Complete technology readiness design review
11. Complete project planning documents for operations, project management, resourcing, technology and IP plan, commercialization plan, and test plan
12. MS3

A process flow chart for the sample Phase 3 process is shown in Figure 5.3.

5.2.1 Documenting the Steps

The more mature an organization is in project management process, the more definition will be provided around each of the steps of Figure 5.3. As a minimum, there should be a document explaining the step including what it is meant to accomplish, how to execute it, and what the expected results are. For those steps that require more prescription, a template for the output can be provided.

A simple write-up may be just a few lines long. If the step is well understood in the organization and the output is simple, this may be sufficient. The template brings the advantage of giving clearer definition of the data required and a consistent output format to simplify review. It also reduces variation from one project team to another. However, a template adds burden as a document that must be maintained by the organization. It also makes the process heavier for the project teams—one more document to learn, fill in, review, update, and so on. Often the optimal solution is to provide documentation for all steps and templates just for the critical ones.

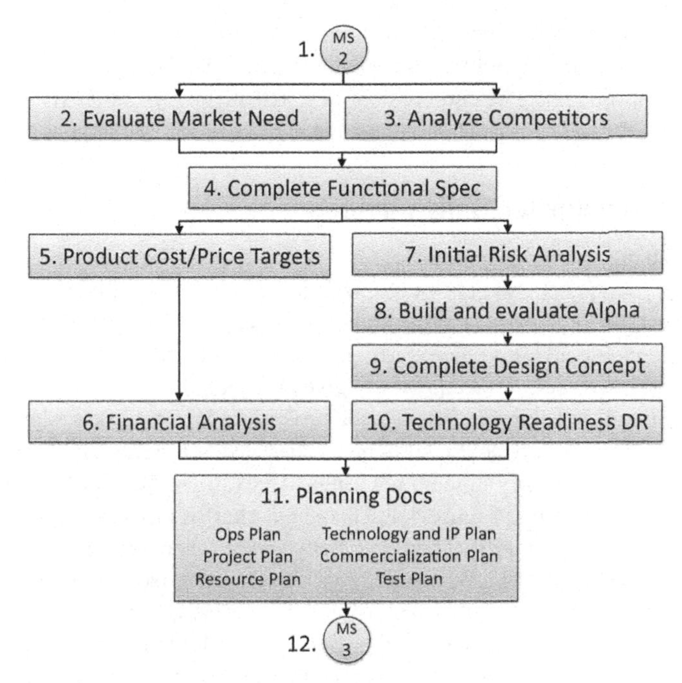

FIGURE 5.3 Sample Phase 3 process chart.

5.2.2 Phased Processes Provide a High-Level View

As with any development process chart, charts for a given phase are a simplification. First, the steps are shown in a rigid sequence: complete initial risk analysis before starting to build alpha units. However, to compress the schedule, the team will usually elect to order at least some of the material they will need for the alpha units before the initial risk analysis is complete. So, one step will usually start before its predecessor finishes.

Further, the Phase–Gate process will define only the major steps that most projects in an organization have in common. There will be a significant number of tasks that are peculiar to each project and so are outside the standard phase definition.

Another simplification is the steps are normally given at a coarse granularity—for example, evaluating market need for this project may have many substeps that are necessary to manage the task on a daily basis:

1. Create a questionnaire.
2. Identify three to five customers who might be willing to take the questionnaire.
3. Select two to three customers.
4. Interview the customers.
5. Analyze and report on the findings.

The granularity of these steps may be too fine to put in a company-wide Phase–Gate process, but the PM will need to manage them to complete the step "Evaluate market need." The PM can create a task set of finer granularity to manage daily work for most of the steps in a standard Phase–Gate process.

5.2.3 Documents for Daily Management

It should be clear that the standard Phase–Gate processes cannot be used by the PM for daily management. They are too coarse and they don't include the steps peculiar to a given project. In order to build a document for daily management, the PM can create a Gantt chart starting with the standard phase definition and then expand it to meet the needs of any given project.

As discussed in Chapters 2 and 3, the process of breaking down tasks in the near term should done weekly, so here the PM needs a lightweight document, something easy to modify in team meetings. An example of a Gantt chart based on Figure 5.3 is shown in Figure 5.4. This Gantt chart is based on Gantt Project, a free Gantt charting tool [9]. There are many such tools including Project Libre [10] and Open Project [11]. Wikipedia provides a comparison of about 200 project management software sets, many of which support Gantt charts [12].

Returning to Figure 5.3, notice that Step 2, "Evaluate market need" is broken down into five substeps. A common enhancement to a Gantt chart is a *timeline* showing major tasks, which is shown just above a fragment of the Gantt chart in Figure 5.5.

For many PMs, a simple spreadsheet-based task list such as shown in Table 5.1 is a better fit for daily management than a Gantt chart. Task lists are easier to modify and so allow the PM to capture new information quickly in team meetings. Start the project with the items from the process chart (Figure 5.3) and add tasks needed for this project that are outside the standard work. As coarse tasks approach, break them down into a granularity of 1 or 2 days. Adjust the details as the task comes into focus. Sort the list so the items due in the near future are positioned at the top of the list. This is a more nimble way to managing daily work—it focuses the team on work in the near future, speeds team meetings, and reduces the workload on the PM. There may still be a Gantt chart or some equivalent for the overall phase at a course level of granularity—perhaps containing just the standard requirements of the phase. This is the larger view that helps the team monitor progress for the phase while placing only a modest burden for maintenance since such a Gantt chart would have perhaps 20 or so items that rarely change. Compare this to a daily-management task list, which might have more than 100 items, many of which will change each week.

To keep the task list nimble, track the minimum amount of information needed to manage the daily work, perhaps just: task name, when due, and who owns it. It is often good to track planned and forecasted completion date. When a task is created, fix the plan and adjust the forecast, which is the current estimated completion date. If delays occur, slip the forecast. This gives a simple history so the team is aware of tasks that are in delay.

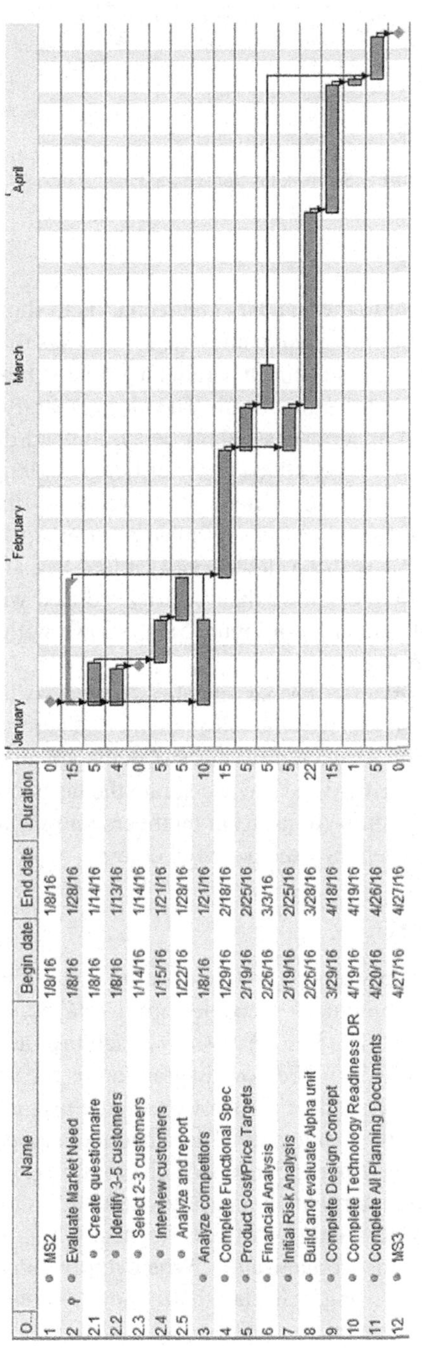
O.. Name Begin date End date Duration
1 MS2 1/8/16 1/8/16 0
2 Evaluate Market Need 1/8/16 1/28/16 15
2.1 Create questionnaire 1/8/16 1/14/16 5
2.2 Identify 3-5 customers 1/8/16 1/13/16 4
2.3 Select 2-3 customers 1/14/16 1/14/16 0
2.4 Interview customers 1/15/16 1/21/16 5
2.5 Analyze and report 1/22/16 1/28/16 5
3 Analyze competitors 1/8/16 1/21/16 10
4 Complete Functional Spec 1/29/16 2/18/16 15
5 Product Cost/Price Targets 2/19/16 2/25/16 5
6 Financial Analysis 2/26/16 3/3/16 5
7 Initial Risk Analysis 2/19/16 2/25/16 5
8 Build and evaluate Alpha unit 2/26/16 3/28/16 22
9 Complete Design Concept 3/29/16 4/18/16 15
10 Complete Technology Readiness DR 4/19/16 4/19/16 1
11 Complete All Planning Documents 4/20/16 4/26/16 5
12 MS3 4/27/16 4/27/16 0
January
February
March
April

FIGURE 5.4 Gantt chart for sample Phase 3 process.

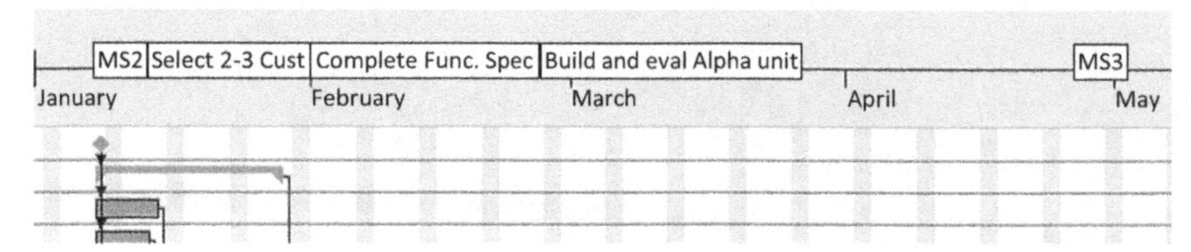

FIGURE 5.5 Timeline for sample Phase 3 process.

5.2.4 Critical Path

A common enhancement to the standard Gantt chart is to show the *critical path*. Developed in the 1950s, the critical path is the set of tasks that define the length of the phase. When two tasks occur in parallel and one is longer than the other, the longer of the two is in the critical path. The idea is that if the noncritical task (the shorter of the two) experiences a short delay, the end date of the phase is unaffected. However, every day added to the critical task lengthens the project. When tasks occur in parallel, the difference between the time to complete the longest task and any of the other tasks is called *float* or *stack*. The critical path then is that path of tasks with no float [13,14]. Float is a an indication of PM discretion; tasks with float can have their priority reduced without affecting the end of the phase. The Gantt chart of Figure 5.3 is shown with the critical path in darker bars in Figure 5.6.

The critical path is effective because it helps the team focus. During schedule reviews, the critical path receives the most attention. There is less tolerance for delays in the critical path. In team meetings and management reviews, resources can be moved from noncritical tasks to accelerate the tasks on the critical path if necessary. And in daily interactions, team members know who is on the critical path so they can help when they see potential delays.

5.2.5 The Pert Chart

The Gantt chart is certainly a popular visualization method in product development today. However, sometimes schedules are displayed with a Pert chart or project network diagram. The Pert chart shows each task as a box (Figure 5.7) in a flow diagram (Figure 5.8). The flow diagram holds the same information as the Gantt chart; the benefit is the Pert chart shows the task precedence relationships more clearly than the Gantt chart. The schedule display in the Gantt chart is probably more intuitive since the right side displays boxes sized in proportion to duration. In the Pert chart, small and large tasks are shown with the same size—duration is indicated only with text. In most cases the PM should use the format preferred in their organization. Whatever the benefits of Pert versus Gantt, the detriments of using a display that is not common in your company will probably outweigh them.

TABLE 5.1 A Simple Task List, an Alternative to the Gantt Chart for Daily Management

#	Name	End Date (Plan)	End Date (Forecast)	Days	Owner	Done
1	MS2	8-Jan	8-Jan	0	Team	Yes
2	Evaluate market need	28-Jan	28-Jan	15	Merle	Yes
2.1	Create questionnaire	14-Jan	14-Jan	5	Merle	Yes
2.2	Identify 3–5 customers	14-Jan	19-Jan	5	Merle	
2.3	Select 2–3 customers	15-Jan	20-Jan	0	Pilar	
2.4	Interview customers	21-Jan	26-Jan	5	Pilar	
2.5	Analyze and report	28-Jan	2-Feb	5	Merle	
3	Analyze competitors	21-Jan	26-Jan	10	Merle	
4	Complete functional spec	18-Feb	23-Feb	15	Greg	
5	Product cost/price targets	25-Feb	8-Mar	5	Greg	
6	Financial analysis	3-Mar	15-Mar	5	Christian	
7	Initial risk analysis	25-Feb	8-Mar	5	Greg	
8	Build and evaluate alpha unit	28-Mar	9-Apr	22	Sanjay	
9	Complete design concept	18-Apr	24-Apr	15	Sanjay	
10	Complete technology readiness DR	19-Apr	25-Apr	1	Greg	
11	Complete all planning documents	26-Apr	2-May	5	Greg	
12	MS3	27-Apr	3-May	0	Team	

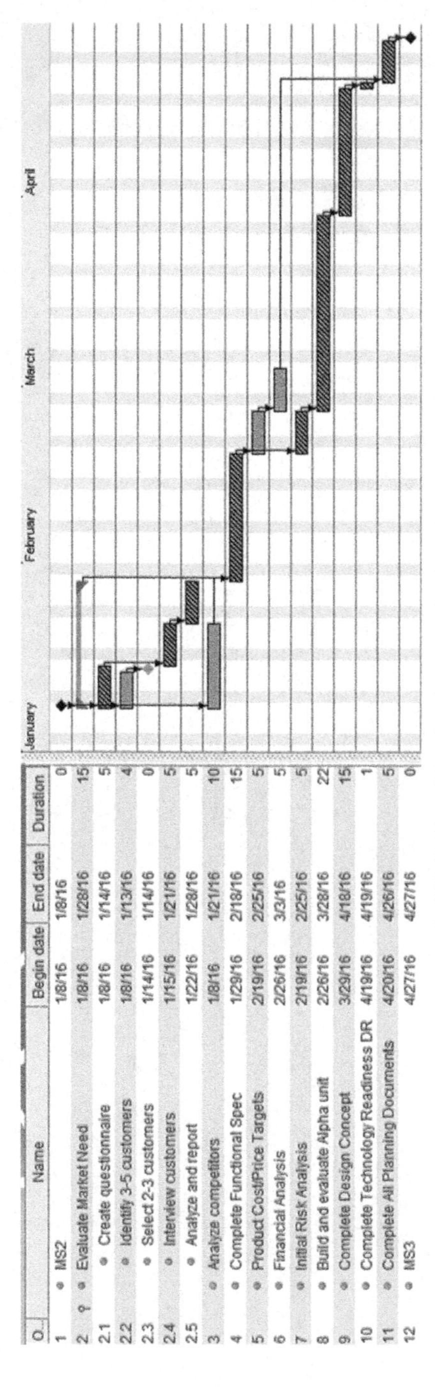

O...	Name	Begin date	End date	Duration
1	MS2	1/8/16	1/8/16	0
2	Evaluate Market Need	1/8/16	1/28/16	15
2.1	Create questionnaire	1/8/16	1/14/16	5
2.2	Identify 3-5 customers	1/8/16	1/13/16	4
2.3	Select 2-3 customers	1/14/16	1/14/16	0
2.4	Interview customers	1/15/16	1/21/16	5
2.5	Analyze and report	1/22/16	1/28/16	5
3	Analyze competitors	1/8/16	1/21/16	10
4	Complete Functional Spec	1/29/16	2/18/16	15
5	Product Cost/Price Targets	2/19/16	2/25/16	5
6	Financial Analysis	2/26/16	3/3/16	5
7	Initial Risk Analysis	2/19/16	2/25/16	5
8	Build and evaluate Alpha unit	2/26/16	3/28/16	22
9	Complete Design Concept	3/29/16	4/18/16	15
10	Complete Technology Readiness DR	4/19/16	4/19/16	1
11	Complete All Planning Documents	4/20/16	4/26/16	5
12	MS3	4/27/16	4/27/16	0

FIGURE 5.6 Critical path for sample Phase 3 process.

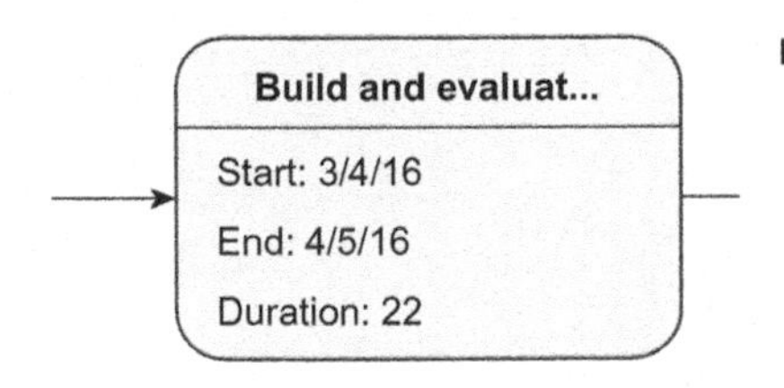

FIGURE 5.7 One task in Pert chart format.

5.3 DEALING WITH SCHEDULE ISSUES

As has been discussed throughout this text, the PM must manage a wide range of facets: budget, resources, risks, change, and many others. However, the dominant measure of project management effectiveness is usually schedule. This is probably for two reasons: first, most problems that occur that are difficult to measure reveal themselves in schedule slip. For example, let's say a new issue has come to light: the team has just learned the shelf life of printer ink is 70% less than the functional specification requires. Understanding the impact of an issue like this is challenging for nontechnical people. However, it's easy to quantify the problem when the PM inserts a 4-month schedule slip to improve the ink chemistry. The second reason is the effects of schedule slip on lifetime profitability of the product are much larger than intuition will lead you to believe. For example, a McKinsey study reported that a 6-month delay in shipment causes a 33% reduction in profit while a budget overrun of 50% causes only a 3.5% reduction [15].

5.3.1 Financial Measurement of Lifetime Profitability of a Product

To investigate the effect of schedule slip, let's start by reviewing three common financial measures of lifetime profitability of a product: net present value (NPV), internal rate of return (IRR), and breakeven time.

Net Present Value

NPV sums investment and sales over years taking into account the time varying value of money. The concept is familiar to anyone who has an interest-bearing bank account: $1000 today is worth more than $1000 1 year from now. The difference between the two is the interest rate.

In product development, the concept is similar. The company invests in a product in the beginning. The goal is for the sales of the product to pay back the investment and, in addition, return a profit. But, while the investment is made today, the return comes from sales in future years where the money is worth less—considerably less in the out years.

Here the value difference in value over 1 year is called the *cost of capital.* The concept is similar to interest, but the cost of capital is typically much higher than normal interest rates. One reason is companies borrow at rates consistent with the inherent risk of an enterprise. Whether a company is borrowing from a

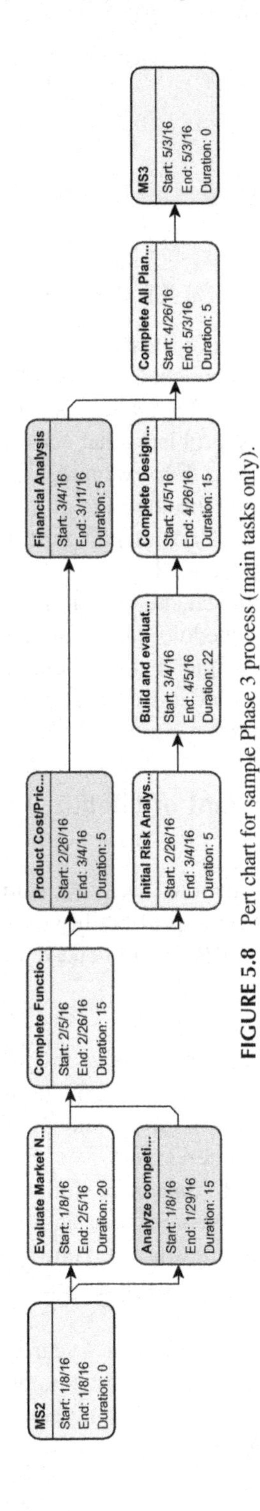

FIGURE 5.8 Pert chart for sample Phase 3 process (main tasks only).

bank, a venture capitalist, or from its stock holders, investors expect considerably more return than is offered by a safe investment like a Treasury bill. Table 5.2 shows how a value of $20k declines over time. With a 10% cost of capital (a value common as of the publication of this book[2]), today's value of $20k is halved if it comes in in 6 years. Figure 5.9 shows the same effect graphically. Note that the cost of capital varies over time and between companies; contact your finance department to find the value used in your company.

This concept is important for project managers to understand. The company invests today when money has its highest value and is repaid in the future when money is worth less. Let's return to Table 5.2 and investigate row 3, where the sales NPV is calculated with a 10% cost of capital. On the far right, the value of 6 years of sales, $20k each year, is $87k. Now, let's say we had to invest $50k today to achieve those 6 years of sales. With an NPV for all sales of $87k, the value of the total project would be $87k − $50k = $37k, assuming a cost of capital of 10%. Any time the NPV is above 0 with the appropriate cost of capital, it represents a profitable project.

Internal Rate of Return

Where NPV sums investments and sales with a specified cost of capital, IRR asks, what is the return rate realized from an investment? Let's take a second example, as shown in Table 5.3. Here a $150k investment produces 5 years of sales that ramp up to $75k/year. The IRR here is equivalent to investing $150k with a 21.7% return. As a comparison, the NPV (10% cost of capital) for this project is shown as $54k. The two measures are closely related. If you modify the cost of capital in the NPV formula until the NPV goes to zero, the result will be the IRR.

Breakeven Time

The third measure to review is the breakeven time: How long does it take to recover the initial investment including the cost of capital? Breakeven is the point in time where NPV is zero. The simplistic payback method presented in Section 2.1.1 just divided peak sales by investment without considering the time varying value of money; this can be misleading especially when the sales are far in the future. The measurement presented here is modestly more complicated to calculate, but with a standard spreadsheet and small investment of time, you'll get a more reliable measure.

5.3.2 Experiment: Effects of a 1-Year Schedule Slip

In this section, we'll use a spreadsheet to simulate the effects of a 1-year slip in product development. There's a lot of detail, but bear with the discussion—it is

2. http://pages.stern.nyu.edu/~adamodar/New_Home_Page/datafile/wacc.htm.

TABLE 5.2 Present Value of $20k over 6 years and with Varying Cost of Capital

Year	1	2	3	4	5	6	Cost of Capital	NPV
NPV at 0%	$20.0k	$20.0k	$20.0k	$20.0k	$20.0k	$20.0k	0%	$120.0k
NPV at 5%	$19.0k	$18.1k	$17.3k	$16.5k	$15.7k	$14.9k	5%	$101.5k
NPV at 10%	$18.2k	$16.5k	$15.0k	$13.7k	$12.4k	$11.3k	10%	$87.1k
NPV at 20%	$16.7k	$13.9k	$11.6k	$9.6k	$8.0k	$6.7k	20%	$66.5k

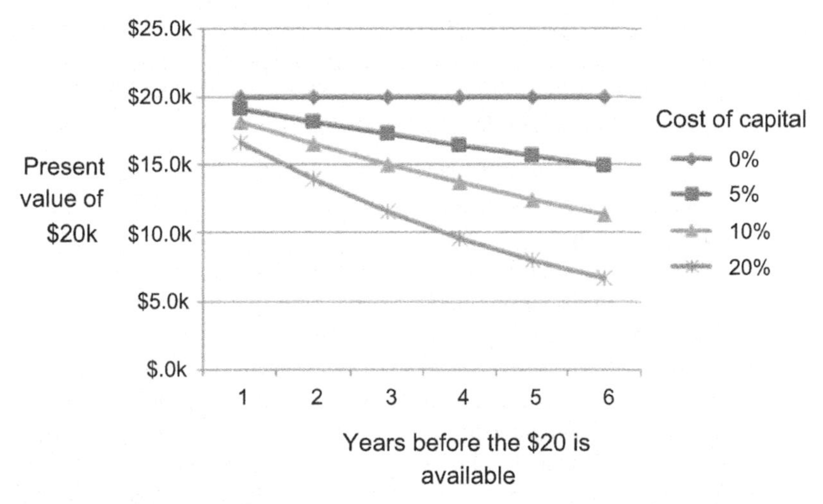

FIGURE 5.9 Present value of $20k over 6 years and with varying cost of capital.

TABLE 5.3 IRR Example

							10%
Y1	Y2	Y3	Y4	Y5	Y6	IRR	NPV
–$150.0k	$15.0k	$50.0k	$75.0k	$75.0k	$75.0k	21.7%	$53.7k

important to understand the factors at work because the detriment of delay is so much larger than is intuitive for most of us. Beyond the cost of capital, there are three main factors that work together to injure financial performance: reduced sales due to late delivery (the ramp-up starts later), increased competitive pressure when the product does arrive (a late start usually reduces peak sales and average margin), and a shorter product life (the factors that end a product's life are usually fixed such as future superior technology). These factors are simulated together in this example.

Let's simulate a development plan for a project that was originally planned for 1 year. Unfortunately, the original supplier of a critical component went out of business late in the program and a new supplier had to be developed. Two options were identified:

- Delay 1 year to allow a new competitor to be developed at minimal cost. No increase in investment is expected, but sales will be delayed 1 year.
- Increase the project investment 50% to incentivize another supplier to accelerate development. This should allow the team to hold the original schedule.

Let's compare three scenarios: the *original* plan before the supplier issue was discovered, the plan where a 1-year delay is accepted but the investment is constant (*slow*), and the plan where investment is increased 50% to hold schedule (*expedited*). We'll use the following assumptions:

- The original investment was $500k. The expedited investment adds $250k.
- The maximum sales for the original plan are $1.2M/year, which will take three years to ramp up to. The first year on the market will produce 1/6th of the maximum, the second will produce 2/3rd of the maximum. The delayed plan will have a similar growth path, but the peak will be reduced 30% due to assumed introduction of competitive products during the second year of development.
- In the 7th year a new technology will displace the product; sales will drop by half. The same thing will happen in year 8. By year 9, sales will go to zero.
- The product will sell at a 50% margin, so half the sales can be considered as return. This is a simplification since margin usually drops over time when competitive pressure forces price downwards.

The margin–investment is graphed in Figure 5.10 for all three scenarios. The first and third, "Original" and "Expedited," are almost identical except for the first year due to the extra $250k required to expedite the development

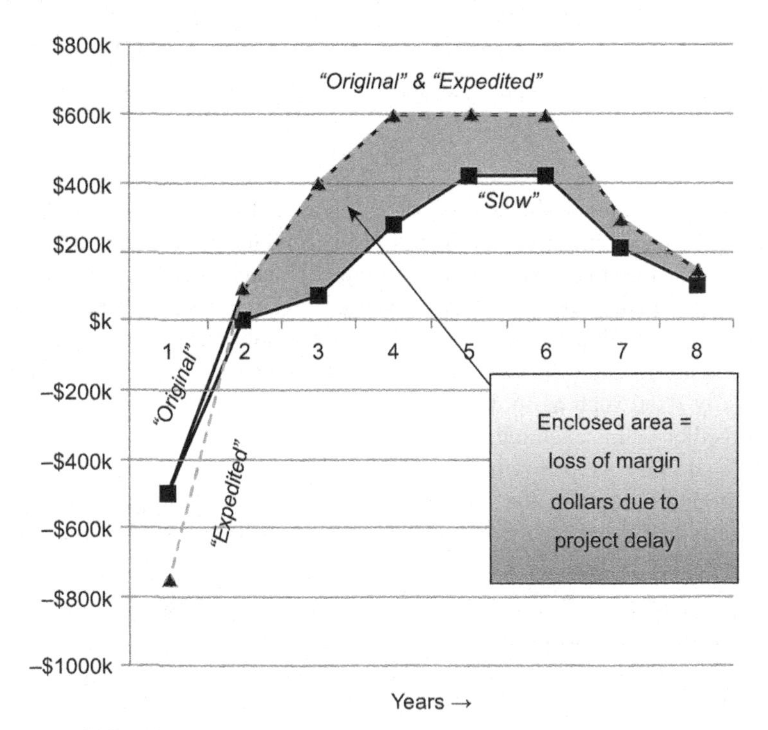

FIGURE 5.10 Cash flow: margin less investment for three scenarios.

with the new supplier (the lines lay atop each other on most of the graph). The sales for the second scenario ("Slow") are pushed to the right 1 year and reduced 30%; the area between the curves after year 2 represents total loss of margin dollars due to the schedule delay.

Analysis for the three scenarios is shown in Table 5.4. There is one column for each of the 8 years being considered with the development years in gray (recall that sales are 0 after year 8). Each scenario has five rows:

1. The investment in development for the first 1 or 2 years, depending on the scenario.
2. The sales for each year following the assumptions above.
3. The margin dollars assuming 50% margin.
4. The sum of investment and margin.
5. The NPV year by year. This row is needed only to calculate the breakeven time, the point where NPV = 0. Since this doesn't occur exactly on a year boundary, this example calculates breakeven time by drawing a straight line across the years where the NPV changes from negative to positive (years 3 and 4 for the first scenario) and calculates the point in the year where the value crosses through zero (3.2 years for the first scenario).

Table 5.4 provides some counterintuitive results. Most startling is the NPV falls 65% ($1274k to $444k) by delaying the project 1 year! The breakeven time slides 18 months (from 3.2 to 4.8 years) due to the 12-month delay. IRR drops by more than half.

If the extra 50% is invested in development costs, the company will enjoy a dramatic improvement. The NPV increases to over $1M, just 18% down from the original plan. The IRR is up 15 points from the "Slow" plan and breakeven time is reduced by more than a year.

The radar chart of Figure 5.11 shows a comparison of the five measures of performance of the three scenarios. You can see the harmful effects of delay—the expedited project outperforms the slow project in every aspect even though development costs were 50% higher. The original scenario sets the standard for comparison; it is the outer pentagon, scoring 100% in all five measures. The scales are set so 100% is the target in each instance and 0% is the poorest result. This results in intuitive measures for four of the measures; the exception is breakeven time, which is displayed as the original plan breakeven time divided by scenario breakeven time (scoring 50% here is equal to doubling the breakeven time).

5.3.3 Unit Price and Unit Cost Variation

The experiment of Section 5.3.2 showed benefits of rapid development assuming costs and pricing remained fixed. However, as Smith and Reinertsen [16] explain, rapid development also leverages the behavior where unit price and cost of new products decline over time. Unit price starts high because when an innovative product is introduced, it has an advantage over the alternatives.

TABLE 5.4 Projections of the Three Scenarios

Year	1	2	3	4	5	6	7	8	Sum	NPV (10% Cost of Capital)	IRR	Break-even Years
Original: investment	–$500k								–$500k			
Original: sales		$200k	$800k	$1200k	$1200k	$1200k	$600k	$300k	$5500k			
Original: margin (50%)		$100k	$400k	$600k	$600k	$600k	$300k	$150k	$2750k			
Original: margin–invest.	–$500k	$100k	$400k	$600k	$600k	$600k	$300k	$150k	$2250k	$1274k	64%	3.2
Original: NPV by year	–$455k	–$372k	–$71k	$338k	$711k	$1050k	$1204k	$1274k				
Slow: investment	–$500k								–$500k			
Slow: sales			$140k	$560k	$840k	$840k	$420k	$210k	$3010k			
Slow: margin (50%)			$70k	$280k	$420k	$420k	$210k	$105k	$1505k			
Slow: margin–invest.	–$500k		$70k	$280k	$420k	$420k	$210k	$105k	$1005k	$444k	29%	4.8
Slow: NPV by year	–$455k	–$455k	–$402k	–$211k	$50k	$287k	$395k	$444k				
Expedited: investment	–$750k								–$750k			
Expedited: sales		$200k	$800k	$1200k	$1200k	$1200k	$600k	$300k	$5500k			
Expedited: margin (50%)		$100k	$400k	$600k	$600k	$600k	$300k	$150k	$2750k			
Expedited: margin–invest.	–$750k	$100k	$400k	$600k	$600k	$600k	$300k	$150k	$2000k	$1046k	44%	3.7
Expedited: NPV by year	–$682k	–$599k	–$299k	$111k	$484k	$822k	$976k	$1046k				

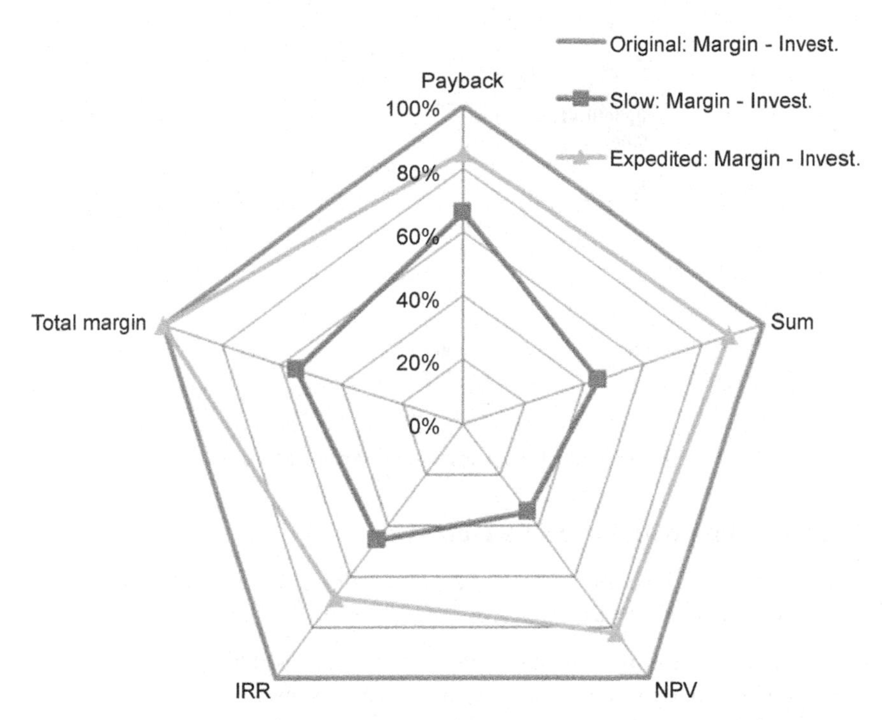

FIGURE 5.11 Comparing performance of the three scenarios, larger shape is better.

Customers are willing to pay a premium to gain advantages from superior technology. At launch, the product is typically available in low volume and so can be priced for those few customers who value it most. As production ramps up, prices are normally reduced to capture a larger number of customers. Also, competitive products will be introduced over time, causing more downward pressure on price. This process continues with prices stabilizing as the competitive field stabilizes. This is shown in Figure 5.12 as "Market price."

Unit cost also declines over time, the primary influence being increasing volumes. Volume allows cost reduction in numerous ways including: the sourcing team can develop lower cost suppliers, tooling can be improved, and process efficiency can increase through continuous improvement. This is also shown in Figure 5.12.

These two effects—declining unit price and cost—give two benefits to companies that introduce new products early, as shown in Figure 5.12. Benefit #1 is the absence of competitive pressure early in a product's life, which allows a pricing premium. Benefit #2 is the extra time the early introducer has to build up volume over later arrivers, thereby reducing its costs in comparison to others. Both of these benefits can yield substantial margin advantage for the company that gets the market first. Had we taken these factors into account, the results of Figure 5.11, which already heavily favored faster development, would have favor that alternative even more.

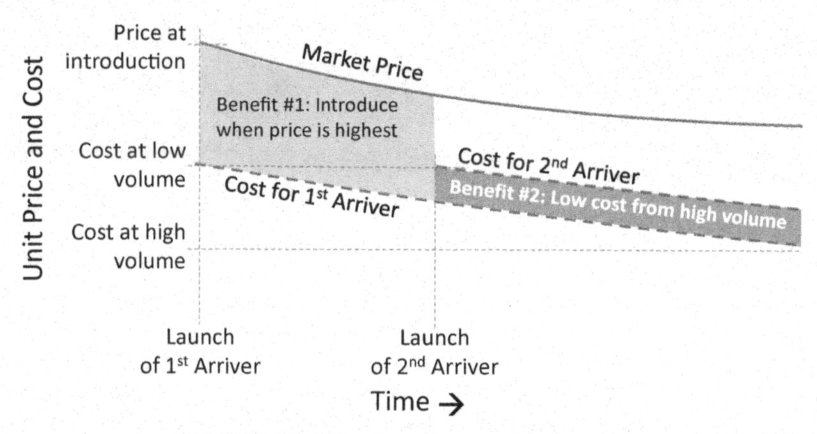

FIGURE 5.12 Unit price and cost variation over time.

5.3.4 Dealing with Schedule Delays

Having established the importance of rapid development, we will now turn our attention to the practical issues of avoiding schedule delays:

- The ability to identify schedule delays as early as possible.
- Taking action to reduce the severity of potential delays.

The Butterfly Curve

The first principle to review is the well-known butterfly curve [17] of Figure 5.13. As this diagram shows, early in a project, the investment is small and the ability to affect the outcome is high. As the project proceeds, the investment increases and the ability to affect the outcome within the allotted time is reduced. The conclusion from the butterfly curve is that it's critical to identify potential schedule slips early, before they start to affect the project. The maturation cycle of schedule slips is:

1. A risk is presented.
2. The risk matures into an issue.
3. The issue absorbs resources and causes delays.

The earlier in the maturation cycle an issue is identified, the greater the team's ability to deal with the issue.

5.3.5 Measuring Delay Directly

Potential delay can be detected in two ways: directly though tracking task completion or, as will be covered in the next section, indirectly through monitoring warning signs. Tracking progress in projects is challenging because of several factors related to measuring task completeness:

- The possibility that all task details were not completed so that a task thought to be complete requires additional unplanned effort.

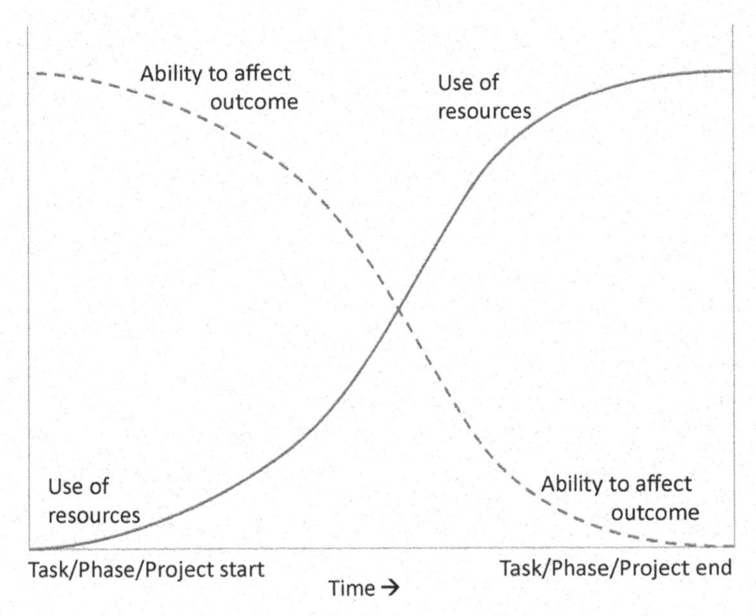

FIGURE 5.13 The butterfly curve.

- The possibility that a poorly executed task will have to be reworked. For example, if the task at hand is to create a production plan, the initial plan might be a rough overview when the task calls for a detailed plan.
- The possibility that the estimate of the task length was incorrect at the outset of the task. For example, if Tucker completed 3 days of a task originally thought to require 5 days, the task would likely be recorded as 60% complete; however, if Tucker were to re-estimate the task now being more familiar with it, 10 days might be more accurate. So, perhaps unknown to the PM, a better estimate of completion is 3 of 10 days or 30%.
- The difficulty in estimating the fraction of completeness for an ongoing task. Partial completion is notoriously difficult to measure because of a negative effect of the well-known Pareto principle: 80% of the (perceived) results come from 20% of the effort. Teams often underestimate the amount of remaining work. In this light, the Pareto principle can be restated as: "The team has 80% of the work remaining even when only 20% of the perceived task remains."[3]

Accurate tracking of project progress requires that the PM is able judge task completeness in the face of these factors. This is a combination of transactional leadership (having strong process, following that process diligently, and reporting results on time) and transformational leadership (staying connected with the

3. Note that this effect is mitigated with proper project granularity—as long as no single task on the critical path is too large, the effect of incorrectly estimating the fraction of work done on the active task will have a modest effect on completion accuracy.

team to understand risks and issues as they arise). We will turn our attention to displaying the results of progress measurement.

Tracking Progress with Deliverables

In many CPM projects, progress is reported by whether key deliverables are finished on time. The key deliverables might be process oriented, such as successful design reviews and FMEAs, or they might be concrete results such as passing a set of tests or getting a positive feedback from a customer evaluating a prototype. The milestones that garner the most interest vary by process, company culture, and an ad hoc reading of recent history. If a project in the recent past experienced costly rework because a vacuum test wasn't properly run, it is likely vacuum tests will receive disproportionate attention.

The problem with tracking deliverables is that they are lagging indicators. The deliverables that are favored by a company to indicate progress usually occur far into a given phase—often near the end. So, by the time the measurement is taken, the project may have encountered unrecoverable schedule delay.

Tracking Progress with the Gantt Chart

The Gantt chart can also be used to display progress. Normally, on a weekly or monthly basis the PM enters progress for each active task and the Gantt chart is augmented with progress lines to display this information. For example, the project of Figure 5.6 is shown in Figure 5.14 with progress as follows:

Tasks 2.1–2.5:	Complete
Task 3:	80% complete
Task 4:	40% complete

The current date is usually shown as a vertical line, as shown below "February" in Figure 5.14. In this way, the Gantt chart that was the outcome of planning becomes the display for tracking progress throughout project execution.

Compared to tracking deliverables, progress tracking provides a leading indicator because the measure is not biased to events that occur late in the phase. The difficulty with progress tracking is that it doesn't provide a measurement that is both intuitive and quantifiable. The reason is that the typical project has many tasks running in parallel and each can be a different level of completeness. Consider the simple example of Figure 5.14: Task 3 is about two weeks late, but Task 4 is about a week ahead. Is the project behind or on time? It requires considerable knowledge of the project to know. Consider a project with many more parallel paths and imagine how difficult it could be to quantify total completion of the project. Progress tracked by the Gantt chart may be so difficult to judge that only the PM has the ability to interpret it. Of course, opaque measurements that are understood by a few people are unlikely to help drive action in an organization. The unfortunate result is progress tracking may give a leading indication of a problem to a handful of people

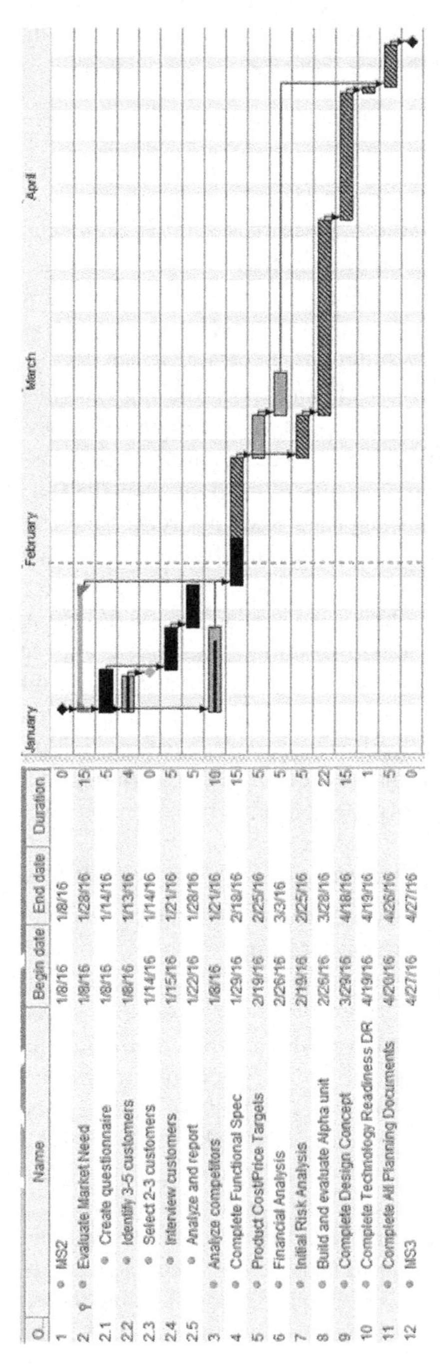

FIGURE 5.14 Critical path with progress tracking (from GanttProject).

most familiar with the project, but decisive action such as adding resources or adjusting project scope may have to wait for a more intuitive (but lagging) measure such as learning key deliverables are delayed.

Tracking Progress with Critical Path Complete

Another progress-tracking technique is monitoring the amount of critical path that is complete against time passed. Here the tasks off the critical path are ignored and the duration of all tasks on the critical path are summed to calculate the total project length; the duration of completed tasks on the critical path are summed to measure progress. This creates a simple metric: critical path days complete, which can be compared to the days elapsed to calculate the variance to schedule.

Table 5.5 is built to track the amount of critical path days executed for the WBS of the project in Figure 5.14. Column 1 shows only the tasks on the critical path, column 2 shows their duration in work days, and column 3 shows the completion date of each task as planned when the project begins (*plan* remains fixed during project execution). The remaining columns to the right allow the PM to enter the amount each critical path task is completed week by week. Note that the total project is planned through May-27, but only the first 10 weeks are shown in the Table 5.5.

With this data available, progress on the critical path can be calculated as shown in Table 5.6. The first row calculates the working days passed based on calendar days, incrementing by 5 days for each week. The second row calculates how many days of the critical path are completed using Table 5.5: in that table, the column under each date is multiplied cell by cell with the duration (column 2)[4]; optionally, this can be converted to percent by dividing by the total critical path, 98 days. The third row of Table 5.6, variation to schedule, is simply row 2 − row 1; this can also be converted to percent by dividing by 98 days.

Visual Progress Measurement: The Schmidt, Run, and Fever Charts

Armed with a simple metric, progress can be plotted in a variety of ways. For example, the Schmidt chart of Figure 5.15 [18] plots percentage of critical path completed against a planned 98-day critical path. The plan is a straight line from start (0 days, 0%) to end (98 days, 100%). A dashed line is drawn above the plan and everything above that is labeled "Continue" with a high likelihood of on-time delivery without further action. Another line can be drawn below the plan; below that the project is behind, with a large risk of a delay; so, the team needs to "Take action." The space in the middle is "Plan"—performance to schedule is marginal so the team should plan actions that can be taken if performance worsens.

The benefit of this approach is that it's intuitive; within a few seconds anyone can see a significant schedule issue was encountered on Feb-26: the team lost ground over the week prior. Of course, understanding the event that caused this and how to recover from it will require analysis of the project, perhaps using

4. In Excel®, this is calculated applying the SUMPRODUCT() function.

TABLE 5.5 Tracking Progress on the Critical Path

Project Planning			Project Execution										
Critical Tasks	Days	Date	Jan-08	Jan-15	Jan-22	Jan-29	Feb-05	Feb-12	Feb-19	Feb-26	Mar-04	Mar-11	Mar-18
Task 2.1/2.2	5	Jan-15		80%	100%	100%	100%	100%	100%	100%	100%	100%	100%
Task 2.4	5	Jan-22			100%	100%	100%	100%	100%	100%	100%	100%	100%
Task 2.5	5	Jan-29				100%	100%	100%	100%	100%	100%	100%	100%
Task 3	10	Feb-12				25%	100%	100%	100%	50%	100%	100%	100%
Task 4	15	Mar-04						50%	75%		100%	100%	100%
Task 5	5	Mar-11										100%	100%
Task 6	5	Mar-18										80%	100%
Task 7	5	Mar-25											50%
Task 8	22	Apr-26											
Task 9	15	May-17											
Task 10	1	May-18											
Task 11	5	May-25											
Total	**98**												

TABLE 5.6 Calculating Progress on the Critical Path

	Schedule Variation										
	Jan-08	**Jan-15**	**Jan-22**	**Jan-29**	**Feb-05**	**Feb-12**	**Feb-19**	**Feb-26**	**Mar-04**	**Mar-11**	**Mar-18**
Working days passed	0	5	10	15	20	25	30	35	40	45	50
X axis—critical path (days)	0	4	10	17.5	25	32.5	36.25	20	40	49	52.5
Y axis—Variation (days)	0	1	0	−2.5	−5	−7.5	−6.25	15	0	−4	−2.5

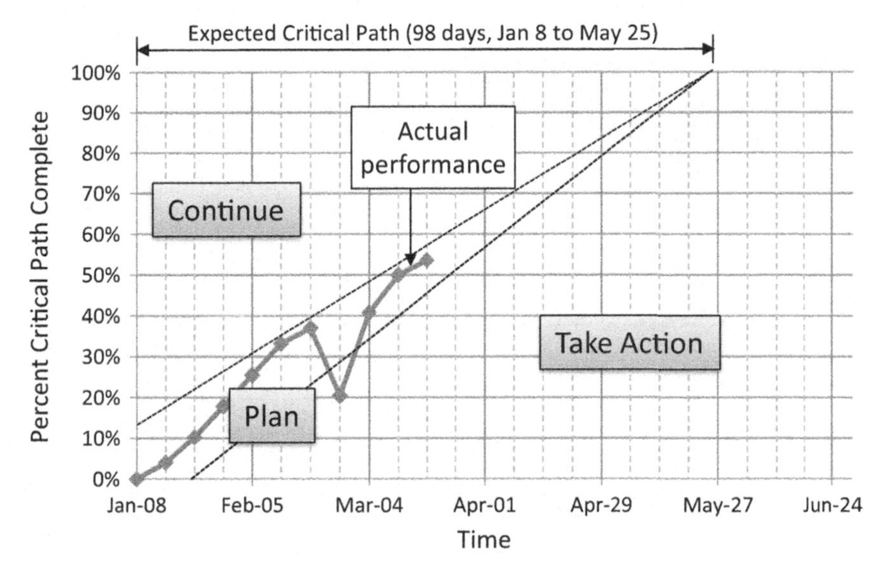

FIGURE 5.15 The Schmidt chart for the project of Table 5.6.

both the Gantt chart and the deliverables list. The primary goal of the progress metric is to indicate if a project needs attention.

Another method of progress tracking, shown in Figure 5.16, uses a format borrowed from quality measurement called a *run chart*. At regular intervals the variation of actual progress to plan can be plotted. The sign here is reversed[5] from the Schmidt chart: positive variance indicates the project team should "Take action," a significant negative variance (say, 1 week) shows the project is ahead so the team should "Continue." One advantage of the run chart is that by plotting variance, it zooms in on the most critical information; by comparison, the Schmidt chart data is crowded in the middle since the upper left and lower right regions are unused in all but the most extreme cases (notice on the Schmidt chart of Figure 5.15 that most of the information appears in a narrow band from (Jan-08, 0%) to (May-27, 100%)).

Another variation on this display is the *fever chart* of Figure 5.17, borrowed from Critical Chain Project Management (Chapter 6) created by Eliyahu Goldratt. In the fever chart, the horizontal axis is the critical path complete rather than time; dates are labeled at the point of measurement. One difference to the run chart is the display for the case where the team has to rework tasks thought complete, the amount of critical path complete is reduced; this is seen clearly in the fever chart when the progress trajectory moves to the left (see

5. Positive variance is favorable for the Schmidt chart but unfavorable for the fever chart as currently defined by CCPM, which will be discussed in Chapter 6. The run chart is not commonly used in project management, so this text will adopt the sign convention of the fever chart. Of course, either convention will work equally well.

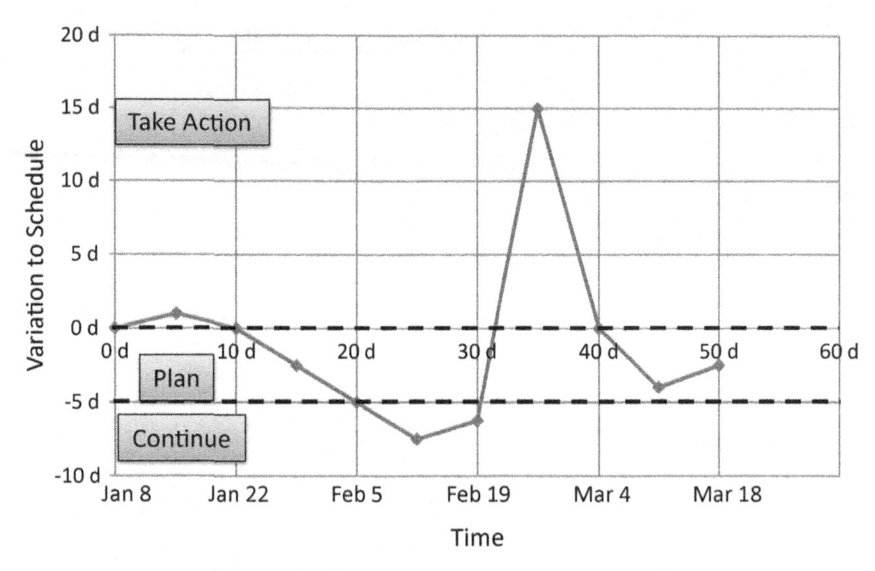

FIGURE 5.16 Run chart for the project of Table 5.6.

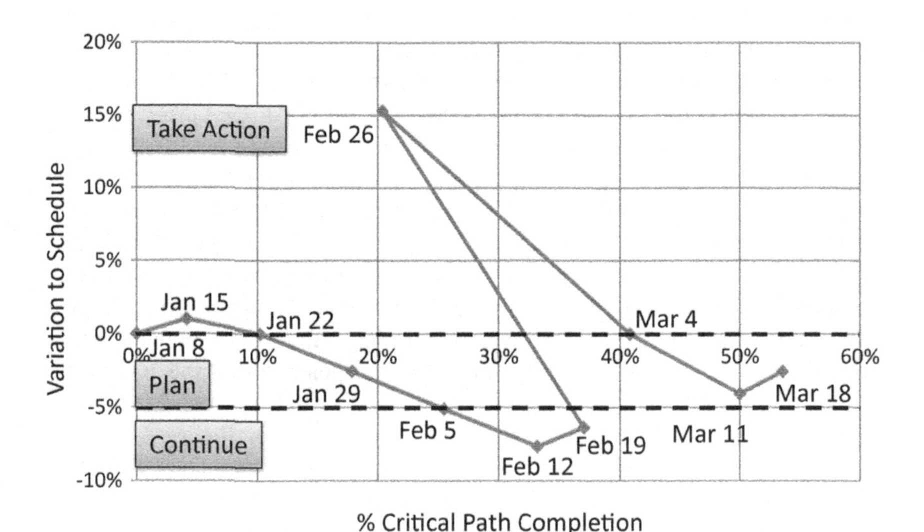

FIGURE 5.17 Fever chart for the project of Table 5.6.

Feb-26 in Figure 5.17). One of the challenges of the fever chart is automatically labeling measurement dates. Many graphing tools do not support this function, so labels may need to be added manually. By comparison, the run chart has time on the horizontal axis and so can be graphed either in dates or percentages with standard spreadsheet charting tools.

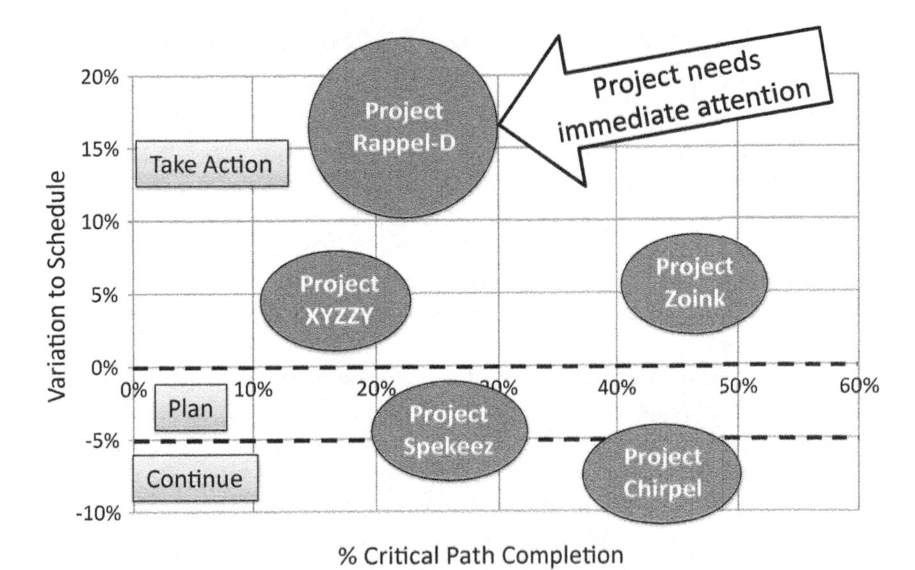

FIGURE 5.18 Fever chart for a CPM portfolio view.

One of the benefits of the % versus % format is the axes are the same for all projects, independent of project length or start date. The fever chart is guaranteed to fit in a width of 100% because the X axis is critical path completed, not time; for the run chart, allow for some amount for schedule overrun. In either case, the %-% normalization supports a snapshot of the organization's entire portfolio that can be plotted on the identical axes, as shown in Figure 5.18. Here all projects are shown in current state so that progress is displayed in an intuitive manner. For example, Project Rappel-D is 15% late less than 25% of the way into the project. On the other hand, Project Chirpel is well ahead and is almost halfway complete. This example also shows the total revenue of the project as being proportionate to the size of the bubble; so, Project Rappel-D is the company's largest project. This fever-chart portfolio view gives a highly visual view of the company's portfolio that any stakeholder can interpret rapidly; the same can be said for a run chart portfolio view; a portfolio view can be done with the Schmidt chart, but data would be crowded along the diagonal. These views provide data for resource discussions—for this example, it seems here that the first step would be to see if any resources from Spekeez or Chirpel (both of which are ahead) could be loaned to Rappel-D as corrective action.

There are good reasons to use both normalized times (that is, in percent) and others to use days on either axis. Different organizations will prefer different measures based on other metrics in their processes. The key point here is that a combination of tracking deliverables and progress gives a richer view of schedule health than either measure can alone. Being able to quantify progress allows the use of intuitive leading indicators that can help drive quick action when delays are encountered.

5.3.6 The Early Warning Signs: Signals of Potential Delay

There are many types of risks to a project, each of which can cause schedule delay. Sometimes, the root cause is known well ahead—even a well-managed risk can mature into an issue that delays the project. But very often the root cause of the delay remains hidden until the damage is done. A supplier who knows they will be late might hide that to prevent you from seeking out one of their competitors. A team member that's in over her head on a technical problem often won't ask for help for any number of reasons: fear of punishment, loss of reputation, or unwarranted confidence that she can fix the problem with just a little more time.

If the PM cannot rely on being informed rapidly about newly discovered root causes, there is an alternative: look for *early warning signs*. As shown in Figure 5.19, the root cause (in that example, a team member lacking skills) may be festering into a large problem that is largely unseen (underground). But the early warning signs (here, unexplained slow progress) can reveal the problem. These signs can help a PM identify there is a problem before the schedule is seriously affected. They are all based on the ability to sense that something isn't "right." Experience is a great help—the more projects you've managed the more you will develop a sense of how things should be. But much of what PMs may lack in experience can be made up by having intimate familiarity with the details of the project and being vigilant.

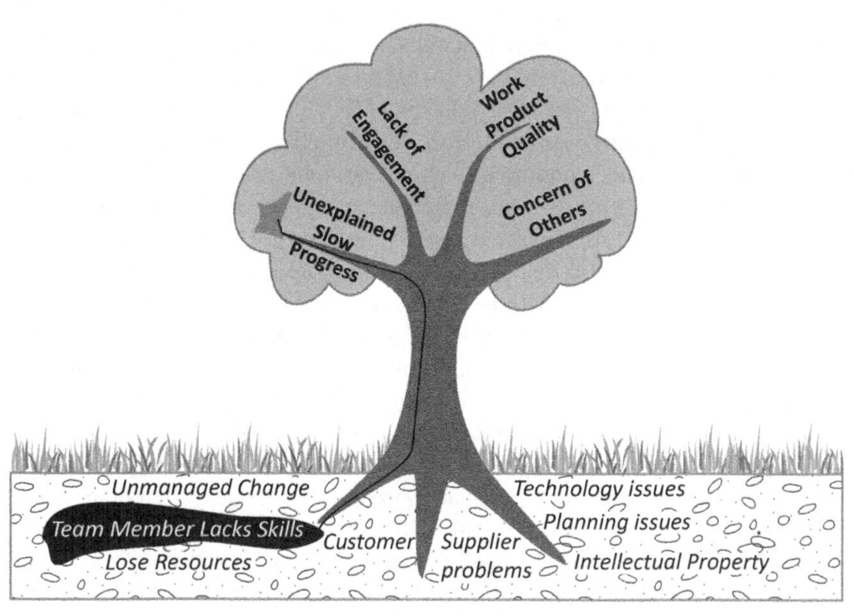

FIGURE 5.19 Early warning signs for schedule delay.

These early warning signs fall into four categories:

1. **Progress is slower than expected.**
   A task, risk, or issue lingers unresolved/incomplete for longer than expected without a convincing explanation of why, for example:
   a. A minor task in a project may be slipping week by week. Any single week of slip might not create concern, but even a small matter slipping across a long period of time should.
   b. A risk remains unresolved longer than expected.
   c. Material not arriving on time for prototypes or other testing.
   d. Long-threaded emails on a stubborn issue without a call to action that resolves or escalates the issue.
2. **Low quality work product.**
   Because work in early phases serves a foundation for later work, poor quality work in one phase will often lead to delay in a later phase. Identifying poor quality work early can prevent delay. For example:
   a. Processes not being followed by the team.
   b. Process steps not being executed diligently. Design reviews, FMEAs, and other review processes are not well attended or the team doesn't diligently search for issues. Items identified at various steps are not reliably transferred to a risk and/or task list.
   c. Reporting is continually behind. Progress is difficult to measure through the "noise" of invalid or out-of-date information.
3. **Signs of faltering engagement.**
   A lack of activity often indicates a task is not being executed well or it may mean no one is working on it at all.
   a. Observing team members are not in places or doing things you would expect. If a big test is planned in a day or two, there should be signs of activity in the lab today.
   b. Lack of questions from customers or suppliers. If someone is using a complex piece of equipment or producing a complex component, there should almost always be questions. If the phone isn't ringing, it may be because no one is working on it.
   c. Lack of feedback from regulators. If the design isn't generating questions, it may be the regulator has pushed the project to the bottom of the pile.
4. **Signs of concern from others.**
   If you learn someone else is not confident that the project is moving the right way, get enough detail that you can follow up on the matter yourself. It may be they see an early warning sign that you missed.
   a. Concerns raised in team meetings. When a team member expresses concern, follow up. It's easy to get frustrated with people expressing negative feeling because it dampens the feel of the meeting. Nevertheless, get enough facts to make an informed determination on the issue.

b. Concerns raised in a hallway conversation.
c. A sponsor, senior manager, or experienced colleague explaining a concern.

Recognizing these warning signs early requires a mix of transformational and transactional skills. Examples of transactional skills include:

- Schedule tasks with the proper granularity, typically a day or two for a project with weekly meetings. Too fine a granularity generates "noise"—so many tasks taking place that it's hard to see the important activities. "A highly detailed schedule requires not only a large amount of effort to establish it in the first place, but also to maintain it" [19]. A coarse granularity obscures problems—it's usually difficult to see delay until tasks slip past the scheduled completion date.
- Define deliverables clearly and review progress. This way you can tell an item is complete because requirements have been met rather than because a predefined period of time has passed.
- Ensure the project schedule is complete. Important deliverables must be assigned or there is a high risk they will be missed.
- Ensure accountability for every task, risk, and issue is clearly communicated.
- Know your processes. If you are unfamiliar with a process step, observe the execution of a similar step in another project. Never been to an FMEA? Attend one for another project so you have a reference.
- Sit in during key process steps like design reviews. Even if you're not technical, you need see the interactions of the team and understand which issues are generating the most concern.
- Manage risks diligently. Review existing risks regularly and thoroughly. Ensure new risks are quickly recorded and tracked. This is discussed in detail in Chapter 9.
- Stay organized. Inaccurate or outdated information hides schedule slips.

Examples of transformational skills that will enable you to understand early warning signs include:

- Listen well, especially when someone expresses concern. It's easy to dismiss a concern in the early phases because there isn't enough damage to see the scale of the problem. But this is the best time to catch a problem.
- Be well connected to your team using transformational skills for people (*connection*) as discussed in Chapter 4. The better your relationships with team members, the more likely they are to be open with you when a problem occurs.
- Stay in contact with key customers, suppliers, and regulators. Participate in meetings. Schedule regular meetings at critical junctures.
- Stay steady in a crisis—if you lose control when people hand you bad news, they'll be less likely to keep you informed in future.
- Reward the people who help spot problem areas early.

5.3.7 Taking Action after Discovering Delay

After discovering an issue that will likely cause a delay, the PM will decide whether or not to take action to maintain the schedule. Of course, not every issue requires attention—slips in tasks off the critical path often can be ignored for a period of time without negative effects. But in those cases where the delivery of important milestones is affected, action is required. That action will likely fall in one of three dimensions: working on the problem yourself, refocusing the team, and increasing the capability/capacity of the team.

Personal Action

Taking personal action is often effective for problems with a limited scope. If a supplier isn't responding to a team member, call the supplier and determine what's wrong. If there's a tough technical problem and you have the expertise, roll up your sleeves. Strong technical ability allows you to help; if you're successful, you've not only sped up the project, but you'll probably have earned respect from your team. But take care not to over-apply this action. It's usually impractical to manage a complex project and take responsibility for significant technical contributions. It can lead to bias ("the problem I'm working on is the important one") or a loss of focus on the project. But if you can help the team without losing focus, jump in.

Refocus the Team

There are multiple alternatives of varying scope for refocusing the team as shown in Figure 5.20. Actions of larger scope are justified for more severe issues. The action of smallest scope is simply refocusing the team on the issues at hand. These are the sorts of things PMs do on a daily basis: stopping by a team member's office to discuss an issue that needs to be accelerated or taking a few minutes in a team meeting to bring consensus to an issue that needs renewed commitment. In these cases, the PM has concluded the team is on the right path but needs a small push to drive the issue to resolution.

When the PM determines adding focus is not sufficient to address the issue, the next step is to shift resources. One well-known technique is to review the critical path and move resources from tasks off that path to the issue of concern. Many times issues that are likely to cause delay are not clear on the schedule, especially in the moments soon after the issue has been discovered. So, deciding which tasks to pull resources from requires more judgment. Perhaps the PM will make the call

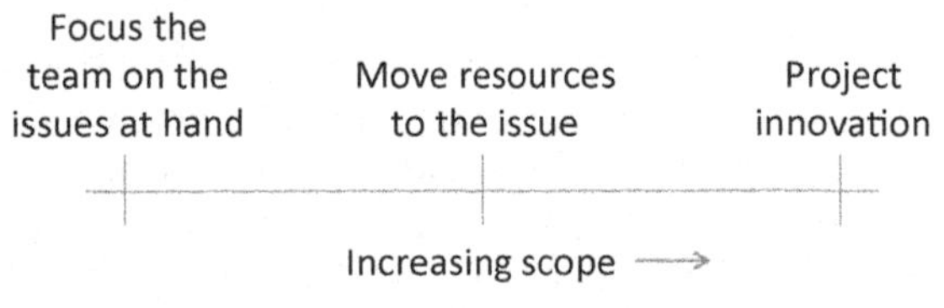

FIGURE 5.20 Alternative methods to refocus the team.

or perhaps he'll call a team meeting to lay out the problem and get consensus on which tasks should loan out resources in order to avert a schedule slip.

The refocusing approach of the largest scope is project innovation, as discussed in Section 3.1.2. Recall that project innovation defines the class of actions where the project is significantly changed in reaction to new information. Here you're reorganizing the project and probably changing team member assignments. Project innovation is a powerful tool—it allows the team to react to a changing situation with the broadest set of alternatives. However, be cautious to use it only when needed. The overuse of project innovation can cause confusion—if the project plan is changing frequently a sense of chaos may drive the team to doubt the PM's abilities.

PMs must understand that refocusing has two sides: increasing the attention some tasks are receiving while reducing the attention on other tasks. It is normally easy to gain management support for increasing attention on problems. For the project sponsor, approving increased effort on a given topic is easy because there's nothing to give up. It's more difficult, often much more difficult, for managers to support reducing attention on another item. It is the reduction that brings risk to the sponsor. So, the team relies on the PM to present the issue in a manner that requires a choice; otherwise, it's too likely to walk away with the unsatisfying decision to increase focus on one topic without the relief of reducing it elsewhere.

Increasing Team Capability/Capacity

The third common action is increasing the resources, or more broadly, the capacity or capability of the team. There are many ways to increase this:

- Negotiate to get more time from a part-time team member—help them shed competing priorities.
- Get temporary help internally or from an outside consultant.
- Add a person to the team.
- Replace one team member with someone more capable.

The first alternative pays off quickly, but any attempt to bring new people into the project almost always takes time to return benefits. Hiring is the slowest—it often takes months to find the right person, bring them into the company, and train them on the products and technologies. This is why adding capacity often doesn't help late in a project. Again, identifying the delay early is critical.

Figure 5.21 shows how these three alternative responses can vary in effectiveness depending on whether they are applied early or late. Your personal help is probably going to be limited to a few occasions, because when you're managing a complex project, you probably won't have the time to make a larger difference. Refocusing produces results quickly, so you can apply it at any point. Adding capacity is usually ineffective late in a project—it takes too long to bring the resource up to speed (apart from getting a larger share of exiting team member time). So, as you select your path forward, bear in mind how long it will take for the action you've chosen to bear fruit.

	Early Action	Late Action
1) PM personal action	Medium	Medium
2) Refocus the team	High	Medium
3) Add capacity/capability	Medium	Low

Success of Correcting Schedule Issues

FIGURE 5.21 Effectiveness of reactions to delay versus when action is taken.

5.4 CPM KEY MEASURES OF EFFECTIVENESS (WITHOUT PHASE–GATE)

This section will provide an evaluation of the CPM method without Phase–Gate according to about a dozen key measures of effectiveness. The following section following will evaluate the differences Phase–Gate can bring when added to CPM. Both are shown in Table 5.7. This table will be expanded in the succeeding three chapters to include Critical Chain (Chapter 6), lean product development (Chapter 7), and Agile (Chapter 8) project management techniques.

Good with high-iteration-cost projects (+)
CPM projects encourage diligent planning before a project starts. The Gantt chart shows projects as requiring long chains of tasks to arrive at a usable result. Extensive planning helps mitigate risks in projects that are expensive to iterate, for example, those requiring new factory equipment and supplier tooling.

Good with low-iteration-cost projects (–)
The lengthy planning that serves high-iteration-cost projects makes the method less flexible for projects with low cost of iteration, such as many software projects.

Process to coordinate varied disciplines (+)
The CPM method is capable of coordinating many functions. The nature of the project team as a temporary group focused on one project allows many people to work together in varying degrees from occasional interaction to full-time core team member.

Mitigates risk before large investment
CPM provides the tools and metrics to allow projects to be structured to maximize review before large investments. However, CPM without Phase–Gate provides no standard work to optimally manage risk. So, the ability to mitigate risk is highly dependent on the skill of the PM and team during planning and the engagement of reviewers.

Provides standard work for planning (–)
CPM provides the WBS (or project task network) as a view of linking the many tasks that make up a project. However, there is little standard work without the Phase–Gate method being added.

TABLE 5.7 Key Measurement of Effectiveness for CPM and Phase–Gate

Measure	CPM	Phase–Gate[a]
Good with high-iteration-cost projects	+	
Good with low-iteration-cost projects	–	
Process to coordinate varied disciplines	+	++
Mitigates risk before large investments		++
Provides standard work for planning	–	++
Clearly defined methodology	+	+
Tools to maintain schedule	– –	
Intuitive, has few adoption barriers	+	+
Well-defined metrics/visualization	+	
Plans shared resources well	–	
Availability of software tools	++	
Sustainable over time	+	++
Low overhead over time	+	–

[a]Phase–Gate is additive to either CPM or CCPM (Chapter 6); it is not a stand-alone method.

Clearly defined methodology (+)
CPM has a clear definition through the many international organizations such as the Project Management Institute that maintain standards. There are, of course, many variations, but the core concepts such as the Iron Triangle, the role of the PM, the Gantt/Pert charts, and critical path are nearly universally practiced.

Tools to maintain schedule (– –)
CPM projects commonly experience delay. In most organizations, missing the schedule by 10% or 15% is considered good and many projects require double the time originally allocated. It's one of the most common complaints about the method.

Intuitive, has few adoption barriers (+)
CPM is an intuitive approach to project management. It's also been the dominant method of project management for decades. People generally understand the approach so it's easy to get individual and organizational buy-in.

Well-defined metrics/visualization (+)
CPM provides Gantt charts, the best known measure of project progress and delivery of work product. Unfortunately, Gantt charts are difficult to

interpret for those outside the project team. Also, Gantt charts created with high granularity often become outdated quickly. As a result, it's common for companies to rely on milestones as more reliable measures of progress; unfortunately, milestones are typically lagging indicators. Often project delays are discovered too late to allow correction. The progress charts presented in this chapter (Schmidt chart, run chart, and fever chart) improve CPM in this respect, but they are not commonly applied.

Plans shared resources well (–)
Because resources are shared between projects, CPM resource planning can be exceedingly complicated. If a team member is on multiple projects, when one project runs into difficulties it often causes problems for the others. PMs often have little warning that a team member will give more focus to another project.

Availability of software tools (++)
There is an enormous array of tools available to support CPM—Gantt charts with a large range of features are provided by high-end tools like Microsoft Project®. At the other end of the spectrum, there are a large number of free and low-cost cloud-based tools. In addition, there are numerous extensions to Excel that provide simple Gantt charts.

Sustainable over time (+)
CPM has good sustainability over time. The method is practiced in many companies so people new to the organization are often able to support it quickly. It is perhaps the most intuitive of all project management methods, which aids full acceptance.

Low overhead over time (+)
The burden of CPM commonly remains fixed over time.

5.4.1 Key Measures of Effectiveness when Phase–Gate is Added to CPM

This section will evaluate how the key measures from the previous section are affected by Phase–Gate being added to CPM or other project management methods.

Good with high-iteration-cost projects (+)
The concept of dividing projects into major milestones with defined approval steps sets the ability of the method to support high-iteration-cost projects well.

Good with low-iteration-cost projects (–)
The concept of dividing projects into major milestones with defined approval steps reduces the method's ability to support low-iteration-cost projects.

Process to coordinate varied disciplines (++)
The Phase–Gate method is capable of coordinating many functions inside and outside of the organization. It expands on CPM in that different functions within the organization can ensure the areas of their focus are properly represented early in the project.

Mitigates risk before large investment (++)
Phase–Gate creates a review process that is structured around decisions to make further investments. Thus, it focuses the team on minimizing risks before large investments. This reduces the likelihood of making large expenditures before they are justified.

Provides standard work for planning (++)
The fundamental contribution of Phase–Gate when added to CPM is to provide standard deliverables for teams to plan out each phase. It ensures key steps are included in planning, even when the team lacks expertise in all areas. This avoids the common problem where projects are staffed in their early phases mostly by design and marketing: initial planning can too easily omit manufacturing, quality, and sourcing steps.

Tools to maintain schedule
Phase–Gate does little to change whether projects meet schedules over the base method.

Intuitive, has few adoption barriers (+)
The Phase–Gate approach is an intuitive extension of CPM (and CCPM). Stakeholders quickly learn the requirements of the phases and so a common language of project completeness quickly forms. The requirements of each phase can be tailored to an organization. People generally understand the need to define the standard work in a major development project so it can be easy to get buy-in.

Well-defined metrics/visualization (+)
The Phase–Gate process supports a rich set of milestones to measure how well the project is producing deliverables. Since the milestones are quickly understood across the organization, they create a common understanding of what is required for a phase to be "done." It does not provide new metrics to measure progress. The deliverables are often a lagging measure of progress because critical deliverables are often loaded near the end of the phases.

Plans shared resources well
Phase–Gate normally offers no tools to improve resource sharing over the base method to which it is added.

Availability of software tools (+)
The Gantt chart, standard flow charts, or deliverables lists are typically able to support Phase–Gate being added to CPM.

Sustainable over time (++)
The standard work of Phase–Gate brings a natural path for improvement within the organization. The phases can begin with a basic set of required deliverables and then be modified over time as the culture of the organization changes.

Low overhead over time (−)
The process can become heavier; over time many new requirements may be added to deal with new issues, but it's difficult to recognize when old requirements are no longer necessary.

5.5 SUMMARY

CPM together with Phase–Gate create a powerful process—a wide range of projects can be executed with it using any number of developers from a few to hundreds. It allows an organization to manage projects from product definition to launch and, sometimes, beyond that. The method has well-known difficulties bringing in projects on time. As discussed in Section 5.3.2, long delays can be devastating to project success—the right product one year late often is no longer the right product. So, the method works but maintaining schedule requires strong leadership from the PM—transactional and transformational; as a corollary, projects with less experienced PMs commonly encounter serious delays. The characteristics of projects and organizations that fit the CPM method or the CPM method with Phase–Gate are shown in Table 5.8.

5.6 LEADERSHIP AND ALTERNATIVE PROJECT MANAGEMENT METHODS

The next three chapters will present four project management methods that compete with the CPM method with Phase–Gate: two generally aimed at hardware projects (Critical Chain Project Management or CCPM, and LPD) and two generally for software (Agile Scrum and Scrumban). Normally CPM with Phase–Gate is the standard against which other methods are compared. That's because the method is so prevalent in industry today. So, when a claim in made that some measure or other improves by some amount, it's usually in comparison to CPM.

TABLE 5.8 Best Fit for CPM

Projects of medium-to-high complexity.
Projects with a high-cost if iteration.
Highly cross-functional projects.
Projects that demand minimal investment risk (Phase–Gate required).
Companies that build deep process (Phase–Gate required).
Projects where schedule slip is acceptable.
Organizations that are less open to nontraditional project management.

But evaluating a project management method is exceedingly complex: there is so much variation between project types, organizations, and project team members that reliable metrics for comparison are yet to be identified. As you read about these methods, bear in mind that the reasons for new-found success are not fully evident. Perhaps the largest factor that escapes notice is the effect of new leadership.

When a new method is brought to a company, simultaneously but often unnoticed, the leadership skills applied to project management also change. As has been discussed throughout this book, leadership skills are a critical part of what makes any project management method work. During a conversion to a new method, often there are a few agents of change—people who perceive the need for the change, lead the organization to accept that need, find a path forward (the new method), and then lead the adoption of that method. These people are, by definition, transformational leaders—they are transforming how the company executes one of its most important functions: new product development. In addition, when a new method is brought into an organization, it often brings with it an improvement in transactional leadership. The stakeholders may be trained in the proper way to work. The new processes are often executed more diligently and senior management is often fully engaged in reviews and approvals during the transition. These changes would benefit any method used to manage projects.

On the other hand, it is certain there are differences in product development project types and these differences allow alternative methods to serve different project types better. For example, a dominant difference between hardware and software development is the cost of iteration. Software often has a low cost of iteration—new versions can be released multiple times per month.[6] This makes practical an iterative approach: get something simple working quickly and then improve it week by week. By contrast, hardware development usually has a high cost of iteration: making substantial changes to a die-casting mold or an assembly line can be very expensive. This normally leads to a demand to minimize iteration, which usually calls for extensive planning, diligent reviews, and thorough approval processes; here CPM or CCPM with Phase–Gate can be a good fit. Agile and Scrumban are oriented to low-cost-of-iteration projects.

So, when improvement comes after a method is introduced, benefits arise from two factors: (1) those that come from superior techniques of the new method, some of which may apply only to a class of projects and (2) those that come from other changes that accompanied the adoption, such as improved transformational and transactional leadership. Unfortunately, these factors are interwoven and it's difficult to isolate the sources of the benefit. The second category is often unnoticed, even at the company where the adoption occurred. Careful reading of testimonials and case studies is required to ensure these factors are properly accounted for.

6. There are examples to the contrary. For example, the cost of iteration for software in highly regulated markets like aerospace and medical can be quite high.

5.7 RECOMMENDED READING

Kendall GI, Austin K. Advanced multiproject management. J. Ross Publishing; 2012.

Lewis JP. Project planning, scheduling, and control. 3rd ed. McGraw-Hill; 2001.

Shina S. Engineering project management for the global high technology industry. McGraw-Hill Professional; 2013.

5.8 REFERENCES

[1] Kendall GI, Austin K. Advanced multiproject management. J. Ross Publishing; 2012. p. 27.

[2] Chichowski M. Introduction to project and process efficiency IEEE benchmark. IEEE, Currently available at: http://www.ieee.de/fileadmin/user_upload/IEEE_PP_Efficiency_Benchmark_Introduction_20140812.pdf.

[3] http://www.isixsigma.com/tools-templates/fmea/quick-guide-failure-mode-and-effects-analysis/.

[4] Agan T. The secret to lean innovation is making learning a priority. Harvard Business Review; January 23, 2014. Currently available at: https://hbr.org/2014/01/the-secret-to-lean-innovation-is-making-learning-a-priority/.

[5] Sobek DK, Ward AC, Liker JK. Toyota's principles of set-based concurrent engineering. Sloan Manage Rev Winter 1999;40(2):67ff.

[6] Ward A, Liker JK, Cristiano JJ, Sobek DK. The second Toyota paradox: how delaying decisions can make better cars faster. Sloan Manage Rev 1995;36:43–61.

[7] http://www.lean.org/Common/LexiconTerm.cfm?TermId=321.

[8] Raudberget D. Practical applications of set-based concurrent engineering in industry. J Mech Eng 2010;56(11):685–95.

[9] http://www.ganttproject.biz/.

[10] http://www.projectlibre.org/.

[11] https://www.openproject.org/.

[12] http://en.wikipedia.org/wiki/Comparison_of_project_management_software.

[13] Lewis JP. Project planning, scheduling, and control. 3rd ed. McGraw-Hill; 2001. p. 253.

[14] Kelley J, Walker M. Critical-path planning and scheduling. In: Proceedings of the Eastern Joint Computer Conference; 1959.

[15] House CH, Price RL. The return map. Tracking product teams. Harvard Business Review; September 2003. p. 42.

[16] Smith PG, Reinertsen DG. Developing products in half the time. Wiley; 1997. p. 9.

[17] Gardiner PD. Project management. A strategic planning approach. Paulgrave MacMillan; 2005. p. 30.

[18] Schmidt MJ. Schedule monitoring of engineering projects. IEEE Trans Eng Manage May 1988;35(2). Updated version available at: https://www.business-case-analysis.com/schmidt-project-progress-chart.html.

[19] Kokoskie G. A comparison of critical chain project management (CCPM) buffer sizing techniques. SYST 798 Research Project. George Mason University; December 2001. p. 6.

Chapter 6

Critical Chain Project Management (CCPM)

In this chapter, we'll discuss critical chain project management or CCPM. The chapter starts with an overview of the method and then relates it to the theory of constraints (TOC), the foundation of the technique. Then a step-by-step process is presented to create a plan for a CCPM project. There is a discussion on the human behaviors that CCPM techniques address. Then there is a discussion of how CCPM provides tools to manage the schedule of projects and project portfolios. The chapter concludes with a comparison of the critical path method (CPM) (Chapter 5) and CCPM.

6.1 AN OVERVIEW OF CRITICAL CHAIN PROJECT MANAGEMENT

CCPM was created by Dr Eliyahu Goldratt in the 1990s in large part because of poor schedule performance of CPM. CCPM changes both the planning and execution of projects to provide more focus on tasks that must be completed efficiently for the project to be done on time. Today, CCPM is applied across a wide range of project types including many examples in product development.

CCPM has delivered impressive results to many organizations. Here are a few testimonials [1]:

- The introduction of CCPM allowed Lucent Technologies Outside Plant Fiber Optic Cable Business Unit to reduce introduction time of new products by 50%.
- Seagate brought its first 15,000 RPM drive to market so rapidly it caused competition to exit the market around 2000.
- Harris Corp. finished a $250M wafer fabrication plant in 13 months against an original projected schedule of 18 months.
- FMC Energy Systems project on-time delivery went from below 50% to above 90% when using CCPM.
- The Boeing T45 training simulator development saw a 20% cost reduction, substantial quality improvement, and 1.5-month reduction in schedule [2,3].

There are many more. Dr Goldratt's website states: "Tens of thousands of people worldwide depend upon Goldratt's business knowledge to successfully

Project Management in Product Development. http://dx.doi.org/10.1016/B978-0-12-802322-8.00006-1

manage their organisations" and then lists numerous cases from BAE systems to Habitat for Humanity [4]. The book *Advanced Multi-Project Management* lists almost 400 examples [5,6] of companies who have adopted the method. CCPM is similar to CPM in that both start with a project network, the diagram of interconnected tasks. Both choose the critical tasks in a similar way, the most significant difference being CCPM uses explicit methods to eliminate resource conflicts where CPM relies on ad hoc methods. At this point, the two methods are similar.

During execution, CCPM then adds many techniques. Some of these techniques are good practices that any project management method can benefit from:

- Increasing efficiency by minimizing multitasking, minimizing procrastination, and ensuring tasks are fully prepared before starting them (called *full kit*).
- Building teamwork by minimization of blame.
- Building a common focus for daily work by focusing on the critical tasks.

The hallmark technique of CCPM is to use a probabilistic method of planning with aggressive duration estimates for each task and then placing a buffer at the end of the project to account for the likelihood that many tasks will not meet those estimates. Project execution focuses on the measurement and management of that buffer. Progress is measured by tracking the critical chain—milestones completion metrics are less important in CCPM than in CPM.

While some authors see CCPM as a breakthrough in project management techniques [7], others characterize it as a collection of known techniques [8,9]. Lechler et al. take up this topic with a convincing argument that the method is unique [10]. There is no doubt that there are many techniques in CCPM that are used in other methods, but CCPM's use of the buffer is unique. Its central position in planning, execution, reporting, and portfolio management differentiate CCPM from other project management methods.

6.2 THE THEORY OF CONSTRAINTS

The critical chain is based on TOC, an approach popularized by Eliyahu Goldratt's book *The Goal* [11]. The basic principle is that there is one constraint that defines the limit of any system. Attempts to improve the system must focus on improving that constraint. Further, people often don't recognize the constraint and so will sometimes attempt to improve a system in ways that don't address it; according to TOC, such attempts will be ineffective.

In *The Goal*, this principle is presented through a novel where a troop of boys on a long hike is slowed by one of boys. The leaders determine they must speed up and then try several ideas that don't address the constraint. They place the slowest boy at the front of the troop where he creates a bottleneck; they place him at the end, where the troop proceeds faster for a time but then must wait for him when he lags too far behind. Success is finally found when they address the constraint, for example, having other boys carry part of the load the slow boy is carrying.

Goldratt asserts that attempts to improve throughput on the factory floor follow a similar pattern. There are many improvements that can be made to manufacturing process, but only those that directly address the dominant constraint are likely to be successful. Similarly in project management, any attempt to improve project performance that does not directly address the dominant constraint will bear little fruit.

Goldratt's book *The Critical Chain* [12] states that the primary measure of project management effectiveness is schedule. (This topic was discussed in detail in Section 5.3.) The constraint for this measurement is the path through the project network that is longer than all the others. He names it the "critical chain," but it's almost identical to the "critical path" used in CPM (the primary difference between the two is the technique used in critical chain to avoid resource conflicts).

An important observation for CCPM and one that extends to lean product development (Chapter 7) is that resources constrain the velocity of project execution. Once the constraint is exceeded, adding more work does not improve performance. Think of a busy highway—allowing more cars to enter slows down the entire system. In fact, a well-tested method of maximizing traffic flow is limiting the cars allowed on the highway.

This concept is extended to project management by limiting the amount of *work in progress* or WIP. Unfortunately, the natural tendency in industry is just the opposite. When project execution slows, fewer projects are completed but the need for new projects continues. When new projects are added without the old projects being completed, the WIP increases; this causes more distractions for the project teams in the form of conflicting priorities and excessive multitasking. Velocity slows and the project queue continues to increase. As the phenomenon worsens, the project team can view their own work as poor and/or see the demands of the organization as unrealistic. Morale falls and this can cause efficiency to decline further. Letting every project "on the highway" usually results in lower productivity.

TOC states that the best way to increase project velocity is to address the constraint, the critical chain. Accordingly, it provides a continuous-improvement process, the *five focusing steps*, as shown in Figure 6.1 [10].

- Identify
  First, find the constraint, the weakest link in the chain. On the production floor, search for the machine or process that limits the assembly cycle. In a project, the constraint is the critical chain of tasks. In a portfolio of projects, the constraint is usually the resource(s) that must be shared the most among the projects.
- Exploit
  Do everything possible to improve performance with the available resources. For one project, this means focusing on the critical tasks, working efficiently, and collaborating well. This is the focus of most of the tools of CCPM: common team understanding of the critical chain, the use of aggressive estimates, and

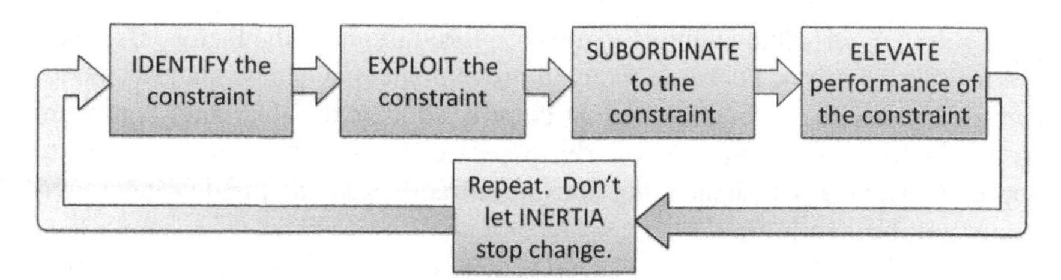

FIGURE 6.1 The five focusing techniques of TOC.

the elimination of multitasking. For the project portfolio, this means managing how the shared resources are used and maximizing their effectiveness.

- Subordinate
  Ensure all resources working on anything but the constraint are available to support those resources on the constraint. For a single project, activities for anything off the critical chain must not distract resources working on the critical chain. In a portfolio of projects, this can mean accepting that noncritical resources may sometimes be allowed to be idle if this ensures the highest utilization of the critical resources.
- Elevate
  After applying the first three steps, if the system performance is still unacceptable, then the limit of the constraint must be increased. The most obvious application of *Elevate* is to add resources to relax the constraint. But there are other alternatives related to increasing efficiency: providing better tools to the team, increasing technical capacity through training and coaching, and borrowing resources from projects that are ahead of schedule.
- Inertia
  If a new constraint is identified in the previous steps, repeat the entire process applied to that constraint. Don't let *Inertia*, the aversion to change, become the constraint.

6.3 BUILDING A CRITICAL CHAIN PROJECT PLAN

A single critical chain project can be planned by first building the task network and then following steps as shown in Figure 6.2. The initial planning of a critical chain task network is similar to a critical path project:

1. Identify the optimal plan granularity, as discussed in Chapters 2 and 5.
2. Connect the tasks in a network that takes into account predecessors as was done in Figures 2.1 and 5.3. The network can be shown as a Gantt chart.
3. Estimate the duration of each task and identify the resources required.

The critical path is then identified using a technique similar to those used to build Figure 5.6. The task network can then be converted to a CCPM plan using the following five steps.

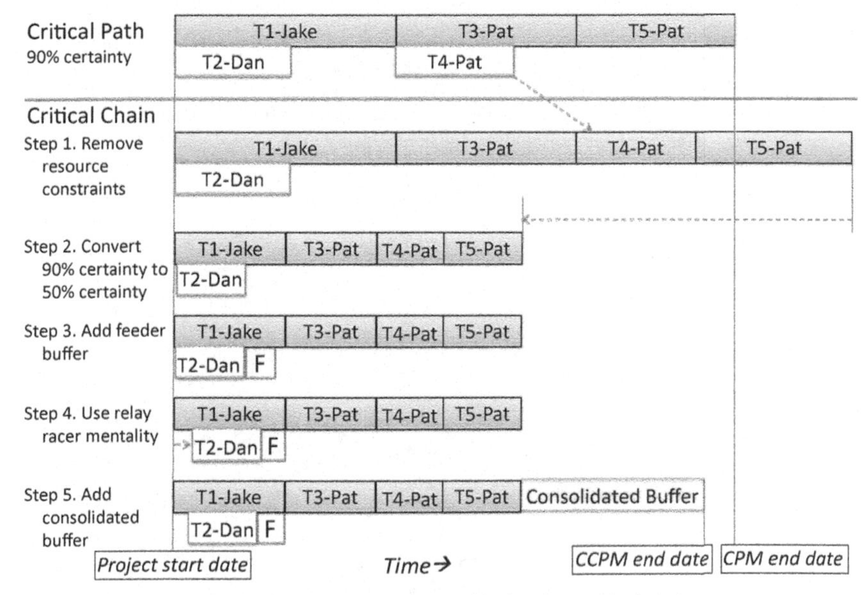

FIGURE 6.2 Five steps from critical path to critical chain.

6.3.1 Step 1. Remove Resource Constraints

All tasks that use the same resource must be staggered so the resource does not need to multitask. This is shown in Figure 6.2 ("Step 1") as putting Task T3 and Task T4 in series since both are owned by Pat. This places T4 in the critical chain although it was not in the critical path.

6.3.2 Step 2. Reduce Task Length by Changing Certainty from 90% to 50%

In critical path methods, normally one question is asked to estimate a task length: how long will it take? In fact, there is no single answer because almost all tasks have significant unknowns. When the question is answered, there is an implicit estimate of certainty. It's rare that a task can be estimated with 100% certainty—no matter how conservative an estimate, it's almost always possible that some unexpected event could occur so the time spent could exceed the estimate. This relationship is shown in Figure 6.3.

CCPM asserts that most people answer with an unnecessarily conservative estimate because they seek to protect the individual task. In CCPM, this mindset is usually called "90%[1] certainty" although the 90% isn't meant to be a precise measure. The point is that when estimating, a significant buffer is added for each task to increase certainty. In fact, when there are many tasks, a great deal

1. Or sometimes 95%.

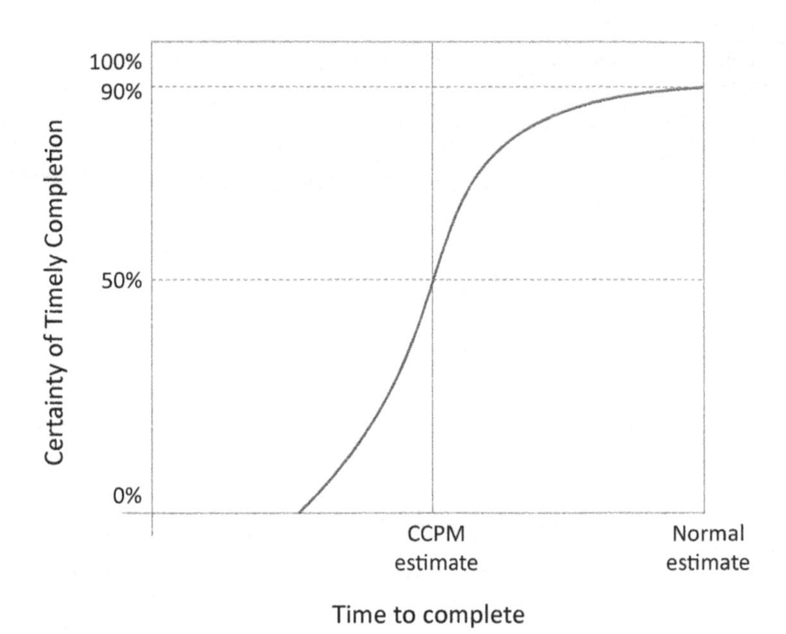

FIGURE 6.3 Estimated time to complete a task versus certainty.

of variation in estimation accuracy can be expected. However, if each task is estimated with a large buffer, the total schedule will be far too conservative. CCPM teaches that tasks should be estimated with 50% certainty—that is, that each task will be as likely to come in on time as to be late. With 50% certainty, there should be as many early tasks as late tasks, so the aggressive estimates can balance, at least in part.

The thinking has an analogy with driving a route with 10 traffic lights. If you estimate the time to get through any one light, you'd need to include the length of a red light because the likelihood that you will arrive at a red light is too high to be ignored. However, the chance that all 10 lights will be red is vanishingly small (at least in the absence of heavy traffic). So planning for 10 full red lights would be excessively conservative. CCPM would recommend taking half the waiting time for each light since the most likely outcome is many lights will be green.

Of course, statistical processes don't reliably produce the mean outcome. Sometimes you might get seven green lights; other times, you might get only three. So, with a 50% assumption, there will be enough variation to create a substantial likelihood that the mean will be exceeded. CCPM addresses this with a consolidated buffer, which will be added to the end of the project as will be discussed in Step 5.

The process to convert 90%-certainty estimates to 50%-certainty could be complex indeed. We rarely know enough about the individual tasks to produce a plot like Figure 6.3 with any accuracy. CCPM gets around this complexity with a startlingly simple assumption: the proper length of task for 50% certainty is about half the length of what people normally estimate (the so-called

90%-certainty estimate). That's precisely the ratio shown in Figure 6.3. There is little objective proof of this ratio; it is counterintuitive to many people and this assumption can be a barrier to adopting the method in some organizations [13].

6.3.3 Step 3. Add Feeder Buffers

Feeder buffers are added where tasks off the critical chain interface to the critical chain. The concept is to make noncritical tasks subordinate to the critical chain; planning must account for uncertainty in the noncritical task and add that uncertainty afterward as a buffer. In Figure 6.2, T2 is off the critical path and so a feeder buffer (marked with an "F") must be added after the task. The feeder buffer is normally assumed to be half of the task length. Feeder buffers are not universally used in CCPM; some proponents teach that they are redundant.

6.3.4 Step 4. Using the Relay-Racer Mentality

When the noncritical tasks are scheduled, they are delayed so they start at the last minute for the task estimation plus its feeder buffer. The concept is that team members should be either working at full speed on a task or not working on it at all. This is analogous to a relay racer who starts running after the runner for the previous leg transfers the baton. The runner is either idle or running as fast as possible—there is nothing in between. As shown in Figure 6.2, T2 is moved to the right so the end of the feeder buffer occurs at the start of T3. The relay-racer mentality keeps focus on the task at hand, limiting WIP in the project and ultimately increasing velocity.

6.3.5 Step 5. Add the Consolidated Buffer

Finally, a consolidated buffer is added at the end of the critical chain. The simplest method is to add a buffer equal to a fixed fraction of the length of the critical chain. In this case, the most popular buffer size is 50%, which was originally suggested by Goldratt; other authors recommend as little as 30% [14].

When the critical path has no resource conflicts (in other words, when Step 1 results in no changes), Steps 2 and 5 combine to provide a CCPM project schedule equal to 75% of the CPM estimate:

$$\text{CCPM estimate} = (\text{CPM estimate})/2 \times (1.5) = 0.75 \text{ CPM estimate.}$$

Any amount added to the critical chain in Step 1 will reduce the difference between the two methods; even so, in most cases, CCPM provides a substantially accelerated schedule. This result is counterintuitive for organizations unfamiliar with the method. Also, using 50% of the critical chain for the consolidated buffer results in one-third of the project being a buffer, which is another counterintuitive trait that is challenged frequently [15]. There is a common misunderstanding that the buffer is a kind of slack added at the end of the project, but expected to be unused unless the project runs into trouble; in fact, in CCPM buffer is part of the schedule, expected to be used to absorb the high likelihood that some 50%-certainty tasks will run long.

There are competing methods for adding the consolidated buffer. An alternative is to add the square root of the sum of the squared critical path tasks, similar to how tolerances in a mechanical assembly are sometimes combined. This normally results in a similar sized buffer. There are more sophisticated methods to determine buffer size, some requiring complex calculations. Geekie [16] and Cox and Schleier [17] describe numerous alternatives.

If you are considering CCPM and are not sure how to size the buffer, it probably matters less than you might think. For many practitioners, the power of the critical chain is not in the mathematical precision of the buffer (or, for that matter, of the schedule as a whole) but rather in the creation of a substantial buffer at the project end. So long as that buffer is large enough to absorb the normal variation of task execution and short enough that the total schedule remains reasonable, you should be able to gain experience with the method. Over time, your estimates will improve and you can then fine-tune the buffer size to the needs of your organization.

6.4 EXECUTION AND HUMAN BEHAVIOR THAT DELAY PROJECTS

A central tenant of CCPM is that there are several common human behaviors that reduce team effectiveness. The approach of CCPM is to create a method that minimizes the effects of these behaviors; this is another hallmark of the method. By comparison, proponents of CPM often give small weight to these behaviors. CCPM asserts that five common tendencies in team members delay projects:

1. The inefficiency of multitasking
2. The student syndrome (a form of procrastination)
3. Parkinson's law ("tasks expand to fill the allowed time")
4. Unintentional disincentives to speeding task execution
5. Protecting individual tasks at the expense of the project

While project managers (PMs) in CPM may recognize some or even all of these problems, there is little in CPM that helps the PM reduce their effects. CCPM builds process to deal with these problems as standard work.

6.4.1 Inefficiency of Multitasking

Multitasking is common in the workplace. Many studies show that people average as little as 3 min on one task before switching to another [18]. CCPM asserts that projects rely too heavily on multitasking, which is a highly inefficient practice. For example, if there are three tasks that each take 1 week, it's too common for a manager to assign all three to one person and allow 3 weeks (or less!). Implicit in this approach is that the person will switch among the tasks effortlessly. In fact, there are three well-known effects of multitasking that are counter to this thinking.

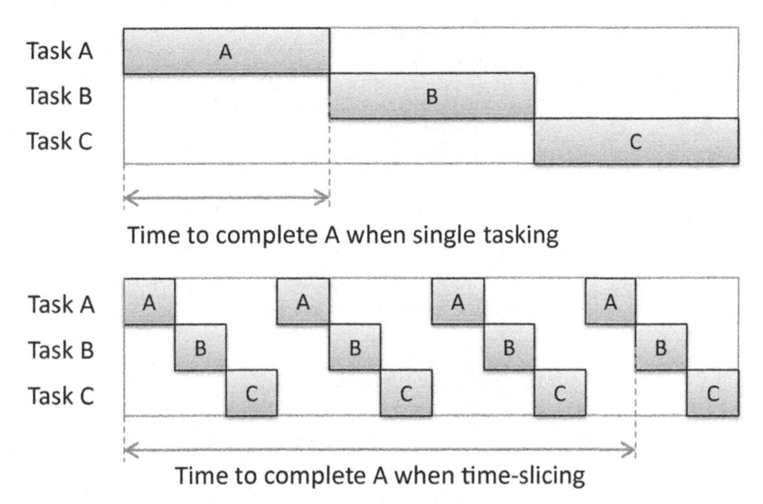

FIGURE 6.4 Lengthening of completion time due to time slicing.

Schedule Delays due to Time Slicing

As shown in Figure 6.4, when three tasks are ideally time sliced, although the three are completed in the same time period, there is a large delay for completion of the first and second tasks. Note that this problem arises even with the assumption that multitasking is a perfectly efficient process, a concept that will be challenged shortly. But even in this most optimistic case, consider the problems that proceed from time slicing. Let's take a common case where the three tasks of Figure 6.4 are in different projects, let's say Task A is in Project 1, Task B is in Project 2, and Task C is in Project 3. By time slicing, Projects 1 and 2 are delayed, but no benefit is provided to Project 3. In general, time slicing creates delay for some tasks that is not balanced by positive effects for others.

Efficiency Reduction due to Context Switching

Although multitasking is often planned as if the process were essentially perfectly efficient, in fact there are several well-known efficiency problems with multitasking. First, changing focus from one task to another, a process called *context switching*, takes a significant effort called a *resumption lag* [19]. If a team member is left to focus on a topic for an extended period of time, she is able to build a mental framework around that task: why the task is important, what the dependencies are, details around various alternatives for technical solutions, and so on. Each time she switches tasks, it takes time to rebuild this framework. The more complex the task, the longer the resumption lag.

Intrusive Thoughts from Suspended Tasks

In addition to the time required to context switch, simply working in a multitasking environment brings inefficiency due to the tendency for people to experience intrusive thoughts from unfinished tasks, sometimes called the Zeigarnik

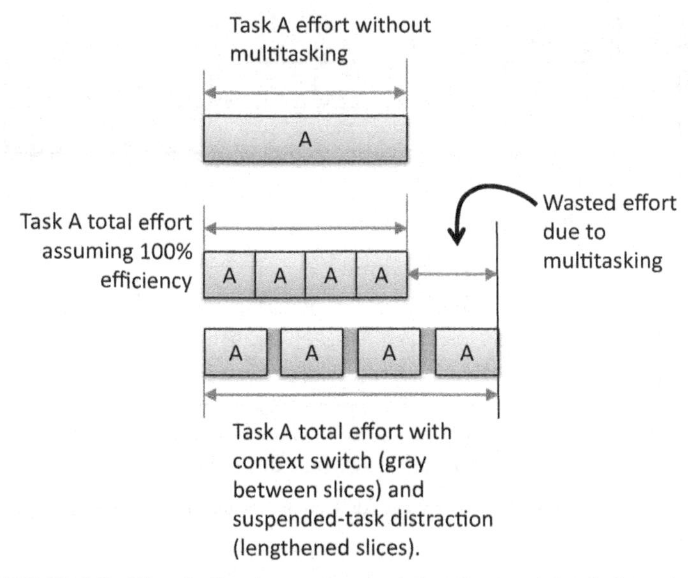

FIGURE 6.5 Wasted effort due to context switch and suspended-task distraction.

effect. Suspended tasks can blur focus and encourage context switching when encountering a barrier in the current task. If you're baffled after 30 min searching for the root causes of a bearing failure and an email pops up related to a suspended task, reading it can prove irresistible. For many multitaskers, no matter what they are working on, there is a sense they should be working on something else.

The effects of multitasking combine to produce wasted effort, as shown in Figure 6.5. So, in addition to the schedule delays caused by time slicing, the total effort to finish a task increases in a multitasking environment.

6.4.2 The Student Syndrome

The student syndrome refers to the way people procrastinate when a deadline is used as the primary driver for completion. It's commonly seen in students, probably because of the nature of modern education where the work product required for a semester (papers, tests, projects, etc.) is neatly scheduled with fine granularity by the teacher or professor. As shown in Figure 6.6, when an assignment is received, the student will invest a small effort. Since the primary motivator is the deadline, the student will, perhaps unconsciously, perform a mental calculation: assuming I work at my maximum rate, what is the "last minute" I can start and still meet the deadline? Any significant work is then delayed until the last minute.

This mindset works reasonably well in school, where requirements are clear and fixed. In the dynamic environment of product development, it undermines

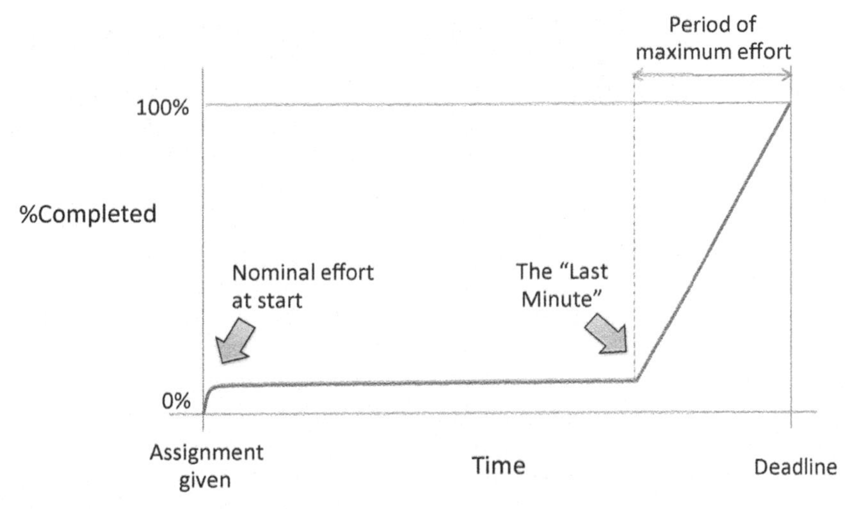

FIGURE 6.6 The student syndrome.

the schedule. When buffer is inserted in anticipation of risks that might be discovered later, the buffer can be wasted—add 5 days to the due date to allow for a risk and the "last minute" simply slides out 5 days. Whatever the buffer, work starts in a frenzy a few days ahead of the deadline; unanticipated issues result in delay no matter how much buffer is added. Not fully cognizant of the phenomenon, some PMs are convinced they are not allowing enough time to deal with the unknowns and so add more schedule buffer. When that doesn't work, the PM can lose trust in the team member; he may then try to micromanage: constantly monitoring the team member's activities. This is another ineffective response—aside from the frustration it causes, it cannot be sustained and it doesn't help the team member grow out of the behavior.

6.4.3 Parkinson's Law

Parkinson's law [20] is commonly stated as "Tasks expand to fill the allotted time." The result is the time allowed for completion becomes a proxy for when the task is done and it's a poor proxy. Probably the best place to see Parkinson's law in your own behavior is when taking a timed test. If you sense you're ahead, you will likely slow down when you encounter a difficult question. If you finish the test early, you're likely to review the harder questions. You use the available time to maximize the test score. A similar phenomenon occurs with meetings—they generally fill the allotted time. This maxim is often quoted derisively, as if people intentionally waste time until the clock runs out. In fact, this result is normally driven by positive behaviors that seek to maximize results. Like a student sitting for a test, most people want to use the time to get the best result possible. You can get more done in a 120-min design review than in 60 min. Both examples have something in common: an expert (either a professor or a design review leader) sets the appropriate time span that others in that process can rely on. When the owner sets the time appropriately, results are usually good.

The problems with Parkinson's law are seen more clearly in day-to-day tasks. A PM and the task owner may make the best estimate they are capable of before the task starts. But once the task begins, the team member may substitute the original estimation as the primary measure of completeness. Tasks that are more difficult to define are more likely to encounter this phenomenon: risk identification, patent searches, and supplier qualification are a few examples. In a practice called *gold plating* in lean product development (Chapter 7), if a task turns out to be easier than was expected, the team member will continue to work on it even after the task is done well enough to meet the needs of the project. As shown in Figure 6.7, there often exist many plausible definitions of *done*—Parkinson's law states that team members will continue working toward an expansive and perhaps even unrealistic definition of *done*. In fact, they will only finish when the time allowed for the task expires.

PMs can respond to Parkinson's law in three ways:

1. Define task deliverables clearly avoiding using time as a criterion.
2. Set the appropriate amount of time for their completion.
3. Create incentives for the team member to complete a task early, which is a specific goal of CCPM.

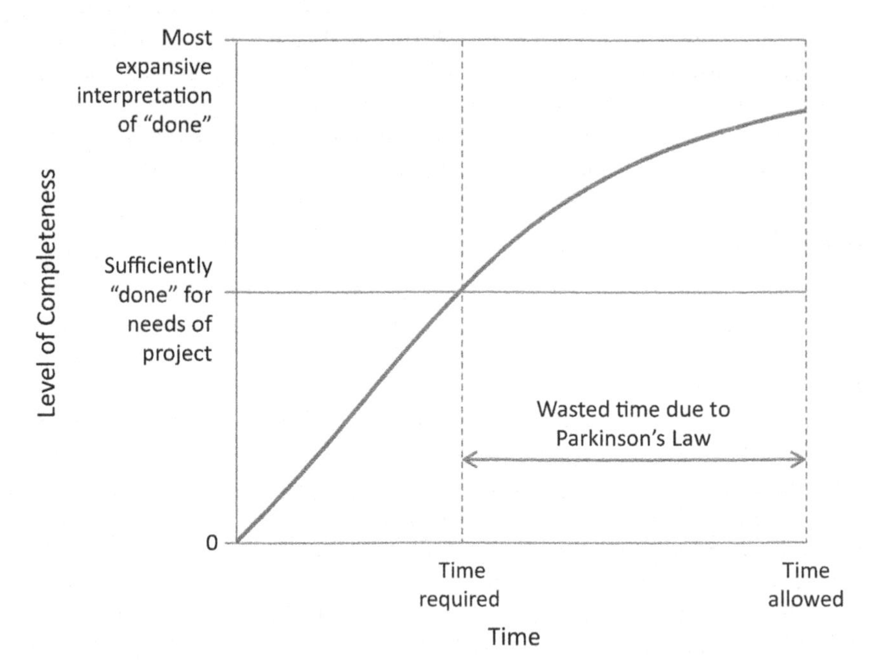

FIGURE 6.7 Parkinson's law.

6.4.4 Disincentives for Early Completion of Milestones

In CPM, the process of setting dates for milestones can inadvertently create disincentives to completing steps early. When a PM uses a conservative estimate but the task completes much more quickly, it can create the appearance of "low balling," the practice of intentionally using longer-than-necessary task estimates to reduce pressure on the project team. Low balling is distained by managers and PMs avoid even the appearance of doing it.

For example, say a custom part from a well-known supplier normally takes 2 weeks to be delivered. However, this supplier also sells to a higher priority customer that from time to time absorbs all their customization capability for a week or two. So most of the time, the supplier can be expected to deliver in 2 weeks, but occasionally it takes 4 weeks. In CPM, the PM will normally allow 4 weeks for the task; in the parlance of CCPM, she is using estimates that have 90% confidence.

However, suppose the project sponsor sees the 4 weeks and suppose also that he has enough experience to know the supplier can normally deliver faster. A tug-of-war ensues where the PM and sponsor dig into their opposing positions. Whoever wins the battle, the PM now has a disincentive to accelerate the task. If it comes in early, she exposes herself to the appearance of low balling and so can undermine her position when a similar conflict arises in the future. None of this is to say the PM will intentionally delay progress; there's no need—Parkinson's law, the student syndrome, and a host of other factors normally prevent unsought progress.

6.4.5 Protecting Tasks at the Expense of the Project

Project performance in CPM is typically based on individual accountability: the PM works with the team to divide up the project into numerous milestones. Tasks leading to the milestones are owned by the various team members. The concept is that if each team member delivers their tasks on time, the project will finish on time. But, as the project proceeds, the critical path encounters barriers so that the team needs to pull together for the project to meet its objectives. As much as the PM might value the team spirit that would cause one team member to pull off their task to contribute to tasks on the critical path, the individual ownership of the task provides a disincentive to take that action. If you're the engineer that owns the tasks around automatic test equipment, you are incentivized to avoid any action that makes the milestone "Automatic test equipment finished" late, whether or not those tasks are on the critical path.

6.4.6 CCPM Responses

In summary, there are at least five sources of delay and inefficiency that come from normal human behavior. CCPM asserts that CPM does little to combat these tendencies. As shown in Table 6.1, CCPM builds process in to reduce the negative effects these behaviors.

TABLE 6.1 CCPM Responses to Human Behavior

Human Tendency	CCPM Response
Overuse of multitasking	Chase multitasking out of projects. Schedule people full time even when shared across multiple projects. Use the relay-runner mindset for task work—either working on a task with full effort or not at all.
The student syndrome	Use aggressive schedule estimates to remove buffer from individual tasks and consolidate it at the end of the project.
Parkinson's law	Use aggressive schedule estimates to minimize the opportunity to "gold plate."
Negative disincentives to complete tasks early	Use remaining buffer to measure project progress. Avoid commitments on individual milestones. Reward team members when buffer is conserved.
Protecting tasks at the expense of the project	Avoid milestones as a measure of progress; focus on the critical chain.

The result of CCPM's focus on the buffer is less to improve mathematical accuracy of estimates and more a way to change the mindset of the project team. According to Loch et al. [21]:

> *Project workers no longer need to protect their own schedule (so they no longer need to low ball), nor can they procrastinate because they impact the overall buffer that everyone… depends on. The entire team "sits in one boat."*

6.5 TRACKING PROGRESS WITH THE FEVER CHART

We now move to the execution phase of the project and the details of creating the most well-known graphical display in CCPM, the *fever chart*. In order to discuss the fever chart and compare it to an equivalent Gantt chart, we will need to create a sample project. This project will have 10 tasks in the critical chain. The task durations are estimated aggressively (50% certainty) and range from 2 to 9 days as shown in columns 1 and 2 of Table 6.2. The critical chain totals to 40 working days. This example will use the simplest buffer calculation: 50% of the critical chain or 20 days. The projected dates in column 3 are calculated by summing the durations in column 2; weekends are avoided using the Excel® function WORKDAYS().

Each week of the project occupies one column in the "Project execution" section of Table 6.2. The project starts on Feb-1. The critical chain is aggressively estimated to complete on Mar-12 and the buffer carries the project to Apr-25. The progress achieved each week for each task is shown in the column below the date. For example, on Feb-15 Tasks 1 and 2 are 100% complete; no other tasks are started.

TABLE 6.2 Ten-Task Critical Chain Totaling 8 Weeks

Project Planning			Project Execution												
Critical Tasks	Days	Date	Feb-01	Feb-08	Feb-15	Feb-22	Feb-29	Mar-07	Mar-14	Mar-21	Mar-28	Apr-04	Apr-11	Apr-18	Apr-25
Task 1	3	Feb-04		100%	100%	100%	100%	100%	100%	100%	100%	100%	100%	100%	100%
Task 2	4	Feb-10			100%	100%	100%	100%	100%	100%	100%	100%	100%	100%	100%
Task 3	9	Feb-23				25%		100%	100%	100%	100%	100%	100%	100%	100%
Task 4	4	Feb-29						50%	100%	100%	100%	100%	100%	100%	100%
Task 5	2	Mar-02							80%	100%	100%	100%	100%	100%	100%
Task 6	3	Mar-07									100%	100%	100%	100%	100%
Task 7	4	Mar-11									100%	100%	100%	100%	100%
Task 8	5	Mar-18										10%	50%	100%	100%
Task 9	2	Mar-22												100%	100%
Task 10	4	Mar-28													100%
Total	**40**														
Buffer	20														

A narrative of the project can be gleaned from Table 6.2:

- Feb-08: The first week of the project shows completion of Task 1, a 3-day task.
- Feb-15: In the second week, the team completes Task 2. Now, 10 days into the project, 7 days of critical chain are completed.
- Feb-22: Task 3 was tougher than expected and the team completed only 25% of the task.
- Feb-29: Bad news. The team recognized the approach to Task 3 was misguided and everything done on it had to be scrapped.

Progress continues as shown in Table 6.2, completing just on time.

6.5.1 Calculating Buffer Penetration

The buffer penetration calculations are shown in Table 6.3. Calculations for each row are:

1. *Days passed* is calendar time: 5 working days each week.
2. *Critical chain completed* uses data from Table 6.2: the completion percentage from the column in the "Project execution" section is multiplied row by row with the task duration in column 2.[2] For example, from Table 6.2, Feb-15 shows 100% of Tasks 1 and 2 (3 and 4 days, respectively) and 0% of all other tasks for a total of 7 days.
3. Chain completion is converted to a percentage by dividing by 40, the length of the critical chain.
4. The buffer penetration is *Days passed* minus *Critical chain completed.*
5. Buffer penetration is converted to a percentage by dividing by 20, the buffer length.

6.5.2 Creating a Fever Chart

A fever chart can be created from Table 6.3. The fever chart typically plots "Critical chain complete" as the X-axis against "Buffer penetration." The fever chart for the sample project is plotted in Figure 6.8. The plot area is divided into three sections:

1. *Continue*: the lower right section where the buffer penetration is low. The project is ahead and the team should continue working as they have been.
2. *Plan*: a section slicing from bottom left to top right. Here, the buffer penetration is marginal; the PM should be leading the team in creating an action plan to be executed if more delays are encountered.
3. *Take action*: The upper left where buffer penetration is unacceptably high. Here, the project has a high likelihood of late delivery and the team must work to recover buffer.

The fever chart can also be used to track usage of active feeder buffers in a similar manner.

2. In Excel®, this is calculated applying the SUMPRODUCT() function.

TABLE 6.3 Buffer Penetration Calculation of Project of Table 6.2. Ten-Task Critical Chain Totaling 8 Weeks

	Buffer Penetration												
	Feb-01	Feb-08	Feb-15	Feb-22	Feb-29	Mar-07	Mar-14	Mar-21	Mar-28	Apr-04	Apr-11	Apr-18	Apr-25
Working days passed	0	5	10	15	20	25	30	35	40	45	50	55	60
Critical chain complete (days)	0	3	7	9.25	7	18	21.6	22	29	29.5	31.5	36	40
X axis—critical chain complete (%)	0%	8%	18%	23%	18%	45%	54%	55%	73%	74%	79%	90%	100%
Buffer penetration (days)	0	2	3	5.75	13	7	8.4	13	11	15.5	18.5	19	20
Y axis—buffer penetration (%)	0%	10%	15%	29%	65%	35%	42%	65%	55%	78%	93%	95%	100%

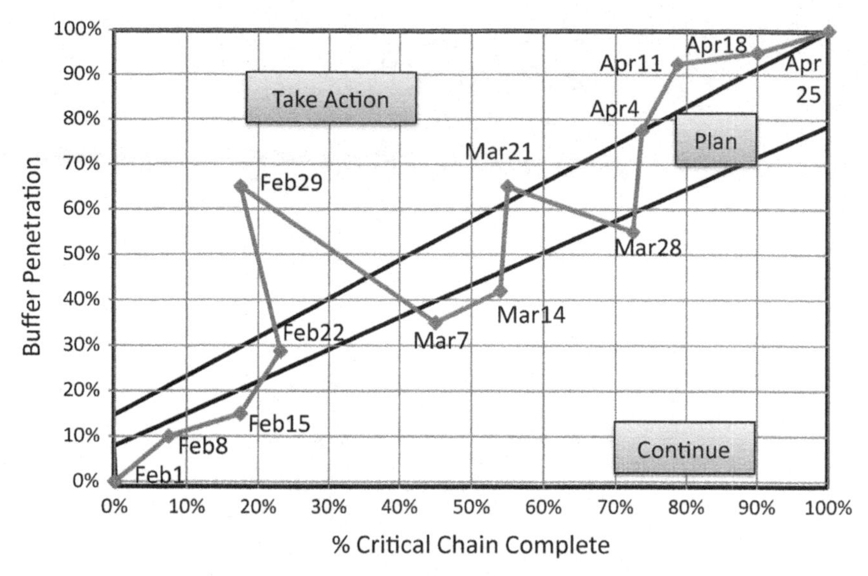

FIGURE 6.8 A fever chart for the example project.

6.5.3 Using the Fever Chart

Now let's imagine what might have happened on Feb-29, when the fever chart would have looked like Figure 6.9. The penetration is well into the *Take action* section of the plot. Further, there is historical data: between Feb-15 and Feb-29, half of the total buffer was consumed without any progress on the critical chain. It would be clear to anyone with the most rudimentary familiarity with CCPM that the project has run into a serious schedule problem. Assuming the fever chart is posted where all stakeholders can view it, the entire team will quickly understand the situation. It's time for the PM to lead the team to execute any or all of the alternatives discussed in Section 5.3.7 to restore the buffer penetration to a reasonable level. If that doesn't result in a credible recovery plan, it will be necessary to escalate to the project sponsor.

CCPM Increases the Focus Compared to CPM

Now let's imagine what the same project would have looked like with CPM. If a CPM project plan were built with the same 10 tasks in the critical path, and with the same start date and end date, the Gantt chart might have looked like Figure 6.10.[3] Here the critical path is shown with hash lines and progress is

3. For this example, each task in the critical path was increased by 50% to spread the buffer between the tasks. This varies the process of Figure 6.2, but was used because it results in the CPM project schedule that is most similar to Figure 6.8. It also assumes there were no resource conflicts so the critical path and critical chain would contain the same tasks.

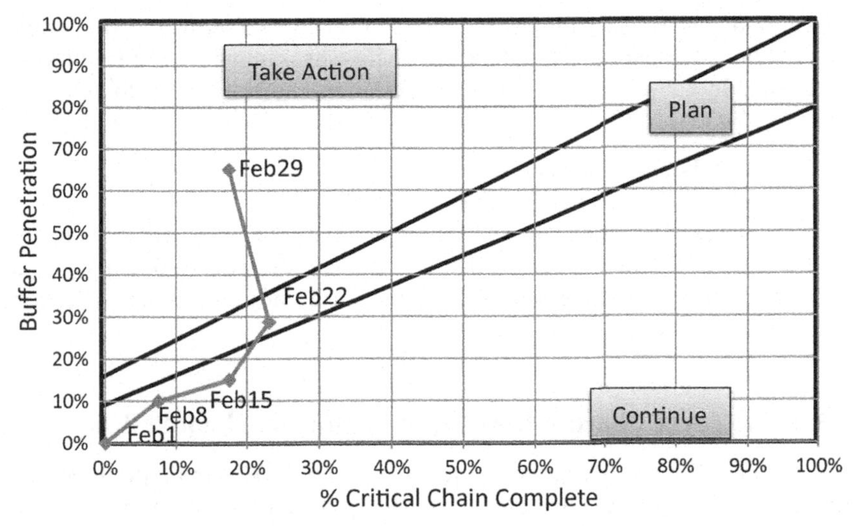

FIGURE 6.9 Fever chart as of Feb-29.

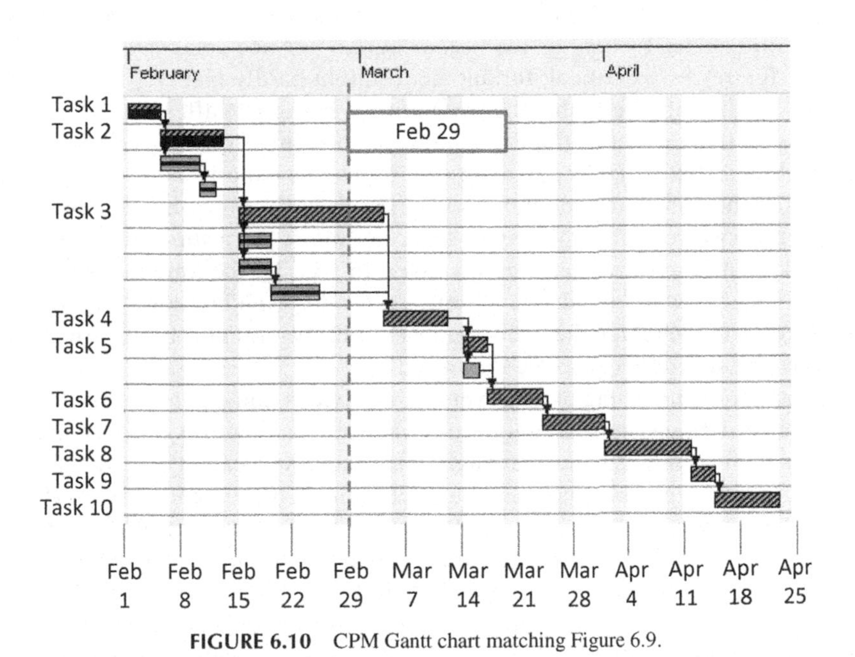

FIGURE 6.10 CPM Gantt chart matching Figure 6.9.

indicated with a black bar in the center of the task rectangle. A few noncritical path items are added between Tasks 2 and 6 as is common in CPM.

So, by Feb-29, when the fever chart showed a serious problem clearly, the Gantt chart shows little urgency. Sure, the project is behind in Task 3, but that's not even due for a week. Besides that, overall progress (that is, considering critical and noncritical path tasks) looks reasonably good. It would be easy for even an experienced PM to conclude there was minimal need for immediate action. In CPM and its accompanying Gantt chart, the need to action is obscured by:

1. The buffer being spread to all tasks.
2. The lack of historical data in a single display.
3. The display of noncritical tasks (all of which are going well in this example).
4. The lack of a progress metric. The Gantt chart shows a 10-day delay in a task, but there's nothing equivalent to the clarity of the *Take action* section of the fever chart.

6.5.4 The Fever Chart: A Leading Indicator of Schedule Risk

For the PM, the fever chart provides a *leading indicator* of progress, an elusive metric in project management. Although it's possible to provide an equivalent display in CPM (see Figure 5.17), it's uncommon in industry today. In a typical CPM project, progress is reported by how late the team is to the milestones favored by the organization—for one company a successful design review is critical; for another it might hardly matter. This indicator is usually *lagging* in that the team reports being late after the milestone is missed. In a perfect world, the team would carefully track the critical path and when a task that was upstream of a milestone experienced a delay, the effect on the milestone would be shown immediately. In reality, that doesn't happen very often. The PM may be reluctant to predict a slip in a milestone 6 weeks out, hoping the team can recover. In fact, by the time the management is made aware of a delay, it's often after the team has done everything they know to overcome the delay. At that point, the slip may be so large that the project schedule cannot be recovered.

The unique reporting of the fever chart provides immediate feedback and it does so without the PM having to have to make a painful decision about whether something is serious enough to escalate. As Kendall and Austin put it, "even when project managers suspect there is a problem … they often find themselves in a conflict—call now for help or wait until disaster is proven!" [22]. In CCPM, the indication of delay is leading so the delay often can be recovered. By contrast, CPM projects are built with no end buffer. Once a date is missed, the entire schedule slips. Parkinson's law and the student syndrome often prevent the team from making up schedule delay. With CCPM, the team can gain confidence because they can recover from even large delays using the end buffer. This improves morale compared to the common experience team members have in CPM projects: once you get behind, you never catch up. People may work on projects in CPM for years without ever working on one that finished on time.

6.6 FULL KITTING

The practice of ensuring everything is needed for a task, project, or project phase is called *full kitting*. This includes activities such as:

- Ensuring requirements are clear and complete.
- Obtaining all approvals.
- Providing all materials needed.

Full kit ensures that once the project/task has started, it can proceed at full speed to completion. This avoids the waste incurred when a project starts and then stops or proceeds with partial effort. Much of full kit is built into Phase–Gate projects by ensuring all tasks in the previous phase are complete and obtaining steering committee approval before proceeding to the next phase. An organization can increase emphasis on full kit by creating a full-kit check list as shown in Figure 6.11 [23].

The creation of the full kit is led by a full kit manager as a highly collaborative process—the entire project team should join in. The concept of full kit also applies to tasks within project. Here, the task owner ensures full kit before starting the task [24].

6.7 CCPM FOR PROJECT PORTFOLIOS

Critical chain also defines process for multiple projects, something that CPM doesn't explicitly address. Two areas of interest are planning multiple projects for which there are resource conflicts and the extension of the fever chart to project portfolios.

6.7.1 Multiproject Planning

CPM doesn't directly address the common problem where there is a single resource used for multiple projects. For example, a company may have one software guru or chemistry genius. Such a situation is common in product development organizations because such people are rare and can add value to many projects. Also there are service organizations that must serve many projects: IT, a certification team, and a test lab are examples. In CCPM, these are called *bottleneck resources*. Conflicts with bottleneck resources are common.

Most CPM organizations deal with these types of resource conflicts on a relatively short time horizon. Because CPM projects are so commonly delayed, it's difficult for the team to predict the needs for shared resources with any accuracy more than a few weeks out. As a result, conflicts are often dealt with almost in real time. A functional leader, project sponsor, or even the steering committee may be called upon to decide which project gets the test lab next week. Clearly such a process is inefficient—it can require meetings with senior managers, effort creating "what if" scenarios to support the decision, and, to any team that loses an expected resource on short notice, delays and frustration.

FULL KIT REQUIREMENTS

Full Kit Manager	**Barry**
Project	**Aardvark**
Planned Start Date	15-Mar
Planned Completion Date	14-Jul

Team	**Material**	**Documents**	**Approvals**
Chemical Engineer	3 Proof of concept units	Market requirements	Budget
Mechanical Engineer	Raw material for stress tests	Proof of concept customer	IT
Manufacturing Engineer		DIN specification	Steering committee
Quality Engineer			
Product Marketing			
Certification Engineer			

FIGURE 6.11 Sample full-kit check list.

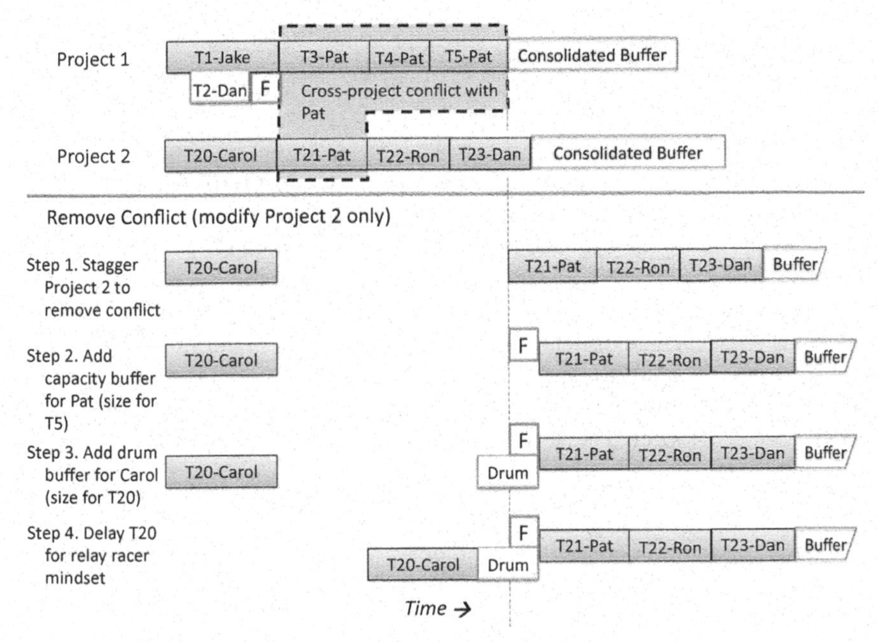

FIGURE 6.12 CCPM staggering in a multiproject environment.

The Drum Resource

CCPM can address cross-project resource conflicts with *pipelining* (also *synchronization*), a process consistent with CCPM processes for single projects. Pipelining recognizes that the constraining resource limits project throughput; in order to protect the integrity of project schedules, CCPM teaches staggering projects to eliminate resource conflicts. To pipeline two projects, CCPM displays both critical chains in a single diagram as shown on the top of Figure 6.12. In this example, Pat is the bottleneck and so is scheduled between projects by modifying Project 2 using the following four steps:

1. Stagger the Project 2 tasks owned by the bottleneck resource. Similar to the process to create the critical chain from the critical path in Figure 6.2, Project 2's use of Pat must be delayed until Pat has finished all duties in Project 1.
2. Add a capacity buffer for the bottleneck resource. This buffer protects Project 2 if the bottleneck resource's task in Project 1 runs over, as it will commonly since CCPM relies on aggressive estimates. In Figure 6.12, this shows as a buffer for Pat with an "F." This buffer is sized to be half of Pat's task in Project 1 and is added into the critical chain of Project 2.
3. Add a drum buffer to ensure non-critical tasks in Project 2 will be ready for the bottleneck resource. In Figure 6.12, this buffer is marked *Drum*, and is sized to 50% of Carol's task T20.

4. Delay the task before the drum buffer to create a relay-runner mindset. In Figure 6.12, the start of Carol's task T20 is delayed so its scheduled end coincides with the beginning of its drum buffer.

This method is workable when there are a few projects that share one or perhaps two resources. Unfortunately, organizations may run dozens of projects in parallel and may need to share design and manufacturing experts, service groups, certification engineers, legal and financial resources, and many other resources. All of this happens in CCPM projects that are in a high state of variability. Remember, CCPM teaches the elimination of milestones (such as "lab testing starts") and acceptance of delays in individual tasks to protect project completion. Of course, shared resources are needed mid-project, so any buffer penetration could push out the need for a shared resource and thus cause conflicts with another project. Raz et al. state:

> *A common criticism of the drum is that it is founded upon a single constraining resource or set of resources. However, multiple resources may be in conflict in different parts of the projects. Further, it requires stable conditions regarding resource conflicts. However, conditions in multi-project organizations are rarely stable [9].*

The Virtual Drum Resource

The drum resource has sometimes been seen as a weakness of CCPM for the reasons expressed above. The virtual drum creates a more workable alternative when many projects are sharing many resources. Stratton states:

> *[T]he instability of the bottleneck resource in project management has more recently been acknowledged by Goldratt (2007). His original guidance (1997) was to plan projects around a "drum" in the form of a resource. This has now been changed to a virtual drum resource [25].*

A virtual resource drum shifts multiproject resource conflict resolution from being dealt with on a case-by-case basis to standard policy. For example, if a particular test lab is frequently used in projects in a given organization and that lab can handle testing three units simultaneously, a policy may be created to limit the number of projects in test phase at any one time to three. If more projects happen to be in the test phase, projects must be put on hold until the lab becomes free. This seems quite similar to how CPM projects deal with multi-project resource conflicts—resolve them real time, delaying the project that is least important to the organization's goals.

6.7.2 Portfolio Fever Chart

The fever chart can be extended to a snapshot portfolio view as shown in Figure 6.13. The projects are each shown in their state of buffer penetration. This provides management with critical information to help restore buffer for

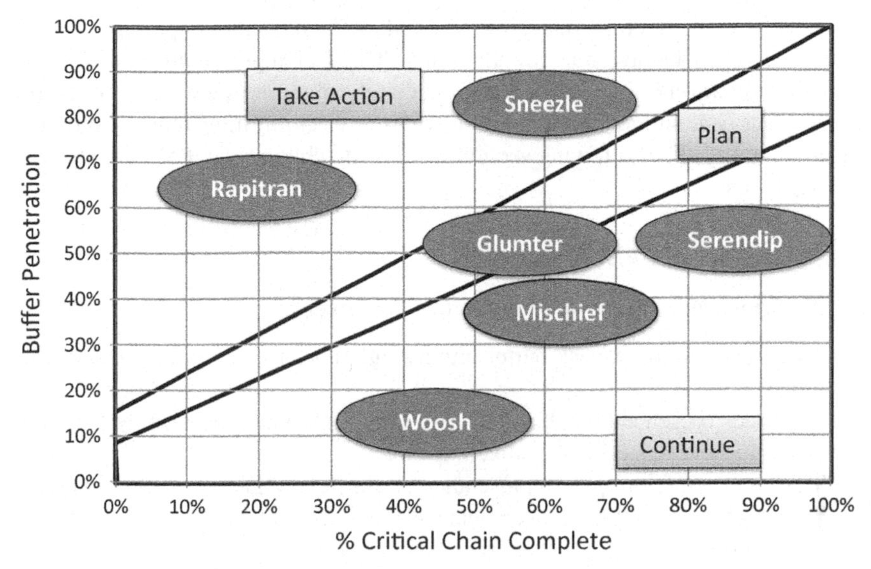

FIGURE 6.13 Fever chart for CCPM portfolio view.

those projects that need it most. For example, a steering committee meeting to review Figure 6.13 would probably start by working to improve Projects Rapitran and Sneezle, both of which are well into the *Take action* section of the graph. If either project needed more resources, the steering committee would look to those projects with the lowest buffer penetration, probably starting with Project Woosh. Again, CCPM's focus on the critical chain using buffer penetration as the key metric, together with the inherent simplicity of the fever chart, provide a visualization of the portfolio that is rarely seen in CPM (Figure 5.18 is the CPM equivalent of a portfolio fever chart, but such a view is uncommon in industry).

Of course, dynamic resource allocation is never simple. There's nothing in the chart that says stealing resources from Project Woosh is the best way to help Project Rapitran recover buffer—that will require analysis and discussion. However, the portfolio fever chart does provide a simple display to identify the problem and get the right discussion started.

6.8 HOW WELL DOES CCPM WORK?

The benefits that have been published for CCPM are convincing. There are hundreds of testimonials with a good share from well-known companies across a wide range of industries. The testimonials follow a similar form: on the first project done after adoption, there are substantial reductions in project schedule and cost in tandem with improved quality and team member morale. These benefits are seen in many types of projects: product development, construction, and sustaining (such as Delta Airline's use with engine repair scheduling).

What are the factors that allow CCPM to bring such impressive initial results? Human behavior is complex and CCPM changes many things in an organization at one time. Some are immediately obvious, such as fever charts, while others are subtle: changing mindset from tracking milestones to a relay-racer mentality. The following section divides the changes CCPM delivers into several types.

6.8.1 Unique CCPM Techniques

CCPM brings a number of techniques that are novel:

- The consolidation of task buffer into a single buffer placed at the end of the project.
- The focus on schedule risk and the use of a single metric to display it: buffer penetration versus critical chain complete.
- The two defined graphic representations: the fever chart of a single project showing historical results for context and the snapshot fever chart of a portfolio using current state of all other projects for context.
- The use of aggressive task duration estimates, cutting initial project schedules by about 25% (first reducing the task durations by 50% and then adding a 50% consolidated buffer).
- A well-defined technique to deal with resource constraints. CPM PMs in general do address resource conflicts albeit with an ad hoc collection of techniques; when this process is applied to CPM, it results in what is often called the "Resource Leveled Critical Path" [9].
- Specified processes for dealing with multiple projects run in parallel including scheduling of capacity-constrained resources.
- Process to protect global optima (such as project launch and portfolio health) over local optima (such as individual tasks in a project or projects in a portfolio), especially the de-emphasis of milestones as progress measures.

6.8.2 Improving the Execution of CPM Techniques

CCPM borrows many techniques from CPM, especially the project network or work breakdown structure (WBS); it is the foundation of both methods. When CCPM is brought to an organization that is executing CPM poorly, training on WBS creation can immediately bring improvement independently of the unique CCPM processes. For example, suppose a CPM organization is planning without full team participation—in fact, in some organizations project plans are simply handed down from a PM or sponsor. This is known to produce poor results. Now let's say that when CCPM is adopted, the department is given training that teaches full-team planning. Of course, the benefit of full-team planning will be realized in the CCPM projects; but, had similar training be provided before CCPM, benefit would probably have extended to CPM projects.

6.8.3 The Use of Lean Techniques

CCPM has many techniques that are common with lean product development especially the focus on the elimination of WIP, the reliance on intuitive visualization (especially the fever chart), and creating a strong team ethic. CCPM teaches that if you reduce schedule, cost and quality will improve—but you need to reduce WIP to reduce schedule. Lean teaches that if you eliminate WIP, schedule, cost, and quality will improve. The desire to eliminate WIP drives good behaviors such as *full kit* and the elimination of multitasking.

6.8.4 Leadership Skills and Senior Management Engagement

When CCPM is adopted by a company, there is always a champion to drive change. It may be a consultant brought in by senior management or it may be a member of the organization, but someone has to convince the organization that there is a better way in order to win over the team to the new methods, and to lead the execution of the first projects. That person must be a strong transformational leader to create new goals in an organization that isn't doing as well as it could. He or she must also be a strong transactional leader to bring success in the execution of the first projects. There is at least some research to support this [26]. As has been covered throughout this text, strong leadership brings advantages independent of the processes used.

One of the common themes in CCPM testimonials is that senior management must be part of the change. Some senior managers are usually part of the decision to switch over to CCPM. Often, they are trained on the method and participate in the early projects. As with leadership skills, senior management engagement will lend some level of benefit to any project management method.

6.8.5 Sample Bias

Some research has suggested that companies that decide to switch over to CCPM are more likely to have poor project management practices [9]. This is intuitive, similar to the fact most doctors see a higher percentage of ill people than are in the general population. As a result, these organizations may have more-than-average room to improve.

6.8.6 Bias Toward Reporting Successes

Most testimonials include the company name and the names of senior leaders. Companies are more likely to sense benefit of publication when it associates them with a success. An article that lauds the once troubled, but now crack, project team as best in class would be much more welcomed than one that details a failed attempt to deploy CCPM.

6.8.7 Summarizing the Effects

This section has presented the many factors that are probably responsible for the repeated successes of CCPM. None of this suggests CCPM is deficient—quite the opposite. The question is: what portion of the benefits is due purely to CCPM techniques and what portion of benefits could be realized with other methods? If your organization is ready to adopt CCPM, the distinction is probably unimportant—if the process works, what does it matter if there might be some other method that got similar results? However, if the method is facing serious barriers to adoption, you probably want to understand how the various parts of the process work so you can select a set of practices that is effective while facing fewer barriers to adoption. For example, many engineers struggle with creating a schedule that is 75% of what the intuitive estimates provide the way CCPM typically does (see Section 6.3.5). If that issue was preventing adoption, you might modify the method, say making your aggressive estimates be 66% of the original and then adding a 50% buffer. This would result in the same project schedule as CPM, similar to what was done in Figure 6.10. That might be an acceptable compromise if it allowed adoption to proceed more rapidly.

An excellent example of this can be seen in a video from 12th Annual TOCICO (Theory of Constraints International Certification Organization) from 2014 [27]. Mark Woeppel of Pinnacle Strategies is a well-known expert in CCPM. He has helped many organizations implement the method successfully. In this presentation, he describes that after 6 months, the client successfully implemented CCPM in part of the organization. However, the remainder of the organization would not adopt it for several reasons, especially the need to change project planning methodologies to support CCPM across the organization. Faced with the possibility of losing the gains hard won over the previous 6 months, he then applied an innovative approach. Because details around the critical chain itself created the resistance to adoption, Woeppel decided to create a solution without it. So, he borrowed many of the CCPM supporting techniques (reduce multitasking, make plans visual, etc.) but without the critical chain itself, the point that is normally described as the core of why CCPM works at all. By setting aside those elements of CCPM that caused resistance, he was able to gain adoption and then bring substantial improvement [28].

6.9 CHALLENGES TO ADOPTING/SUSTAINING CCPM

For all the testimonials to the method's virtues, is has been adopted by only a minority of companies. Barriers to adoption include that it is new (compared to CPM), it has numerous counterintuitive assumptions, and it requires specialized software tools.

CCPM does present numerous barriers to adoption. It has two assumptions that stand out: the task length estimate with 50% certainty is half that with

90% certainty (the "normal" estimate), and the buffer should be half of the critical chain [10]. These two combine to provide project schedules that are reduced by 25% with no obvious reason, something project team members can find objectionable.

CCPM is normally adopted wholesale—it requires a great deal of training and new software tools. Typically, consultants must help companies through the transition. Even with tools and training, buffer management is still difficult. According to one paper:

> *To maintain a balanced perspective, however, we must also point out that a number of firms failed to implement CC and complained about the complexity of the CC approach in changing behaviors and expectations and managing the extra complexity of buffer management [10].*

There is a lack of software tools that support CCPM compared to CPM or Agile. A few packages to consider are:

- http://www.pd-trak.com/criticalchain.htm
- https://www.prochain.com/
- http://www.stottlerhenke.com/products/aurora/
- www.exepron.com

6.10 CCPM KEY MEASURES OF EFFECTIVENESS

This section will provide an evaluation of the CCPM method without Phase–Gate according to more than a dozen key measures of effectiveness. These ratings are added to Table 5.7 to create Table 6.4. Please refer to Section 5.4 for details on the evaluation for CPM and Phase–Gate. (The assumption here is Phase–Gate makes roughly the same changes in CPM and CCPM.)

Good with high-iteration-cost projects (+)
Like CPM, CCPM projects encourage extensive planning before a project starts.

Good with low-iteration-cost projects (−)
Also like CPM, the lengthy planning that serves high-iteration-cost projects makes the method less flexible for projects with low cost of iteration.

Process to coordinate varied disciplines (+)
Like CPM, CCPM is capable of coordinating many functions.

Mitigates risk before large investment
Like CPM, CCPM provides the basic tools and metrics to support review. Also like CPM, CCPM without Phase–Gate provides no standard work to optimally manage risk, so this is dependent on the individuals planning and reviewing.

TABLE 6.4 Key Measurement of Effectiveness for CPM and CCPM

Measure	CPM	CCPM	Phase–Gate
Good with high-iteration-cost projects	+	+	
Good with low-iteration-cost projects	–	–	
Process to coordinate varied disciplines	+	+	++
Mitigates risk before large investments			++
Provides standard work for planning	–	–	++
Clearly defined methodology	+	+	+
Tools to maintain schedule	– –	++	
Intuitive, has few adoption barriers	+	– –	+
Well-defined metrics/ visualization	+	++	
Plans shared resources well	–	+	
Availability of software tools	++	–	
Sustainable over time	+	+	++
Low overhead over time	+	+	–

Provides standard work for planning (–)
Also similar to CPM, CCPM provides the WBS but there is little standard work without the Phase–Gate method being added.

Clearly defined methodology (+)
CCPM has a clear definition, beginning with Eli Goldratt's *Critical Chain* and continuing with most of its key concepts widely accepted by practitioners. The fever chart, 50% certainty in task planning, buffer management, relay-racer mentality, and many other core concepts are almost universal in CCPM organizations.

Tools to maintain schedule (++)
CCPM provides numerous tools and techniques to maintain schedule such as consolidated buffers, relay-racer mentality, fever charts, and aggressive estimates. The many testimonials bear out the method's ability to improve timely completion of projects.

Intuitive, has few adoption barriers (– –)
CCPM has numerous counterintuitive assumptions such as reducing schedule time by 25% over CPM, having a large buffer at the project end, and use of complex buffers to manage shared resources. It's also instituted quickly with a large set of changes coming simultaneously.

Well-defined metrics/visualization (++)
CCPM leverages Gantt charts and contributes buffer penetration, a leading indicator of progress. It also provides the fever charts for single projects and for project portfolios. Fever charts have an equivalent in CPM, but they are not often used (see Figures 5.17 and 5.18).

Plans shared resources well (+)
CCPM does provide methods to help manage sharing resources across multiple projects, but they can be complex to manage.

Availability of software tools (–)
There are many fewer tools available to support CCPM than CPM. This can be a major barrier to adopting the method, especially for companies that need project management software to tie into their enterprise.

Sustainable over time (+)
While the method is used by a minority of companies, testimonials demonstrate that CCPM is sustainable over time.

Low overhead over time (+)
The burden of CCPM commonly remains fixed over time. When combined with Phase–Gate, it is common that Phase–Gate processes will get heavier.

6.11 SUMMARY

CCPM has an impressive weight of evidence to show its effectiveness in reducing delay for high-cost-of-iteration projects. As the same time, there are significant barriers to adoption. CCPM is really a collection of methods and all need not be adopted simultaneously. Organizations that recognize a need to convert and are open to wholesale change will adopt the method in a single step. Other organizations may see benefit, but be closed to some of the more counterintuitive elements of CCPM; those organizations can consider adopting those parts that would be immediately acceptable, for example the aggressive reduction of multitasking or relay-racer mentality. Presumably, if the method is adopted only

TABLE 6.5 Best Fit for CCPM Projects

Best Fit—CCPM and CCPM with Phase–Gate
Projects of medium-to-high complexity. Projects with a high cost of iteration. Highly cross-functional projects. Projects where meeting schedule is critical. Organizations open to nontraditional project management.

in part, when those parts do bring benefit, the organization will be more open to adopting a larger portion later. The places where CCPM (accompanied by Phase–Gate) fit best are shown in Table 6.5.

6.12 RECOMMENDED READING

Leach L. Critical chain project management. 3rd ed. Artech House; 2014.

Newbold R. Project management in the fast lane: applying the theory of constraints (The CRC Press Series on Constraints Management). CRC Press; 1998.

Woeppel M. Projects in less time: a synopsis of critical chain. Pinnacle Strategies; 2005.

6.13 REFERENCES

[1] Roberts R. Critical chain project management: an introduction. Stottler Henke Associates. Presentation available at: http://www.stottlerhenke.com/product/wp-content/uploads/2013/09/2010-12-09-Critical-Chain-Project-Management-Introduction-FINAL.pptx.

[2] Phillip S. (Boeing) T45 Undergraduate Military Flight Office Ground based training system. Presentation available at: http://www.realization.com/results/engineer-to-order-manufacturing#BoeingT45-full.

[3] http://www.realization.com/pdf/casestudy/Case-Study-Airline-Boeing.pdf.

[4] http://www.goldratt.co.uk/Successes/pm.html.

[5] Kendall GI, Austin K. Advanced multi-project management. J. Ross Publishing; 2012.

[6] http://www.chaine-critique.com/fr/Les-resultats-de-la-Chaine-Critique-33.html.

[7] Steyn H. Project management applications of the theory of constraints beyond critical chain scheduling. Int J Proj Manage 2002;20:75–80.

[8] Maylor H. Another silver bullet? A review of the theory of constraints approach to project management. In: Van Dierdonck R, Verbeke A, editors. Proc. 7th Internat. Annual Eur. Oper., Management Assoc. Conf. Gent (Belgium): Academic Press Scientific Publishers. p. 4–7., 2000

[9] Raz T, Barnes R, Dvir D. A critical look at critical chain project management. December 2001.

[10] Lechler TG, Ronen B, Stohr E. Critical chain: a new project management paradigm or old wine in new bottles? Eng Manage J December 2015;17(4).

[11] Goldratt E. The goal: a process of ongoing improvement. North River Press; 1984.

[12] Goldratt E. Critical chain. The North River Press; 1997.

[13] Herroelen W, Leus R. On the merits and pitfalls of critical chain scheduling. J Oper Manage 2001;19:559–77.

[14] Loch C, De Meyer A, Pich M. Managing the unknown: a new approach to managing high uncertainty and risk in projects. John Wiley and Sons; 2006. p. 56.
[15] Critical chain project management theory and practice, Roy Stratton. In: POMS 20th annual conference Orlando, Florida, U.S.A. May 1–4, 2009.
[16] Geekie A. Buffer sizing for the critical chain project management method [Master's thesis]. Department of Engineering and Technology Management, Faculty of Engineering, University of Pretoria.
[17] Cox J, Schleier J. Theory of constraints handbook. McGraw-Hill Professional; 2010.
[18] Matthews T, Czerwinski M, Robertson G, Tan D. Clipping lists and change borders: improving multitasking efficiency with peripheral information design. Montréal (Québec, Canada): CHI; 2006.
[19] Salvucci DD, Bogunovich P. Multitasking and monotasking: the effects of mental workload on deferred task interruptions. CHI; 2010.
[20] Parkinson CN. Parkinson's law. The Economist; November 1955. Available at: http://www.berglas.org/Articles/parkinsons_law.pdf.
[21] Loch C, De Meyer A, Pich M. Managing the unknown: a new approach to managing high uncertainty and risk in projects. John Wiley and Sons; 2006. p. 57.
[22] Kendall GI, Austin KM. Advanced multi-project management. J. Ross Publishing; 2013. p. 27, p. xiv.
[23] Kendall GI, Austin KM. Advanced multi-project management. J. Ross Publishing; 2013. p. 259.
[24] Huang C-L, Li R-K, Chan Chung Y-C. A study of using critical chain project management method for multi-project management improvement. Int J Acad Res in Econ Manage Sci May 2013;2(3). Available at: http://www.hrmars.com/admin/pics/1833.pdf.
[25] Stratton R. Critical chain project management theory and practice. Currently available at: https://www.ntu.ac.uk/nbs/document_uploads/94592.pdf.
[26] Verhoef M. Critical chain project management [Master of Project Management thesis]. Utrecht (The Netherlands): HU University of Applied Sciences Utrecht, 2013 p. 26.
[27] Woeppel M. Deconstructing critical chain: a failed implementation leads to an evolution. TOCICO; 2014. Currently available at: http://vimeo.com/107942371.
[28] Based on a conversation between Mr. Woeppel and the author. November 2014.

Chapter 7

Lean Product Development

This chapter will present lean product development (LPD). LPD is derived from lean manufacturing: those principles that today are so well accepted on the factory floor are applied to product development. The chapter will start with an introduction to lean manufacturing, which was created by Toyota starting in the 1950s and has been widely adopted in the West starting in the 1990s. Today, these principles are almost universally accepted in production processes. The chapter then applies lean thinking to product development to build LPD techniques. It concludes by comparing LPD to critical path management (CPM) and critical chain project management (CCPM) from Chapters 5 and 6.

When applying lean principles, be wary of the differences between production and product development processes. The fundamental difference is that production processes are repetitive: able to make the same product again and again, always seeking to reduce cost, shorten delivery, and improve quality. By contrast, a large portion of product development processes are nonrepetitive since they create a product that does not exist. Of course there are repetitive steps within those processes such as releasing drawings or deploying a software test; those repetitive actions are the most natural fit for lean production thinking. So, lean manufacturing tools bring the opportunity to improve, but users must focus on the concepts that are most relevant to the domain of product development. In this chapter, seven lean tools are presented as examples of that conversion to demonstrate the approach.

LPD provides a focused way of thinking and brings a number of powerful tools, but it is not a step-by-step sequence. It is evolutionary—usually adoption begins by focusing on a limited space: work on a limited problem set with a limited number of people. For example, a company could start LPD by deploying a handful of lean techniques to improve their design process for electronic projects. Performance is measured after deployment to validate the improvement. Once the improvements are validated and stabilized, they can be spread to other parts of the organization and new areas for improvement can be selected. The cycle continues indefinitely. Step by step, lean thinking changes the organization's product development tools, the processes, and, ultimately, the thinking of all involved.

Because LPD is evolutionary, it improves on organizational processes as it finds them. LPD doesn't create a start-to-finish product development process the way Phase–Gate with CPM or CCPM does. And, where revolutionary methods like CCPM and Agile usually start with a jarring wave of change, LPD seeks improvement through a gentle but relentless stream of change. This brings LPD both its greatest strengths and weaknesses. On the one hand, there is usually

Project Management in Product Development. http://dx.doi.org/10.1016/B978-0-12-802322-8.00007-3

minimal resistance to the initial adoption of LPD because the first changes are modest and intuitive, at least for those organizations that accept lean manufacturing. On the other hand, there is no end to LPD because lean thinking is never satisfied—there is always something else to improve. Where revolutionary methods require the greatest commitment at the outset, the commitment required for LPD starts small and increases over time; accordingly, the temptation to return to business-as-usual and leave the lean conversion half done is ever present.

7.1 AN INTRODUCTION TO LEAN THINKING

The adoption of lean manufacturing in the West is often credited to James Womack and Dan Jones. In their book *Lean Thinking* from 1996 they stated five foundational characteristics of lean [1]:

1. Understanding Value
   Value is anything a customer is willing to pay for. A lean organization should clearly state and validate the value created for the customer by the production process. In lean thinking, every activity undertaken by an organization should create or support the creation of value.
2. Value Stream Mapping
   A production system should be viewed as a stream where value is added step-by-step until the final product or service is complete. That stream should be meticulously understood and mapped.
3. Flow
   All value-add steps flow toward the customer and that flow should be fluid. Every component and activity should flow smoothly. Bottlenecks interrupt flow and should be removed or minimized. Work in progress (WIP), the accumulation of partially completed products, reveals bottlenecks and so should be minimized.
4. Pull
   Value should be pulled from the customer, not pushed by the company. The delivery of a product can be viewed as the customer pulling value from the system such as when a burger is removed from a chute at a fast-food restaurant; the empty chute signals the worker to produce another burger. Making a burger might empty the sliced-tomato bin, which would signal a worker to refill that bin. And so on. The entire value stream is arranged this way. Each downstream process pulls assemblies and components from upstream processes and in doing so signals those processes to produce more.
5. Continuous Improvement
   In lean thinking, no production system is ever finished. The many processes must be perfected again and again. This is continuous improvement–relentless activity to move the system step by step toward perfection: always improving, never being satisfied.

For those companies that are diligent, *lean production* will become a complete system to manage every aspect of the business: R&D, ops, sourcing, customer service, sales, and so on. Properly executed, lean production reduces waste of every type; effort, physical space, capital, and time are all reduced compared to traditional management systems. Lean production is best known from Toyota; their implementation is called the *Toyota Production System.* It started after the Second World War and is widely credited with making Toyota one of the most successful companies in the world.

7.1.1 Toyota Production System

The Toyota Production System (TPS) is attributed to Taiichi Ohno, the head of production at Toyota after the Second World War. It is based on a simple premise: increase value and reduce waste; then all performance indicators will improve: cost will be reduced, quality defects will decrease, and delivery time will shorten. TPS is a system of continuous improvement based on standard work augmented by Kaizen ("improvement") to bring sustainable improvements. TPS is said to rest on the twin pillars of *just in time* and *jidoka*, which are discussed below. The foundational process of TPS is the *Deming cycle*: plan, do, check, and act (PDCA).

TPS was made famous in the West in the 1990s due in large part to the book *The Machine That Changed the World* [2]. It is now widely accepted as being far superior to mass production systems of the early twentieth century. Today, companies in virtually every industry have converted to lean production: construction, hospitals, manufacturing, and many more.

Jidoka or *autonomation* guides the development of machines and processes to detect defects quickly and then stop production until the conditions that caused those defects have been corrected. This stands in stark contrast to mass production systems, which build subassemblies in batches and then test to detect defects. In mass production, by the time a defect is discovered, it may be present in a great number of subassemblies, causing much more rework than if the defect were discovered earlier. Jidoka improves quality and minimizes the waste of rework at each point in the process.

Just-in-time (JIT) delivers just what is needed to the input of a process just when it is needed. Delivering too much or too soon creates the waste of overproduction. Delivering too little or too late interrupts flow, thereby creating the waste of waiting.

Plan-Do-Check-Act (PDCA)

Plan-do-check-act, often called the Deming cycle, is the template process in lean manufacturing. This was brought to Japan by the American W. Edwards Deming in the 1950s. The PDCA cycle is never-ending as shown in Figure 7.1.

Plan
Evaluate a process to determine:

1. the goals,
2. the gaps in current performance versus those goals, and
3. the changes needed to close the gaps.

Do
Adopt the changes developed in the plan.

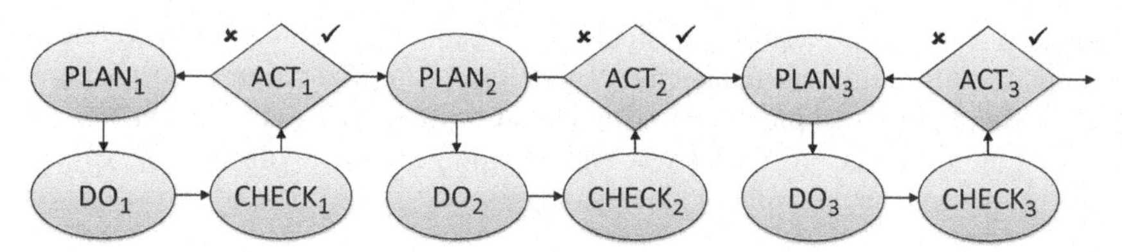

FIGURE 7.1 The never-ending cycle of PDCA.

Check
Evaluate the performance of the process after the changes have been implemented.

Act
Decide if the new performance is acceptable. If not, repeat the PDCA cycle until it is. If so, move the changes to standard work to sustain the improvements. Then identify the next area to improve and repeat the PDCA cycle.

7.1.2 Waste

Waste is defined as any unneeded activity that doesn't create value for the customer. Value is defined as anything customers are willing to pay for in a product or service. Waste prevents the organization from operating at its full effectiveness. According to lean thinking, reducing waste will improve every aspect of a company's operation: profitability, quality, worker morale, customer satisfaction, and so on.

The Eight Wastes

Most presentations of lean manufacturing present the eight wastes [3], thought by many to be a complete listing of the primary types of waste present in a production system (Figure 7.2 and Table 7.1). They are axiomatic—stated without

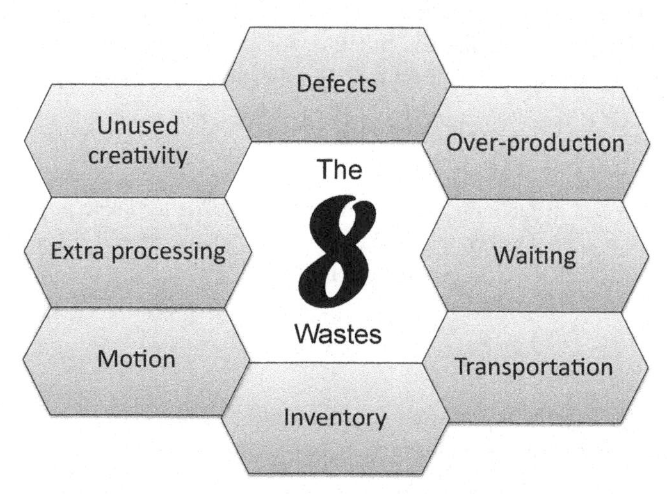

FIGURE 7.2 The eight wastes of the Toyota production system.

TABLE 7.1 The Eight Wastes of Production Systems

Type of Waste	Description
Defects	Defects include anything that causes a customer to reject or be dissatisfied with a product. Defects have high costs such as the time to rework an assembly, scrap material, and the effort an organization must expend responding to disappointed customers.
Overproduction	Overproduction is producing more than is needed or producing before there is a need. It is the goal of lean production to produce just what is need just when it's needed. Overproduction often leads to two other wastes: increased inventory and excessive WIP.
Waiting	Waiting is waste created when production is idle. Waiting can be caused by many factors, for example, having insufficient material, nonoperational equipment, or an incomplete manufacturing team.
Transportation	Transportation is waste generated by moving goods more than is necessary, for example, carting subassemblies between two processes that are further apart than they need to be.
Inventory	Inventory is the storage of goods produced or purchased before they are needed.
Motion	Wasted motion is created when workers must move more than is necessary. An example is when a worker must reach down to pick up an item that could be placed within immediate reach (sometimes called "point of use") or a worker having to walk more than is necessary. (*Motion* is wasted movement of people; *transportation* is wasted movement of product.)
Extra processing	Extra processing is created anytime effort is spent to produce a feature or function that the customer doesn't value. It is sometimes called *gold plating*.
Underused talent	Underused talent comes anytime creativity that could improve the production process is not applied to the system. This waste includes ignoring ideas or not engaging workers to create new ideas. Creativity can come from anyone associated with the manufacturing process, from factory workers to company presidents.

proof. However, they have such wide acceptance from organizations that are outstanding at production processes, they have great credibility. Understanding the eight wastes and acting to reduce them can transform an organization. When searching for waste, everyone is speaking the same language. After waste is identified, there is common purpose to vigorously eliminate it.

While the eight wastes are normally applied to manufacturing processes, there are direct corollaries to product development, as will be discussed in Section 7.2.1. There is variation in the number of axiomatic wastes. While these eight are probably the most commonly cited, sometimes only seven wastes are listed ("Unused Creativity" is not always included). On the other hand, Tapping lists 12 wastes adding overburden, unevenness, environmental waste, and social waste [4].

7.1.3 Continuous Improvement

Continuous improvement is both a learning mindset and the actions that mindset creates to improve production, either by increasing value or reducing waste. As shown in Figure 7.3, the continuous improvement cycle begins where the organization finds itself at the outset, whether weak or strong. It starts as a transformational process, finding new areas to improve and spurring creativity to generate ideas on how to create the improvement. It ends as a transactional process; diligently moving improvements to standard work, measuring their value, and filtering out those ideas that turn up short.

The centerpiece of the continuous improvement cycle is the Kaizen event. A Kaizen event is typically a multiday day event where a cross-functional team meets to take up a topic. It seeks to create ideas for improvement and run simple experiments to validate the effectiveness of the ideas. After the Kaizen, the ideas are moved into standard work, a step that can require months—changing process, measuring improvement, and so on. When the measured improvement meets the goals from the Kaizen, the organization then moves to the next improvement in an ever repeating cycle. A single iteration will advance an organization one step in its journey toward lean. That step will be small [5]—substantial gain is achieved after many cycles.

Error Proofing

One goal of continuous improvement is to move processes toward being error proof, often called *poke-yoke* from Japanese (usually pronounced "pohkah-yohkah"). Error proof means errors are prevented or they are so obvious they will be corrected rather than becoming defects that affect the customer [6]. The seatbelt is often given as an example of error proofing because it's so difficult to engage incorrectly. *Poke-yoke* is a target for each design element, each assembly step, and each process. While a handful of exceptional designs might achieve this lofty status, in general it is a goal to strive for, understanding it will rarely be achieved in full.

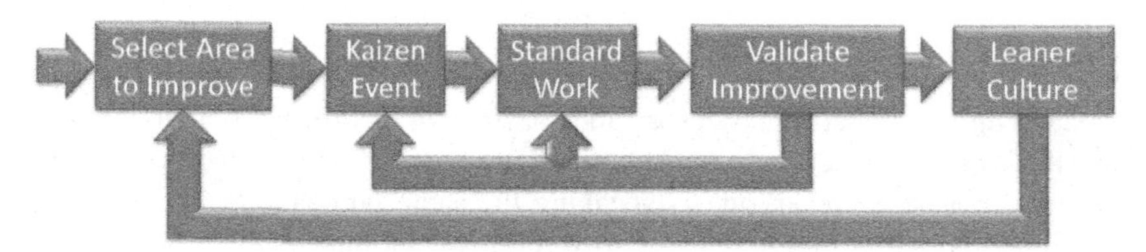

FIGURE 7.3 Driving continuous improvement with Kaizen.

Poke-yoke has special meaning for product design and process development. When a defect occurs in a production process, *poke-yoke* takes the focus away from the worker. Unfortunately, it's too common in western mentality to blame workers—"if they would try harder, our defects would decline." In lean thinking, when a defect occurs, the question is "why isn't the design closer to *poke-yoke*? If it were, there would be fewer defects." If a screw is left out of an assembly, don't blame the worker. Instead, ask why there are so many screws.

7.1.4 Kaizen Events

Small steps are the heart of Kaizen. ...Slowly...you'll cultivate an appetite for continued success and lay down a permanent new route to change [7].

A Kaizen event is one of any of a number of cross-functional multiday meetings targeted at identifying specific ways to make the next increment of continuous improvement. Kaizen events (or just Kaizens) fall into two classes: the broader *flow* Kaizen, which focuses on a value stream, and a narrower *process* or *design* Kaizen, which focuses on a single element in the value stream. While Kaizens are traditionally focused on the factory floor, they can target improvement in customer service, sales, design, finance, HR, logistics, or any other part of a business. A Kaizen is typically 3–5 days long [8].
Ron Taylor lists a framework of six steps for a Kaizen event [9]:

1. The event is initiated by defining one or more goals in an area to improve.
2. A cross-functional Kaizen team is selected.
3. A facilitator starts the event by training the team on the process that will govern the event.
4. The team builds consensus on the problem as it exists, often called the *current state.*
5. A measurable problem statement is created by the team.
6. The team works to create and validate solutions.

The Kaizen process empowers the Kaizen team to solve a problem. If the team follows the process and stays focused on the defined problem, they can make decisions that are usually reserved for managers in traditional organizations. And, ideally, there is no rank in a Kaizen—an assembly worker and the vice president of engineering each get one vote. Of course, that ideal may not always be fully met—subordinates are often deferential; still, there is no event that levels out the org chart the way a Kaizen can.

Kaizen events are initiated by a leader who selects an area that needs improvement and then builds consensus to support a Kaizen. Building that consensus is often no simple task, partly because Kaizens are expensive: they occupy numerous people for the better part of week and may require travel for some. Beyond cost, Kaizens require a commitment from the organization to deploy the changes the Kaizen team creates. For those management teams who are most comfortable prescribing solutions from the top, this requires sharing power with a team of people, all or most of whom are well beneath them on the organizational chart. So, lean change must come from both the bottom and the top of the organization. From the bottom because Kaizen events bring improvement by empowering the people that work in the area daily; from the top because the output of the Kaizen will only be effective if it is supported by the management of the company.

Kaizen Team

Members of the Kaizen team are selected with four primary criteria [10]:

1. **Technical knowledge in the area where the Kaizen is focused**
   The Kaizen result will only be as good as the knowledge represented by the team.
2. **Enthusiasm for change**
   Some people are fast adopters; others are averse to change. The purpose of Kaizen is to bring change, so the team should be biased for change. Unfortunately, sometimes those most knowledgeable are also those most averse to change; in such cases, the knowledge factor normally wins so long as the bulk of the Kaizen team is oriented to change.
3. **Confidence that improvement is possible**
   A Kaizen team must believe real improvement can come about from the event. This moves beyond simple enthusiasm for change; it must include confidence in the organization's ability to implement the Kaizen output. That confidence is partly innate to the individual (self-efficacy from Chapter 4) but also influenced by the organization's history—it will be hard to find a person confident the Kaizen will make things better if the management consistently rejects Kaizen-team recommendations.
4. **Ability to focus until the end**
   Kaizens run on a compressed schedule so team members must be able to focus on critical elements and complete their work on time.

The leader who initiated the Kaizen may continue to lead through the Kaizen or may delegate to someone else in the organization to lead from that point on. Aside from a leader, the Kaizen team will need a facilitator to manage the meeting, ensuring all participants are engaged and following process. The facilitator is the Kaizen process owner and should not be invested in any particular solution.

A Kaizen is an egalitarian event. Decisions are made by the team consensus. Neither the Kaizen leader, nor the facilitator, nor the highest ranking member of the team dictates the outcome of the Kaizen—that responsibility belongs to the Kaizen team as a whole. Since the facilitator manages the event, she should not participate in the event activities such as contributing solutions, voting, and ranking. This allows her to lead the Kaizen without having even the appearance of bias for the outcome.

Standard Kaizen Events

Kaizens can be tailor made for any area to improve. However, there are many well-known standard Kaizen types. Some that will be most interesting to project managers (PMs) are the following.

Toyota 3P Process

The Production Preparation Process or 3P is one of the most transformative Kaizens [11]. Where most Kaizens accept the design or process as it is, the 3P seeks to improve by changing the design of the product or the processes used to produce it. To a westerner, the 3P can feel awkward. The basic premise is to isolate the functions of components in a design or process and then find equivalents in nature [12]. The hinge of a car door might find a parallel in skeletal movement. The team then can sift through other natural equivalents that accomplish the same thing, finding the best fit in nature for the Kaizen problem statement. Of course, the more mechanical the problems being addressed, the better the nature equivalents fit. The many ideas from team members are melded together into potential solutions in a process that eliminates personal ownership within the Kaizen. The team evaluates the solutions and picks a few to proceed to rapid evaluation [13].

3P Kaizens require doing, not just thinking, so it's typical to build a quick proof-of-concept (POC) unit to demonstrate the idea, even if it's made from paper towels and duct tape. The process is called rapid evaluation and it can be surprisingly effective. The team proceeds through cycles of creativity, building quick POCs and applying evaluation filters until the strongest ideas emerge. In the end, the winning solution is usually a team idea because the many contributions from individuals meld together to create a single solution. A 3P with a knowledgeable cross-functional team, a well-defined problem, and an effective facilitator will usually generate strong, creative solutions.

Value Stream Mapping Kaizen

> *A value stream is the sequence of activities an organization undertakes to deliver on a customer request [14].*

Value stream mapping or VSM Kaizen charts a process to understand which actions add value versus which do not (often referred to as *value- and non-value add*). The VSM starts with the current state and then seeks to create a future state to increase the value-adding steps (to increase value) and reduce the nonvalue-adding steps (to reduce waste). One premise of lean is only 5% of what any employee does adds value—the VSM identifies those steps and seeks to amplify them. Often much of the learning in a VSM is coming to consensus and writing on paper the current state, a process that can take a lot longer than anyone in the Kaizen expects at the outset. You'll need a team that together has full knowledge of the real process, which is what people actually do rather than what they are "supposed" to do. When the team maps the actual steps accurately, much of the waste is immediately apparent and the path to improve is often clear from that point.

Single Minute Exchange of Dies (SMED)

The SMED tool seeks to reduce the time required to convert from producing one part to another. SMED is a critical element of single-piece flow because a production system able to change over quickly can produce a greater variety of products in one process. This stands opposed to production systems that have lengthy changeover times, perhaps 24 hours or perhaps more. These systems require large batches between changeovers to justify the time to convert the line. As conversion time is reduced, batch sizes can become smaller and remain economical to produce. As the time for the exchange approaches "single minute" (literally, 1–9 minutes) it becomes so small that single-piece flow can be realized.

Six Sigma

Six Sigma is a process of identifying defects and then systematically eliminating them, starting with those that occur most often. The goal of Six Sigma is to design products and processes with variation so small that defects occur at a rate no larger than three or four occurrences per million opportunities. It should be pointed out that Six Sigma is commonly thought of as a whole philosophy unto itself rather than a part of lean manufacturing. However, it's frequently used in lean organizations and so is discussed here.

The name Six Sigma derives from the standard of deviation (or "sigma") of the *normal* (Gaussian) variation of a process. In a Six Sigma process, one sigma of variation is 1/6th of what is needed to generate a defect. For example, let's say a casting is being produced and the length of the part should be 100 mm ± 0.6 mm. This product would meet Six Sigma if:

1. The process produced a product length that averaged 100 mm,
2. The process variation had a normal or Gaussian distribution, and
3. One standard of deviation of the process variation is less than or equal to 1/6th of the tolerance; in this case, sigma ≤ 0.6 mm/6 = 0.1 mm.

In Six Sigma, process capability is measured in terms of defects per million opportunities or DPMO. An opportunity is an identified failure mode so that one component can have many opportunities to produce defects. Another common measure of quality defects is DPM, the defective products per million samples. DPM is independent of the number of opportunities: either a part is defect free or it is not. DPMO is DPM scaled down by the opportunities for a failure. So for our casting above, if there were 10 identified failure modes, a process capable of a DPMO of 500 could produce a product with a DPM of 5000. Table 7.2 shows the DPMO of processes that are capable of 3-, 4-, 5-, and 6-Sigma.

TABLE 7.2 Process Capability versus DPMO

Process Capability	DPMO
3-Sigma (fair)	66,807
4-Sigma (good)	6210
5-Sigma (excellent)	233
6-Sigma (outstanding)	3.4

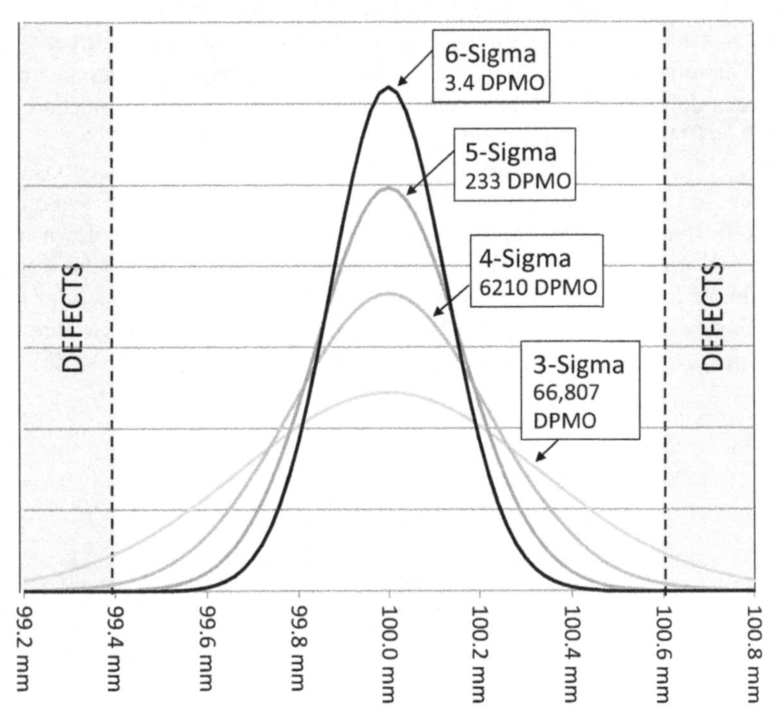

FIGURE 7.4 Gaussian (normal) distribution for 3-, 4-, 5-, and 6-Sigma process capability.

Six Sigma guides product and process designers to reduce variation. As the standard of deviation of the produced parts drops in relationship to the upper and lower limits, the DPMO drops. Figure 7.4 shows the normal (Gaussian) distribution for the length of our example casting for fair quality (3-Sigma), good quality (4-Sigma), excellent quality (5-Sigma), and outstanding quality (6-Sigma). The journey to 6-Sigma is a long one because there are usually many processes generating defects and within each process, multiple root causes for those defects. It takes time to bring the many processes into Six Sigma control.

Suggested Steps

This section has presented a short introduction to Kaizen events. Entire books are written on Kaizens and a complete discussion of Kaizen events is beyond the scope of this text. If you have further interest, see the recommended reading list at the end of this chapter.

7.1.5 WIP Control

Work in progress or WIP is the accumulation of assemblies between process steps. If there are two production cells where one feeds another, when the upstream cell produces goods faster than the downstream cell can consume, then WIP, the goods between the two cells, will accumulate. The effect is undesirable, a kind of inventory on the factory floor—it consumes resources to pay for the material and processing. It also allows defects to accumulate that could have been detected quickly in a downstream step were the part not accumulated in WIP. WIP brings no value to the customer.

The TPS controls WIP primarily by blocking production of new goods when excessive WIP has accumulated. TPS prescribes a *pull system* where goods are produced from the upstream process only as they are needed for the downstream process. A small amount of WIP may be required in order to prevent waiting. As a simple example, let's say an upstream process requires 20 min to produce an assembly as shown in Figure 7.5. That process feeds three unsynchronized downstream processes, each of which consumes one of these assemblies in

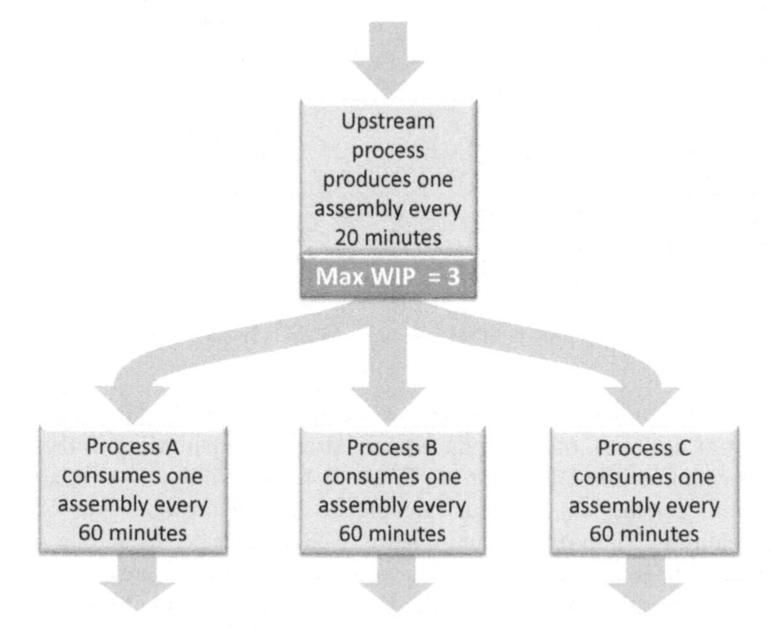

FIGURE 7.5 Simple WIP control scheme.

60 min. Since the upstream process is three times faster, it is able to keep up with the three downstream processes over a long period of time. However, a WIP of three assemblies might be required for the condition where all three downstream processes require an assembly in a short span of time (something possible since the downstream processes are unsynchronized).

7.1.6 Kanban Inventory Management

Kanban is a pull-system for material used in a production system. Two or more bins of equal size are used to store parts for a process. These bins define the WIP for this process. A simple example of a Kanban system is to store material in two bins. Suppose a part had a lead time of 7 days and in that time the process consumes a maximum of 50 pieces. In that case, a simple Kanban system would have two bins each sized to hold 50 pieces. Parts are consumed from one bin until it is empty. The empty bin is used to signal an order for the next 50 pieces. The order is placed and the remaining bin is used to provide parts while that order is being filled. When a two-bin system is properly sized, the WIP is limited to about one bin.

The name Kanban is Japanese for *sign board* and originated from the Kanban systems in the 1950s where a cardboard card was kept in the parts bin; it had written on it all the information necessary to place the order to refill the bin. When the bin's parts were consumed, the card signaled the need to refill the bin, either as a purchase order to an external supplier or as a production order to an upstream process. Today Kanban has a broader meaning and there may be no physical sign board, but the original approach of using an empty container to signal a need to refill remains.

7.1.7 Go to Gemba

Gemba or "place where the action is" is typically the production area being considered. Jim Womack talks about his repeated experience of being brought to a well-appointed conference room to discuss a problem in a factory. He immediately advises his clients to escape that sterile environment and "go to Gemba" [15]. Solving problems requires a complete understanding of the current condition. While managers might prefer talking about factory issues in the comfort of a conference room, the best solutions are most likely to be understood by going to Gemba, which could be an assembly line, the customer service department, a test lab, or a customer site.

7.2 LEAN PRODUCT DEVELOPMENT

LPD applies the concepts of lean production to product development. Every activity of the product development team should be applied to tasks that will provide features or performance that the customer will value; each should have minimal waste. There are many parallels between manufacturing and product development. Here the design team members are the workers whose product is the information necessary to build and support a product: drawings, specifications, software, and analysis. Waste is still any unneeded activity that does not generate value for the customer. WIP is now partially completed projects. Continuous improvement here is applied to design processes rather than production facilities. The eight wastes have their corollaries as well.

While there are many parallels, there are stark differences between product development and production systems. The most obvious example is that in product development, it's rare for the same activity to occur twice. By its nature, new product development involves many steps that are occurring for the first time: a new technology, a new algorithm, or a new mounting configuration. Also, information is normally stored in computers; it is much more difficult to "see" than physical WIP spilling out onto the factory floor. In lean manufacturing, variation is the enemy: process designs aspire to *poke-yoke* processes that provide the same results no matter who performs the work. By contrast, product development seeks the best output of every team member accepting a wide range of capabilities from technology gurus to first timers. An informed reading of LPD accepts both the corollaries of and the differences between the two endeavors.

7.2.1 The Eight Wastes of Product Development

The eight wastes of production systems have corollaries in LPD [16–18]. They are not as universal as those from the TPS and a couple are perhaps a bit strained; nevertheless, the equivalence between the two demonstrates ways to apply lean thinking to product development (Table 7.3).

7.2.2 Kaizens for Product Development

Kaizen events can be applied to product development. Many factory floor Kaizens can be used, but the 3P Kaizen translates directly since one of its functions is to improve the design of products. The more concrete the design activity, the easier to fit it into a 3P; mechanical design is a direct fit while a 3P for software architecture is a stretch for most people. However, the core of the 3P can be applied to problem solving in such diverse areas as order entry, user interfaces, configuring a product, and project management. The main steps are:

1. Create a clear definition of the problem with measurable goals for the Kaizen to achieve.
2. Visualize the current state (varies widely according to the problem being addressed).
3. Allow each team member to provide solutions independently in silent brainstorming. In a 3P, the initial ideas are recorded by individuals without discussion. This allows all people to contribute at the maximum of their ability and levels the playing field for people who are less outgoing. Often 100 or more ideas can be generated in 10 or 15 min.
4. Ideas are quickly presented to the team by the creators, usually in less than one minute per idea.

TABLE 7.3 The Eight Wastes of Product Development

Defects	Mistakes, errors, and design flaws that require rework, generate scrap, or cause customer rejection.
Overproduction	Effort applied to a feature or performance characteristic that does not bring value to the customer.
Waiting	Any time project resources are ready to work on a task but are delayed. Delay can come from a prior step being incomplete, the team missing one or more members needed for a task, or the team being idled until an approval is received.
Transportation	The waste of knowledge transfer, for example, training a new team member who is replacing someone moved to another project.
Inventory	Partial designs that are shelved for later use; partially complete projects.
Motion	Product development tools and processes that don't meet needs (requiring excessive manual work) or are overly complex (required excessive effort to learn or use).
Extra processing	Gold plating—continuing to apply design effort after the feature or performance characteristic has met customer needs.
Underused talent	Providing an environment or process system that doesn't obtain the maximum creativity of every project team member.

5. The many ideas are grouped together to form a single category of solution. Typically about 20 categories will be created. The formation of categories merges the ideas of the individuals into groups so that the original ideas are no longer discernible. This eliminates the common problem in traditional brainstorming where the individual ownership for ideas persists, making conflict more likely when ideas are selected for further investigation.
6. The team then evaluates each category. The means of evaluation vary widely—it can be a simple vote or a complex evaluation of many factors: cost, size, reliability, and so on.
7. Those categories that score well are moved to the next stage: rapid evaluation. The group is broken into teams to build simple proof-of-concept units that demonstrate the performance of an approach. It might be a crude user interface, a plywood mold, or hacked-up circuit board. Typically only a few hours are allowed, so the teams must focus on the core of the idea.
8. The rapid evaluation units are then evaluated by the team as a whole. The two or three best are accepted and another cycle of rapid evaluation is used to create a single solution.
9. The team develops a plan of all activities that need to take place after the Kaizen for this implementation.

These steps taken together form the "plan" step of Deming (PDCA) cycle. Over the succeeding weeks or months, the plan must be executed ("do"), the results "checked," and then the organization must "act": if the results meet the customer need, they must be stabilized; otherwise, the team may return for another 3P. The 3P is one of the most powerful problem-solving tools a PM has access to. It is effective on a wide range of design and process issues.

7.3 SEVEN LEAN TECHNIQUES FOR PRODUCT DEVELOPMENT

Lean product development applies the principles of lean production to the many activities required to create a product. The focus in both is to increase value and reduce waste. Like lean production, LPD seeks continuous improvement relying on Kaizens to make progress and stabilizing those improvements over time with process.

In traditional project management, the Iron Triangle (see Figure 1.1) shows that three characteristics of a project must be balanced: project cost, schedule, and deliverables (generally, features, performance, and product cost). Improving one requires compromising the others: need more features? Be prepared to trade off project cost and schedule to get them. LPD also views these dimensions as being interrelated, but seeks improvement by removing waste and increasing value at every step of the process. This improves one or more of the three dimensions without compromising the others.

LPD, like lean techniques in general, is a collection of methods that are created to increase value and reduce waste. They are borrowed from lean production systems, repurposed from the factory floor to the product development team. LPD is a way of thinking and unlike CPM, CPPM, and Agile, is not defined by any set of methods. True, there are lean methods, but those methods are tailored to the problem at hand, the customers being served, and the culture of the organization. Unlike CPM or CCPM, lean has no widely accepted process base, so users will need to develop that base. One option is start with a Phase–Gate process and then use lean thinking to continuously improve; that is the viewpoint of this chapter. But lean thinking can be applied to virtually any current state and any organization whether large or small, whether focused on consumers, industry, or government, or whether the culture is new to lean or has been using lean for years.

The remainder of this section will discuss seven of the most common lean tools for product development: visual workflow management, go to Gemba, WIP control, minimal batch size, Kanban project management, lean innovation, and Obeya. As shown in Figure 7.6, LPD requires the mindset of Kaizen and continuous improvement to drive the selection of tools: which ones to use and how to tailor them to the immediate purpose. LPD is evolutionary—start small and grow incrementally, always sustaining what works and trimming what doesn't.

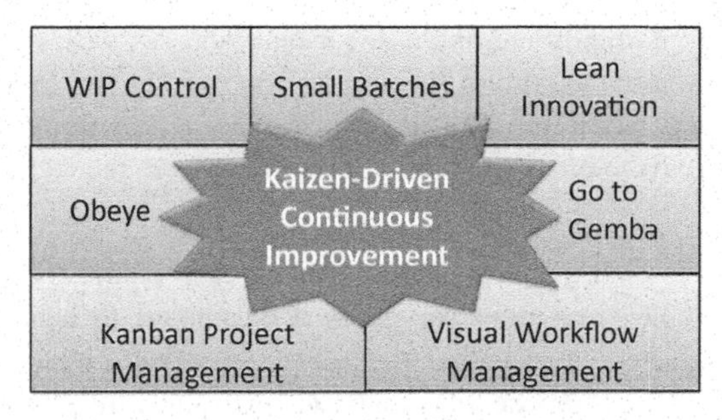

FIGURE 7.6 Lean tools for product development.

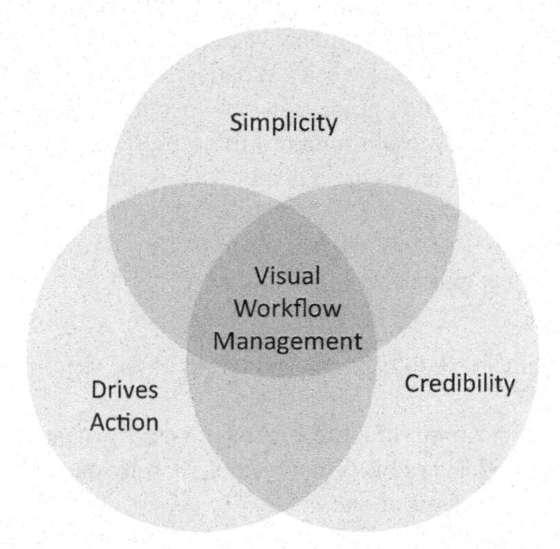

FIGURE 7.7 Requirements of visual workflow management.

7.3.1 Visual Workflow Management

Visual workflow management creates charts and other graphics that show where the team is, where they are going, and how they will get there. The critical components of visual management are simplicity, credibility, and driving good action as shown in Figure 7.7.

- **Simplicity**
  Visual management must be easy to understand. A person from outside the project should be able to view the information and understand the current state, the desired future state, and how well the team is doing to fill any gaps. It should comfortably fit on a single piece of paper or a single screen shot with most of the story told graphically (comfortably includes an 8.5″ × 11″ sheet with a font no smaller than 10 or 12 points). There should be a bias for concrete data, avoiding abstraction. And use formats common in your organization. If

your company normally charts engineering resources with bar charts, don't switch to a radar chart unless there's a strong need. The less the company leadership has to wade through new graphs and terminology, the more they will comprehend the issue; the more they comprehend, the more they can help.

- **Credibility**
  The data that is used to create visual management charts should be detailed and reliable. The chart then serves as a dashboard to a larger and more complex data set. The way that data is processed to create the dashboard should be intuitive. Suppose your team wants to show the number of features complete in a specification with 200 features. The visual display might be key performance indicators (KPIs) such as features completed this month, features remaining, and so on. A senior manager may once in a while "pressure test" the display: "10 features this week? What were they? Have they all passed acceptance testing?" At that point, you can open the full data set, filter so the 10 are visible and review them for a few minutes. If you try to generate visual data manually, the risk of having a display that is misleading or out of date is so great, you can be sure it will happen at some point (probably during one of those pressure tests!). If you pass one or two close reviews, you'll earn trust. If you don't, expect to have to work hard to regain it.
- **Drives actions**
  Visual management should be designed to drive action. Driving action requires:
  - Choosing the most important few areas to visually manage. If product cost and design schedule are the critical items, visually manage them. The team can't focus on 25 items at one time.
  - Quantifying performance: software defects, product material cost, assembly drawing backlog, or % test plan complete. Develop plans with measurable results.
  - Clarifying accountability: create countermeasures when gaps exist. Clearly show who is leading and when results are expected.

Visual Management Example

Let's build up an example of visual management. Suppose one of the most important goals of Project Floink, a second generation automatic welder, is the reduction in parts cost from $1247 (cost at project start) to $850 (target cost). Figure 7.8 shows an example visual management board.

Now, let's walk through the criteria for visual management given above and see if this visualization meets the requirements:

- Simplicity in the graphics is enhanced by having minimal text, using concrete measures ($), clarifying actual, plan, and gap, and having a clear start and target cost. Top issues and countermeasures are shown at a high level. The entire chart fits easily on a single page.

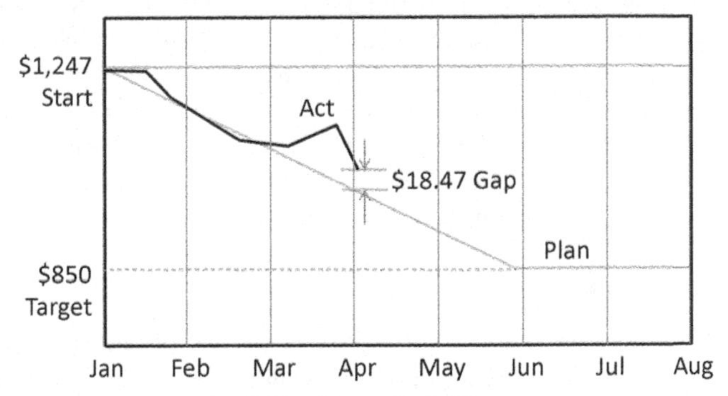

Top Cost Items Report: April (Total Gap: $18.47)

Top issues

New gluing machine takes 6 min. longer than expected (+$8.57)
Rolled steel costs increased due to commodity price (+$6.25)
Unsuccessful negotiating redux in monitor software (+$2.50)

Mitigation:

May 17-19: Glue machine 3P Kaizen for cost redux (-$8.50 target, Gretchen)
May 30: Reduce sheet metal piece count from 7 to 4 (-$7.00 target, Brandon)

FIGURE 7.8 Sample visual management display for product cost.

- Credibility here requires that the current cost is built up from a bill of materials and assembly process definition that can be made available for any stakeholder to review. Ideally the bill of materials would be accessed directly from the visual chart, either through a link to an material requirements planning system or cut-and-paste to a hidden sheet in a spreadsheet file.
- Does this drive good action? First, the focus is on one of the project's most important goals: cost reduction. Second, plan, actual, and gap are shown. Root cause is shown for most of the gap with mitigation that has clear targets, owners, and dates.

Consider how powerful a tool such a display would be if it was updated regularly—perhaps printed weekly or even displayed continually on a large monitor and updated daily. Everyone on the team would know where the gaps are and who's leading. The team will be more likely to pull together. And when someone acts out of step with the goal, say adding a questionable feature that widens the gap by $8.00, the PM (or Brandon or Gretchen) is likely to show up at that person's desk a few minutes after the change is published. Compare this to the PM keeping such a spreadsheet on a PC and reviewing it from time to time.

Like so many concepts in lean thinking, the visualization is flexible. It can be transformative if right-thinking is applied: the right areas to focus, the right level of detail, and reliable updating. At the other extreme, if the wrong area is chosen to focus on, or the detail is too fine to see the issue or too coarse to drive action, or if the chart is constantly out of date or riddled with errors, it will bring little value. This is an illustration of what is best and worst about lean. Lean is a mindset with few prescribed tools and techniques. A skilled practitioner can bend lean techniques to fit almost any issue in any project. At the same time, it's easy to create expensive "fake" lean work that drives no improvement—events that are not finished, Kaizens poorly defined or missing needed talent, and widescreen displays showing out-of-date information.

7.3.2 Go to Gemba

Going to Gemba is a key concept in lean production. It's just as true for product development, although the place "where the action is" is often quite different (Figure 7.9). Here are several examples of Gemba for product developers.

Product Point of Use

Lean thinking teaches two principles above all others: know the value you bring to customers and eliminate waste. The best way to learn the value of your planned product to the customer is to observe the customer using similar products in the most real-world conditions possible. You won't find that in your office. You won't even find it during a customer visit if you stay in a conference room. Whatever you developing, observe people using something similar in the environment where they are meant to use it. The world of product development is full of stories of developers going to Gemba and learning a great deal in a few hours about products they worked on for years: interviewing users, observing users, and using the products themselves to do what users normally do. I recall one of the first such stories I heard when a woman became CEO of a razor company and required her mostly

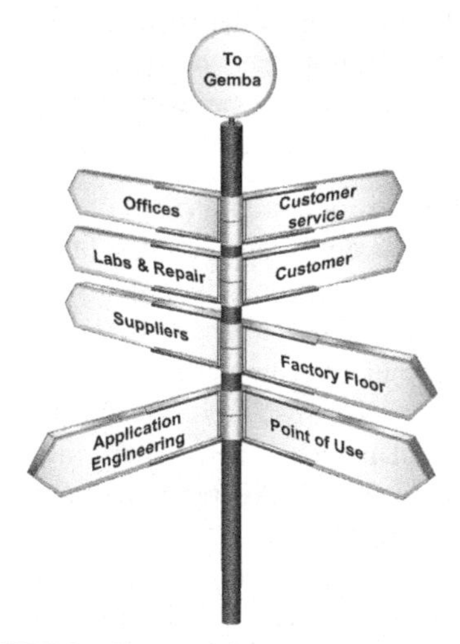

FIGURE 7.9 The many Gembas the team can go to.

male staff to shave their legs with the company's products. You can only imagine how much they learned at Gemba (their bathroom in this case) in a short time.

In their book, *Lean Product and Process Development*, Ward and Sobek provide an outstanding illustration of going to Gemba. The company Menlo created "High Tech Anthropologists" (HTAs) who go to Gemba for days to watch end users in their native environments [19]. They seek to deeply understand the full context in which their products are used. HTAs work to identify *pain points* for their users—shortcomings of their products that indicate a likely place they can add value. They watch users and they interview them. They return several times as the team evolves from understanding how the product is used to identifying problems users are enduring to creating solutions that increase the value they bring. Menlo created a wide range of standard work to guide teams through all phases of learning. In this example, the company did more than send developers to Gemba—they created an entire culture of Gemba through process and extensive engagement.

Factory Floor

In any design-for-manufacturing activity, spending time in the manufacturing area will improve your ability to create designs that eliminate waste. An engineer in his office might imagine quality defects arise from assembly workers that don't apply themselves properly. But spending a few hours building the product can open his eyes to how difficult it is to produce a product, one after the other, when the design has manufacturing shortcomings. Deep understanding of the processes and equipment used by the factory and the supply chain allows designs that can be built faster and with fewer defects.

Laboratory

Spending time in the laboratory while tests are being executed will be a benefit to those designing the tests. Tests that seem straightforward in the conference room may have defects that are quickly revealed in the test environment. Similar thinking applies to other service groups: repair, customer service, and applications. When the people outside your group are experiencing a problem, go to their work area so you'll understand better what they are dealing with.

Engineering Offices

Engineering managers can be too comfortable with meetings in their office or in conference rooms. Spending time with the engineer who has encountered a roadblock can bring breakthroughs. I recall once several years ago where a mechanical engineer was stuck for several days designing a spring—it didn't seem possible to meet the spring's complex requirements of axial flexibility, torsional stiffness, and long life. We discussed the problem many times in a sterile environment after we encountered the roadblock. However, when I went to Gemba—in this case, the engineer's office—and sat with him for a couple of hours the issues became clear. The issue that had blocked progress for more than a week was resolved in an hour or so.

Supplier Factory

You can also go to Gemba at the supplier site. When there is a stubborn defect with a supplied component, be prepared to get off the phone and travel to Gemba. A few hours at a supplier site might bring more value than weeks of long-distance talking. As with all examples of "go to Gemba" you will understand complex problems much better if you are present to see the problem, ask questions, and work the alternatives.

7.3.3 WIP Control in Product Development

The reduction of WIP is central to lean thinking—in the factory or with knowledge work such as product development. In product development WIP includes:

- Too many projects active at one time (organizational WIP).
- Excessively complex product iterations, either too many features or too large a performance increase (project team WIP).
- Multitasking (personal WIP).

Any open work is WIP and, according to lean thinking, WIP brings waste. At the organizational level, too many projects being open prevent management from engaging deeply in all projects. This increases the work required for review, approvals, and reprioritization. All of this adds delay. If the general manager is receiving capital requests from three projects and she's not deeply familiar with any of them, the process to get approval will be slow. Aside from the fact that she'll have three times the work of approving one project, she'll need more meetings to bring her up to speed; she'll have more concerns about unknown risks. This will require more preparation from the PM, more analysis and, ultimately, more delay.

The same is true for project team WIP. If the team is building too complex a product in a single iteration, they will go longer before getting feedback from the customer. A simple example: let's say the team is developing a lab instrument that needs to communicate via four different protocols. If the product doesn't release until all four are complete, any defects in the communication architecture won't be found until much of the work is complete. Had the product been released with one protocol, the team could receive feedback much sooner. Let's take the worst case where initial feedback reveals that the architecture does have defects that affect all protocols. Now, development on new features must stop while the team reworks all four protocols.

WIP on a personal level is damaging as well. An engineer split among five projects will work less efficiently. In Chapter 6, we discussed the problems of multitasking from a point of view of reduced velocity. CCPM demands the relentless minimization of multitasking. Lean thinking takes a similarly dim view of multitasking because of the damage it does to flow. The longer it takes an engineer to complete a task, the longer it takes to find defects.

So, WIP brings waste at many levels. Like an overloaded highway, WIP chokes off flow. Market data that drove the project is getting more out of date. Opportunities with customers are evaporating as the competition passes by. The technology in the new product is aging before it even gets to market. And defects go longer without detection, forcing more rework, leaving less time to add value for the customer. And the defects we're discussing are not simply coding errors or design flaws. Almost every function in a company is at one point or other involved with product development. That means defects can be generated anywhere in the organization: marketing, sales, customer service, ops, sourcing, finance, R&D, test, certification, management, and more.

WIP in knowledge work is more difficult to see than WIP in a factory. Factory WIP can be seen with a walkthrough of production. Buggies full of subassemblies parked waiting for processing tell the story quickly. WIP in knowledge work is information hidden away on computers or taxing the mental capacity of the product development team. It may be completely unnoticed.

The Math of WIP

Donald Reinertsen has developed a mathematical view of WIP in product development [20]. He seeks to convert WIP to economic loss to help demonstrate the harm it causes. His first step is to use queuing theory to predict the performance of a highly loaded system. The example is the simplest of all queues, the so-called M/M/1 queue. This is a queue with no limits where only one event can be processed at a time. The events arrive randomly but once processing begins, it has fixed time to completion. An example would be a simple communication protocol where all packets take a fixed time to process, let's say 20 ms. The packets arrive according to a Gaussian distribution. Of course, there's a wide difference between a simple M/M/1 queue and complex WIP in a product development system; nevertheless the example is instructive because the math is simple enough to get a closed form solution, and yet it demonstrates behavior seen in common queues such as highways, factory floors, and product development teams.

The closed form for the M/M/1 queue is given by Little's formula [21]:

$$L_q = \frac{\rho^2}{1-\rho}$$

where L_q is the expected time to process a packet and ρ is the loading of the network. Figure 7.10 graphs the result. When you attempt to run the system at 85% capacity, the average size of the queue is about five. Said another way, it takes six times as long[1] to process the event at an 85% load than it does with a load under 40% where L_q is close to 0.

1. The six times is five times waiting in the queue for five events and then one time for processing time, versus no time in the queue at low L_q.

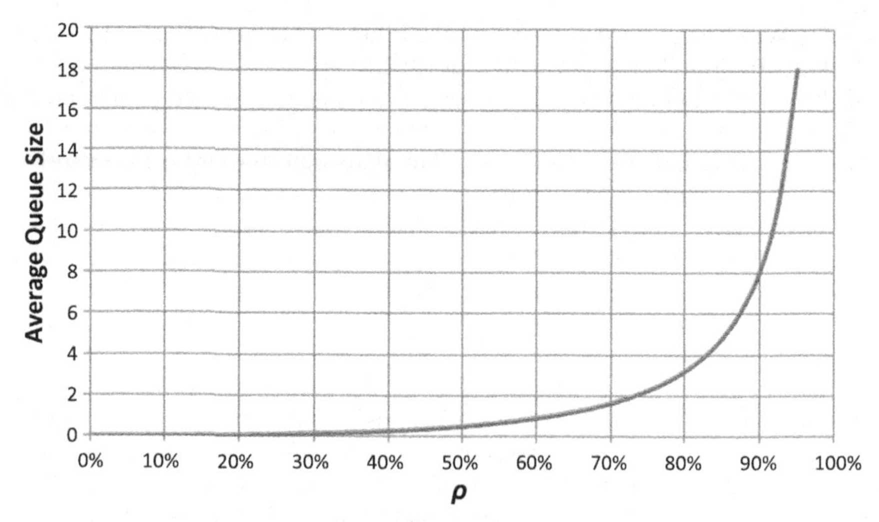

FIGURE 7.10 Little's formula: queue size versus ρ.

Anyone who has spent time sitting in traffic will recognize Figure 7.10. A highway can be busy but moving fast; then if it gets a little busier progress stops for everyone. Let's say at 80% capacity, traffic moves along nicely and your 15-mile commute takes 20 min. But add a few more cars and that same commute can take 90 min. A simple example: if 8000 vehicles on a road average 20 min for 15 miles, but 2,000 addition cars increase that to 90 min, we can say the cost to the 8000 cars of adding 2000 more is:

$$(90-20) \times 8000 = 560{,}000\,\text{car-minutes} = 9333\,\text{car-hours}$$

Assuming one or two people per car, that amounts to about 15,000 h of waste.

Now apply this idea to product development. When the system is operating near capacity everything takes much longer to accomplish because of waste–waste that is hard to even see, much less measure. The waste comes in the form of:

- Defects occur throughout the organization more often (due to a less effective team) and take longer to correct (due to a longer time of detection).
- Overproduction of features and performance the market doesn't value because the marketing work is out of date.
- Waiting for approvals, reviews, and resource allocation decisions.
- Transportation of knowledge between team members because projects are longer and team composition changes more than is necessary.
- Underused talent due to low morale from teams that don't "win" and are overloaded with rework of every stripe.

The Cost of WIP

In Reinertsen's construction, Little's formula can then be used to estimate the cost of overcapacity. The formula is mirrored horizontally so the horizontal axis becomes excess capacity. This is shown in Figure 7.11. The vertical scaling

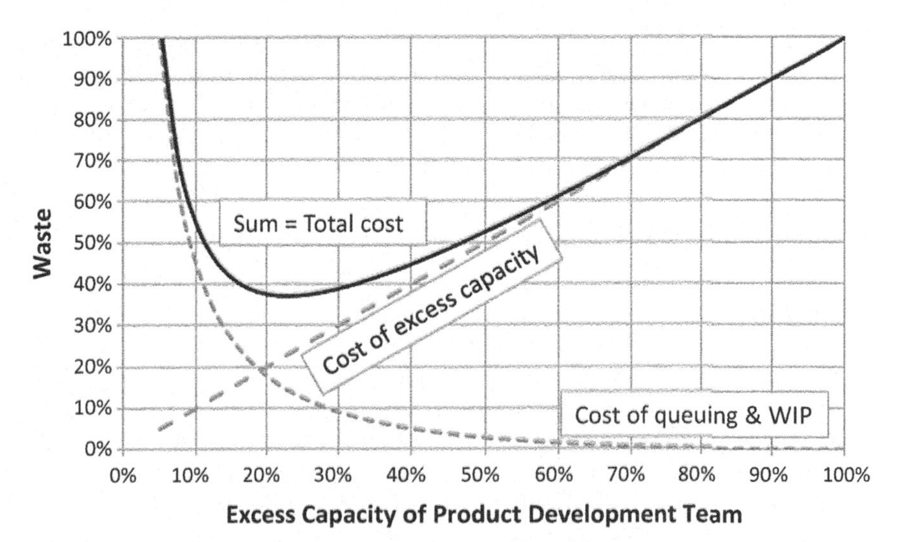

FIGURE 7.11 Total waste versus excess capacity.

is subjective; here it's scaled as 95% usage (5% excess capacity) is twice as expensive as the baseline. This seems intuitive, but is much less severe than Little's formula, which yielded almost 20 times for the same comparison. Then another cost line is added: the cost of idle resources. This is, of course, a straight line. If project resources are 100% idle, the extra cost is 100%; if they are not idle, there is no waste due to idleness.

The final step is to add the two costs together to obtain the total cost of excess resources. By considering both effects, we see the optimal cost is not obtained when resources are 100% utilized, but rather at something like 15% or 20% excess capacity. The finding is analogous to a highway—total volume of cars per hour is optimized in a similar range. Further, a common rule of thumb on the factory floor is resources should be about 85% utilized for optimal results. While the analysis is helpful conceptually, few organizations can measure the cost of overused resources. So acceptance of the concept may have to come as much from judgment as measurement.

It's interesting to note how CPM, CCPM, and LPD come to similar conclusions from different directions. The discussion in Section 5.3 showed how delay caused great harm to lifetime profitability of a project. CCPM assumes schedule is the critical constraint and all effort should be made to reduce delay. Lean sees WIP as the villain and delay as one of many negative effects. Each arrives at a similar conclusion: avoid delay. But, while they share a goal, the techniques they teach to accomplish it often differ.

Minimizing WIP

In LPD, minimizing WIP is central to improving performance. Reinertsen wrote, "Queues are the root cause of the majority of economic waste in product development" [22]. The most common way of having excessive WIP in a product development organization is having too many projects running simultaneously. The Toyota Production System normally limits WIP by blocking entry of the upstream process. This solution is commonly applied to highways, using traffic lights at entrance ramps to limit entering traffic.

One of the difficulties of applying a pure blocking strategy to project management is tasks in a project have much more variability than work on the factory floor.

- Duration
  Project tasks range in duration from more than a week to less than an hour. Such variation is rarely seen in a production system.
- Value
  Project tasks also vary more in value. A delay in a task on the critical path is likely to delay a project. A delay in other tasks may cause no harm. In a production system, value of the product is not used as part of the blocking mechanism; the unit in the downstream process blocks the upstream process independent of the relative value of the finished unit. In a project, tasks are prioritized to avoid low value tasks blocking high value tasks.
- Workers
  There is variation in workers in all knowledge work: IT, plant engineers, and product development. On the factory floor, the assumption is any trained worker is capable of executing a process properly. On the project team, there is large variability in skill sets and variability of capability within those skills; one goal in product development is to obtain the largest contribution possible from each person.

The variability of product development imposes a need to find alternatives to deal with WIP [20] beyond simple blocking. For example, few PMs would knowingly allow a task on the critical path of an important project to be blocked by a task that brought relatively low value such as processing ordinary engineering change orders (ECOs). The alternatives for WIP reduction in product development fall into two categories: increase capacity and reduce loading as shown in Figure 7.12.

Capacity can be increased through adding new hires and consultants; this is effective for all WIP; WIP isolated to a part of the organization can be addressed by borrowing internal resources thereby shifting resources to the group with the greatest need. This is facilitated by staffing the department with overlapping skills. It can also be facilitated by staffing with so-called *T-shaped resources*: people who are deep in a skill (the vertical bar in "T") and able to coordinate with others with different skills sets (the horizontal bar in "T").

Loading can be reduced in several ways. The simplest method is to block entry such as CCPM does with its severe restrictions on multitasking. An alternative to the on/off limits of blocking is to adjust criteria for accepting new work as WIP increases. When WIP is near zero, projects with relatively

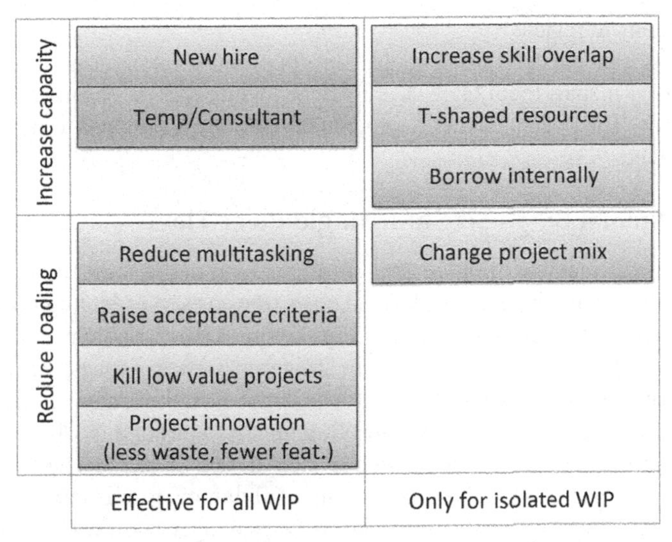

FIGURE 7.12 Ways to reduce WIP in product development.

low value can be accepted; when WIP is high, only the highest value projects are started. Another common practice is to kill projects of low value during periods of high WIP. Project innovation is another alternative, most commonly by removing lower-value features and functions from ongoing projects to speed the projects to completion. If the WIP is isolated to a group and the department overall has excess capacity, another alternative is to change the mix of project work and so shift loading from the group with high WIP to other parts of the department.

WIP product development is hard to measure, but the best places to look for it are [23]:

- Marketing
  Too many projects being evaluated for potential development. Too many other duties to be able to apply much effort to product development (e.g., product line management, marketing communications).
- Analysis
  Need for too many instances of analysis by any one technical expert.
- CAD
  Insufficient CAD resources supporting many projects.
- Purchasing
  Production purchasing will prioritize purchasing activities for the factory. Purchasing requests from product development can back up.
- Prototyping, testing, and tooling
  Specialized resources are often required for prototype assembly, test labs, and tool fabrication. These resources are often organized in service groups that serve many projects.

- Management review
  Projects can queue waiting to get an opportunity for management review. Normally management reviews are batched with reviews scheduled at regular intervals. The queue can delay projects if approval is required to move to the next stage.

7.3.4 Minimal Batch Size and Single-Piece Flow

Batching is the practice of processing multiple items at the same time. Batching is economical when there are high fixed costs (the costs of executing the process independent of the number of items) in relation to the variable costs (the cost added by each item being processed). The term *costs* is expansive, including anything that needs a resource that must be conserved: time, capital equipment, power, floor space, or even mindshare. Batching is all around us. As I write this I'm in a Boeing 777 several miles above the Atlantic Ocean batched with hundreds of other passengers. The fixed costs of flying over an ocean are so high and the incremental cost of adding one more passenger so low, batching in air travel is an economic necessity for most of us.

In product development, there are many processes that have a high fixed cost. Regulatory certification is expensive and so discourages incremental changes that require recertification. The cost of tooling molds and dies is another expensive fixed cost in product development as is setting up production cells. These fixed costs make it more economical to develop many features and functions in a single product before tooling or building production cells; this is another reason to batch product development.

The most common form of batching in product development is the creation of many features and functions in one iteration and bundling them together for the next step. In the process of Figure 7.13, many features are waiting in design like a rat swallowed by snake, while little activity occurs in the other steps.

If batching creates large economic advantages, it's counterintuitive that lean thinking teaches to avoid batching. The lean ideal is the absence of batching: *single-piece flow*. For product development, this ideal is embodied in the smooth flow of Figure 7.14, with each feature and function flowing smoothly through the process.

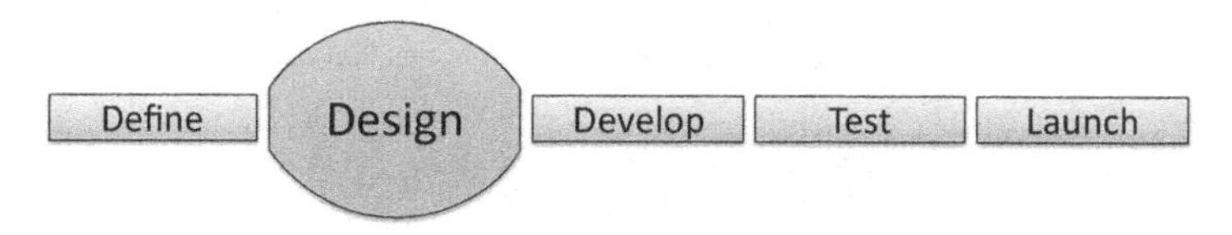

FIGURE 7.13 Product development process with batching.

FIGURE 7.14 Product development with single-piece flow.

The justification of single-piece flow is that, while the economic advantages of batching are relatively easy to calculate in most cases, the costs of batching are large and hidden. As a result, people have historically overestimated the benefits of large batch sizes in product development processes. Single-piece flow is an ideal to strive for understanding that some amount of batching will remain for virtually all product development. Even for uncertified software, the product development class with one of the lowest iteration costs, there is some level of batching owing to the costs associated with releasing a version of firmware to the market. So, the ideal may be single-piece flow, but there is little expectation of reaching that goal in every aspect of product development. Lean thinking teaches to diligently search for batching and when it's identified, vigorously work to minimize the batch size. Lean thinking also reveals the high costs of batching, flushing out its hidden penalties so fully informed decisions can be made.

The costs of batching are in the creation of waste. Some of the most common examples are as follows.

Slower Detection of Defects

Batching slows the detection of defects by delaying downstream processes, one of the most reliable ways to detect defects of upstream processes. Referring to Figure 7.14, each step in the process can detect defects in any step to its left. Product developers often uncover that feature definitions are incomplete or conflicting after they start the design process. During development, it's common to find the design is unable to meet the goals of the product. Of course, the entire goal of testing is to find defects.

Batching slows detection because the upstream process step is executed many times before the downstream process is executed once. If a marketer defines 50 functions, each with the same definition defect that will be detected in Design or Test, 49 more defects will be created than would have been in single-piece flow. Waste in hardware projects is created not only by wasted design effort but also in wasted cost of prototype parts. If an error is detected after fabrication of many components, all those components may become scrap. Had the error been detected quickly, fewer components would have been fabricated.

Another waste caused by slower detection occurs when a feature must be abandoned due to information learned in the test phase. This causes waste because any feature or function dependent on the abandoned feature would be unneeded. For example, suppose a light sensor is originally rated up to 200 °C. The sensor and all its supporting components must then be designed for 200 °C. However, suppose that during testing, the core sensing technology is found to be unreliable above 175 °C. Assuming the product still had substantial value in the market rated at 175 °C, all effort expended for the supporting components to raise their maximum temperature above 175°C would be waste.

Slower detection of defects means defects are likely to accumulate in larger number, which is another type of WIP. More bugs in software means more bug reports. More quality defects in products at launch means more root-cause/

countermeasure actions queueing up. And a larger number of defects implies greater communication about those defects until they are resolved: customers, end users, and distributors will all want to be informed of the progress resolving any defects that affect them.

Loss of Urgency

When batch sizes are large, the urgency the team feels diminishes. If a software developer has 1 week to fix five software defects, he is likely to feel urgency to reduce the count to 0. If there are 250 defects present, he's unlikely to feel the same urgency to reduce the count to 245. The same is true with any defect WIP—if 60 mechanical drawings have defects, curing one defect doesn't seem to contribute much to progress, so why hurry? Of course there are tools PMs can use to create urgency such as creating a plan to correct 10 drawings a week and then measuring to plan. But building and managing those tools takes effort–effort that is waste compared to the high urgency that occurs naturally from low WIP.

More Difficult to Find Root Cause

Batching also makes it more difficult to determine root cause. This can be seen in software between Development and Test. When software exceptions are found in Test, if they are due to multiple dependent errors, it is almost always more difficult—sometimes much more difficult—to understand the root causes. Were the software defects presented to Test one at a time, the root cause would be easier to determine. The same effect occurs in hardware. If a base product and many options are designed simultaneously, a dimensional defect in the base product can translate into a similar defect in a bolt-on option.

Because batching increases the time between when a design is complete and when a defect in the design is detected, there are several wastes. First, the developer must recall the details of the design. If only a few days have passed, those details will be fresh in her mind—she is more likely to know root cause intuitively. But if defects are found 6 weeks later, she may have to spend hours reviewing mechanical drawings, electrical specifications, or code segments just to reclaim the familiarity she had with the design when it was being created.

Feedback from Customers

Batching features in products also slows the feedback path from the customers and end users. If a base product or even a proof-of-concept can be delivered to a customer to evaluate, defects can be detected early in the design process so they can be corrected quickly and avoided in future. Consider the diagram of Figure 7.15. The assumption is that despite the best efforts of a diligent team, there will remain some number of defects only a customer or end user will find. These are usually *soft* defects—such as not meeting ease-of-use expectations, not providing functions in the expected combination, or not supporting how a

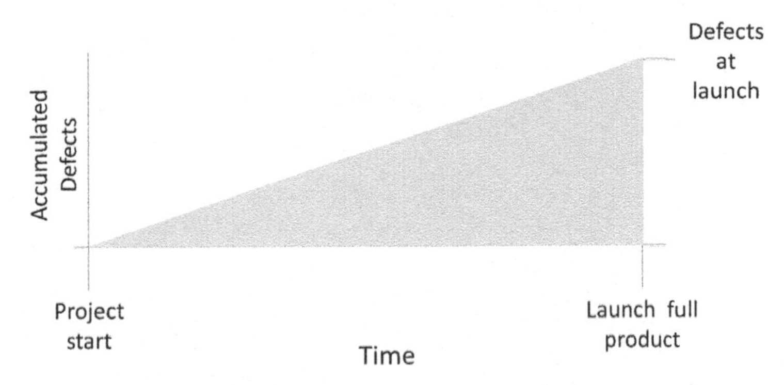

FIGURE 7.15 Customer-detected defects with large batch size.

FIGURE 7.16 Including customer feedback as part of the design process.

customer expects to use a product whether or not it's in the specification. I recall one customer that expected their maintenance workers to stand on the ends of mounted motors to work on machines—this requirement never made it to a specification, but it was a customer need. These sorts of defects are exceedingly difficult to find early in a design; in most cases some will be found by the customer. When the customer's first view is the full product, these sorts of defect will accumulate during the entire design cycle.

So, how can we use smaller batch size to reduce the number of total customer-detected defects? First, we must change thinking about the design process of Figure 7.14 to include customer feedback. Given that the only way to find all defects reliably is to get the customer's feedback, including this feedback in the process is the intuitive response. So, Figure 7.14 must be augmented to include customer feedback as shown in Figure 7.16.

Now the question becomes how to reduce the batch size. A simplistic answer, such as develop one feature at a time, is almost always impractical because of the fixed costs associated with product development. However, there are many well-known ways to obtain customer feedback beyond releasing the full product. Examples include presenting mock-ups and storyboards to customers very early in the design process. Soon after the design is completed, proof-of-concept units can be supplied to key customers. After that, low-feature prototypes can be provided on a selected basis. Then the base product can be released and the process ends in the release of the full product. At each interface with the customer, there is an opportunity to detect defects. Of course, defects are unlikely to be driven to zero because of these actions, but the total number of defects could be much lower at the launch of the full product as shown in Figure 7.17. Traditional project management teaches that this can lower efficiency—all the

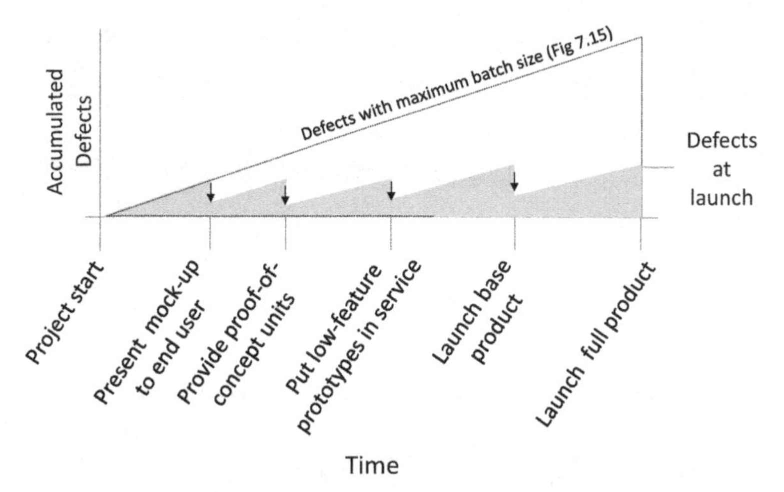

FIGURE 7.17 Customer-detected defects with minimum batch size.

effort expected to get early feedback is time that could have been applied to the final product, accelerating the launch. Lean thinking teaches that the removal of waste brought about by the thoughtful reduction of batch size will more than pay for the additional effort.

One point many readers will have noticed is that the steps presented to reduce batch size are also used in traditional project management, albeit to a lesser degree. That's true—lean thinking didn't invent proof-of-concept units or incremental-feature launch. However, lean thinking provides an overall reference frame to understand why these steps improve the project. In traditional project management, these actions stand alone, often as things that ought to be done but are not, at least to the extent they should be. Lean thinking provides a framework to understand why they work, which then informs PMs of what to look for in the development process that shows the batch is not right-sized. This is true in many parts of lean thinking—the focus on understanding value and reducing waste provides a framework to understand the entire development process, which then informs how the process can be improved. Traditional project management presents similar concepts, but often as prescriptions; however, giving the *what* without the *why* tells only half the story, which is probably why these types of steps are often poorly executed in traditionally managed projects.

Tom Gilb takes this principle to its extreme but logical conclusion: no single step in a project should be allowed more than 2% of the total project or 2% of the budget to deliver value. For a 1-year project, that means delivering value every week! This approach is common in software projects as will be covered in Chapter 8, but Gilb suggests the approach applies to hardware projects as well. He suggests a process called *Decomposition* with numerous concepts that allow the project to be divided into small slices. For example, one suggestion to reveal the smallest increments of value is to deliver value only to a narrow range of customers that can be defined in terms of experience level (novice or expert), geographical location, or as needing support for only a narrow use case of the product. He points out that product developers often react by stating that decomposing their project into such small increments is impossible. However, he makes a convincing case that it is almost always possible and gives a number of examples where it has been used successfully [24].

Reducing Changeover Time

If the primary reason for large batch sizes is to reduce the effect of fixed costs, then it must follow that an effective method to support smaller batch sizes is to reduce fixed costs. This is the principle of SMED in Section 7.1.3 and it can be applied to product development. The example of releasing products in increments was presented above. There are several other methods, which follow.

Using Software Models

Software models are commonly used across the spectrum of product development. One reason is to reduce change over time—software models of a tire can be changed in a fraction of a second. This concept is expanded in *hardware-in-the-loop* (HIL) [25], a technique where a physical controller operates on simulated hardware. Such capability at one time was so complex and expensive, only the most well-funded organizations could afford it; now it's so accessible that universities use it in student competitions [26]. The concept is reversed in *rapid control prototyping* (RCP) [27]—here the controller is run in simulation and the hardware is physical. These both present opportunities to reduce changeover time during development.

Automated Software Testing

Automated software testing reduces set-up costs for different software tests. This allows a full battery of tests to run immediately after changes are made to software, allowing rapid detection of software defects. This method is particularly effective for defects that affect seemingly unrelated areas of code (a common problem for software developers) through shared memory, changed computational timing, or other types of resource sharing. Automated software testing often requires a large investment for the many types of hardware that are relevant to the test regimen and for staff to write tests and manage conversion of exceptions (tests that fail) to confirmed defects.

Automated Changeover of Hardware for Testing

Automated change of hardware reduces set-up costs for tests. Tests can be varied rapidly and during times when staff are not present, both of which reduce the costs of running tests. Since testing is one of the most reliable means of detecting failure, this reduction in set-up allows more testing (testing in more combinations and running tests longer) and thus allows the team to be more likely to find defects.

Fast Prototyping

Fast prototyping is an important investment in hardware design allowing design ideas to be rapidly built and tested for design defects as well as be rapidly deployed to key customers for the identification of design and definition defects. Software designers have enjoyed the benefit of fast prototyping for many years: software is famously easy to build up in various combinations for broad testing internal and external to the development team. Similarly electronic designs have long been able to go from design to prototype in a few days based on rapid prototype PCB houses. Recent advances in 3D printing have brought similar capability to mechanical engineers. Some companies augment these capabilities with prototype machine shops and dedicated machinists to bring new designs to life in days.

7.3.5 Kanban Project Management

Kanban project management (KPM) is a simple visual management tool for projects of low or medium complexity. It is similar to the Kanban inventory management system described in Section 7.1.6. In KPM, each task is listed separately, sometimes written on a sticky note and other times keying into software that mimics a white board. The Kanban board is divided into four or five columns with the tasks starting on the left in the least mature state and flowing to the right as they mature. Figure 7.18 shows an example KPM board based

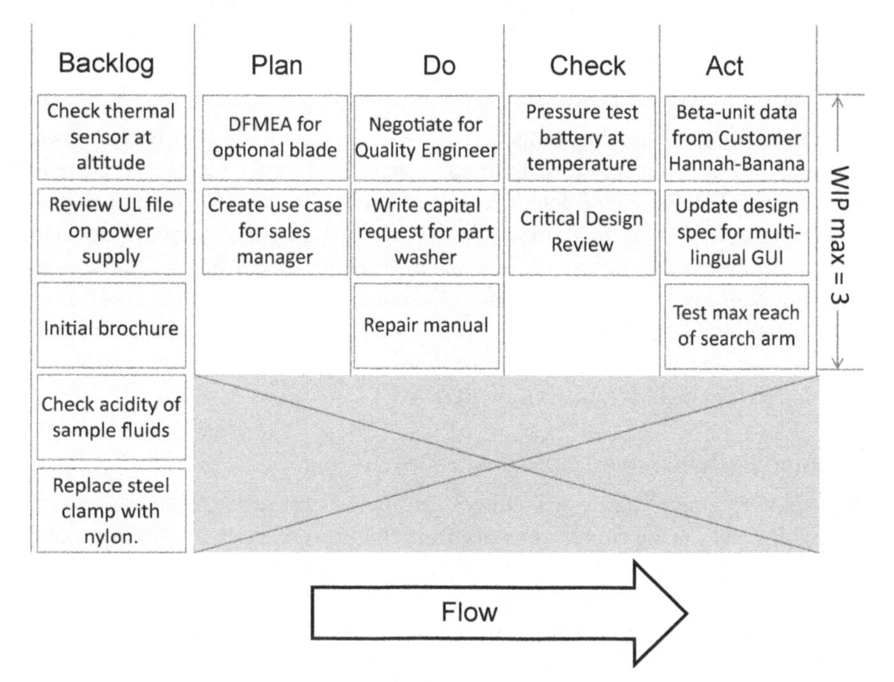

FIGURE 7.18 Kanban project management based on PDCA.

on the Deming cycle (PDCA). However, Kanban categories can be tailored to a specific need. For software development, the columns might be: backlog, specify, code, unit test, functional test, and document.

KPM is a pull system. Except for Backlog, columns are limited typically to about three items. In the example above, Plan has two items, so it can pull an item from Backlog; similarly, Check can pull from Do. However, Act and Do are at their maximum, so no further tasks should be pulled in. In this way, the Kanban management system provides a highly visual means to manage WIP, keeping the team from having too many tasks in process at one time.

KPM brings immediate and obvious benefits. First, it displays and limits multitasking in an intuitive way. Anyone with 30 seconds of training can tell when there are too many tasks in a column. Kanban boards are simple to use—where Gantt and fever charts require training and experience, a team with a white board, a stack of sticky notes, and a marker can start a Kanban board in a few minutes.

Kanban boards are also excellent tools for communicating the current project status to any stakeholder close to the project (they are generally too detailed for those who review the project on rare occasions). Everyone on the team can see what everyone else is working on. Priorities are made clear in a single simple graphic.

Kanban boards also support continuous improvement. Bottlenecks become apparent quickly and the team can improve process to remove them. Tasks that don't flow are easily spotted; the team can get to root cause on the issue and improve process. Tracking Kanban boards with measurements on process provides further illumination—for example, the team might track days-in-queue to see where bottlenecks form most often. The Kanban board is evolutionary—it starts with a small change to process and grows over time.

Kanban boards are easy to use, but one common flaw must be avoided to prevent WIP [28]. Do not separate columns of work with columns that hold WIP. For example, in the Kanban board of Figure 7.18, adding "Hold" columns between, Plan and Do, Do and Act, or Act and Check. Such columns create WIP and prevent any real flow across the Kanban board. To be clear, Backlog is not WIP because work on the task has not started. The purpose of Kanban management is to create flow and reduce WIP; columns for WIP thwart that purpose.

KPM Limitations

KPM has its limitations, the most obvious being the inability to link tasks together in complex ways. If a project has 50 tasks that have numerous predecessor/successor relationships, managing the project through Kanban can be impractical. You could place all 50 tasks in an enormous backlog column, but to pull the tasks out of backlog you would have to ensure all of a task's predecessors where complete. Unfortunately, the Kanban board doesn't show those relationships.

Another challenge with KPM is managing to a schedule. The Kanban board creates flow and manages WIP, but doesn't manage velocity or schedule well. So, the PM can see the current state, but cannot directly predict completion times of complex projects or intermediate milestones where handoffs need to occur. For example, suppose a set of chemical tests needs to be complete before designing the disposable packaging. The predecessor is primarily a task for chemists and the successor is for mechanical engineers. If the PM has to pull the mechanical engineers into the project full time when the chemistry is sufficiently complete, that event has to be scheduled well in advance. The standard Kanban board doesn't provide tools for this.

The net result of KPM is that it's a powerful tool to manage tasks on a short time horizon. By itself, it can be used to manage relatively simple projects, for example customization and cost reduction; it's not normally capable of managing complex projects alone. However, it can be combined with other methods that can fill that gap. For example, long-term issues of a large product-development project could be managed with CPM, but the Kanban board can be used to manage weekly tasks. So, the larger project might have a Gantt chart or deliverables list scheduled out for months, but there might be a Kanban board that gets updated for weekly tasks. This hybrid approach is used by Agile Scrum, as will be discussed in the next chapter.

How to Start with KPM

Perhaps the best advice for those who want to try KPM is to try it first as a board for managing personal daily work [28]. I took that advice about 6 months ago and never looked back. I created the Kanban board shown in Figure 7.19 with four sheets of printer paper and a stack of sticky notes; it's taped up beside my desk so I never go more than a few minutes without seeing it. (I tried it on a web-based

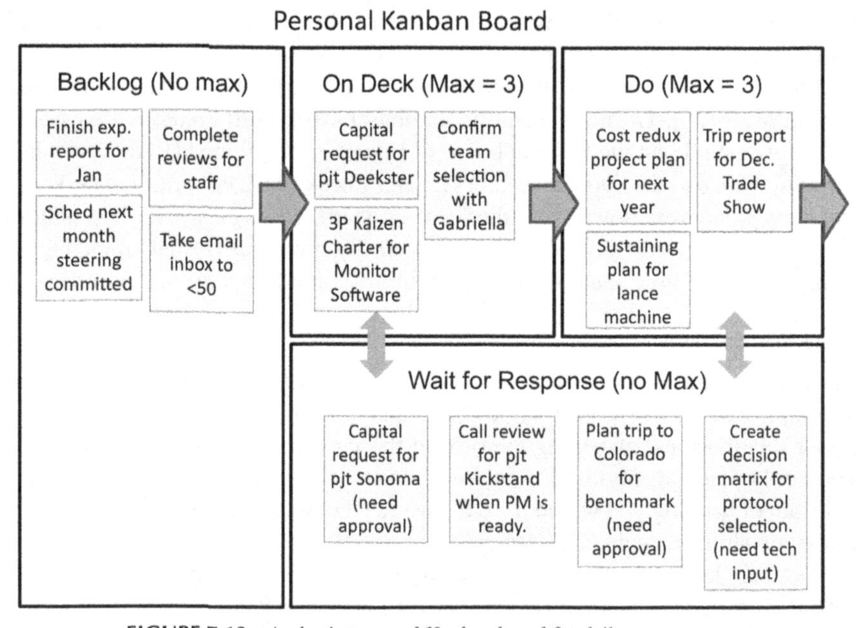

FIGURE 7.19 Author's personal Kanban board for daily management.

Kanban, but prefer the visibility paper provides). It immediately brought order to my workday. It also helped me cut down on multitasking. When On deck or Do are overloaded, I return a sticky to Backlog. One personal improvement came when I recognized I was much more efficient at email when I attack it in uninterrupted blocks of time. So now I'm more likely to process 20 or 30 emails in 30min rather than trying to do a few here and there. I learned that each time I found myself with 10minutes free, I was drawn to email when there were more important things I needed to do. The Kanban board made that obvious.

One final point: I added a Kanban category *Wait for response* that I recognize could become a kind of WIP column of the nature warned against above. However, I found I needed it since my Kanban typically flows in a matter of days, but people often take a week or two to respond. In my mind, it's more backlog than WIP because I'm not expending any effort until I hear back or until too much time has passed (when I feel too much time has gone by without a response, I move the sticky back to Do or On deck and then follow up on or escalate the issue).

After about 3months experimenting, we started to use a Kanban board to address new issues in the management team. As a management team, we now use this board to manage any issue that affects customers but can't be resolved with standard work. We use a simple PowerPoint[2] format to manage a weekly call. This also has worked well as marketing, operations, and sales leaders meet and get fast visibility on important and fast-changing situations. It also lets us track progress and reliably monitor items sitting in backlog.

Both of these examples played to the strengths of KPM: numerous unrelated tasks of modest complexity that are important but need not be executed on a strict time schedule. Here the Kanban board creates visibility of prioritization and flow; it reduces multitasking. These examples also show the flexibility of KPM—creating column headings that are tailored to the needs. No special software or training was required; just get started.

7.3.6 Lean Innovation

Another concept in LPD is lean innovation, an expansion on WIP reduction. Lean innovation espouses an iterative approach to creating new product definitions. It begins with the *minimum viable product* (MVP)—the most basic variation of the product that brings real value to a customer—and then aggressively tests that version in real-world applications. Iteration continues until the core product is competitive in the market place. Lean innovation eschews the classic product development method of creating a comprehensive product specification—this isolates the team from the customer experience. As Tom Agan [29] put it, "lean innovation is not a better innovation process; rather it's a more efficient learning process."

---

2. We could also have used a Kanban software application, of which there are many. I evaluated Trello and Wrike, both of which were easy to use and also sufficient for our needs.

One of Agan's findings was that learning was the largest factor in reliably producing revenue from new products. Lean innovation creates a better learning environment through quick iteration and trial in the most real-life situations possible. In this way, lean innovation captures knowledge more quickly than traditional approaches. Agan also found that companies who required debriefings of successes/failures doubled revenue from product launches; if a third party facilitated, revenue increased again. Lean innovation and the MVP are covered in more detail as a means of reducing risk in Section 9.3.3.

7.3.7 Obeya

Obeya (or Obeye or Oobeya), Japanese for "the big room," is a space to display information on every aspect of the project: design, manufacturing, quality, sales, marketing, sourcing, finance, and management. Obeya is credited in achieving a shorter time to market with reduced cost and improved quality. Obeya, like its predecessor the "war room," is a space dedicated to visual management [30]. Activities, issues, and deliverables are displayed in an intuitive, graphical format to facilitate frequent discussions [31] and disseminate information rapidly and reliably to all stakeholders. A schematic of an example Obeya room is depicted in Figure 7.20.

One view of the Obeya expands the information display to a colocation of workspace. The entire team works in the Obeya room and key topics they work

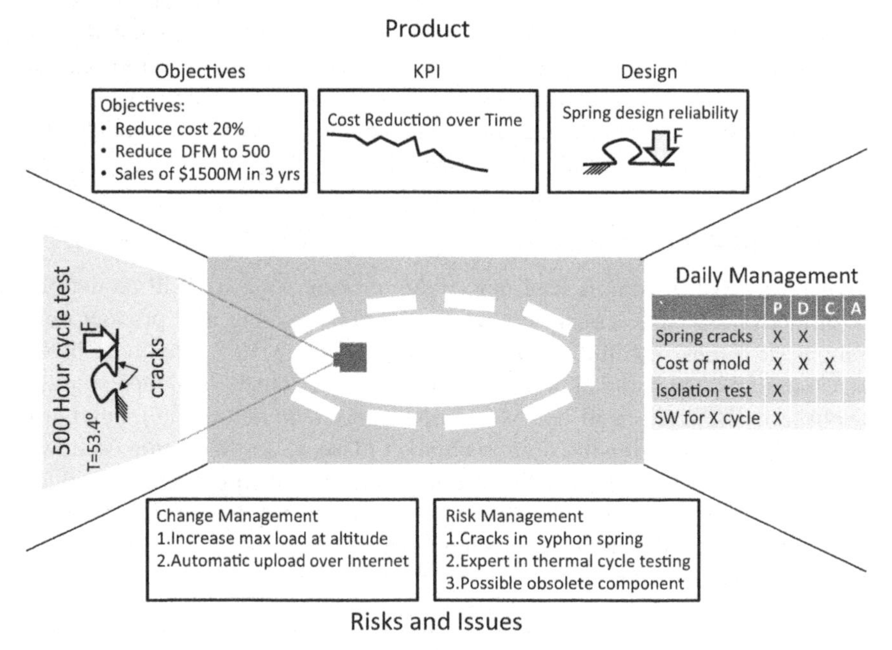

FIGURE 7.20 Sample Obeya room.

on are displayed on the walls of their workspace. In traditional cross-functional projects, the PM might visit team members in their offices or rely on infrequent meetings. By contrast, Obeya allows the broad team to gather frequently in an area rich with information that can help them understand current state and make decisions. This facilitates better communication of progress and clearer information for decisions [30,32,33]. This Obeya room is almost a continuous Kaizen.

The Obeya room walls are continuously covered with key information as shown in Figure 7.20. The information can be divided into categories such as [34]:

- **Project objectives**
  - Project goals
  - Key product specifications, acceptance criteria
  - Progress tracking
  - Value stream maps
  - Design of critical components
- **Project schedule**
  - Gantt charts/Task lists
  - Progress charts (Schmidt, run, or fever charts)
  - Deliverables lists with progress toward completion
- **Resources**
  - Team members and service groups
  - Resource utilization risks
  - Performance to budget
- **Information for management decisions**
  - Changes being managed
  - Key risks and issues
  - Upcoming approvals

With Obeya, all stakeholders are better informed because they have easy access to all high-level project information. This facilitates team alignment and builds stronger team spirit. It also supports rapid management decisions [35]. It supports pushing more decisions down to the project team [36], an important factor in lean thinking.

Simply having the information available on intranets is insufficient to achieve the advantages of an Obeya room. First, those less familiar with the project such as managers will not know where to find the critical information for the current state of the project. Second, because the information is not continuously displayed, it doesn't become a focusing tool for the team. For example, if removing material cost from a PCB is the primary objective for the team for the next 2 weeks, the PM might display a graph depicting estimated cost versus time, updated daily to show improvement. A continuous display will focus the team much more than the same information stored away on a network drive or SharePoint® server, seen only by those informed and proactive enough to review it.

Types of Obeya

There are many types of Obeya and each Obeya room can be tailored to the company culture, the available space, the geographical makeup of the project team, and the needs of the project. The Obeya discussed above is focused on a single project. An Obeya room can be built for any purpose from managing a project to managing a business. They can be electronic (e.g., using the iObeya board [37]) to support dispersed teams. They can have dozens of flat screen monitors to display information that changes by the minute.

In summary, the project Obeya room provides the following advantages:

- Clarity on key project activities to help coordinate all stakeholders.
- Simplify communication.
- Clarity on areas that require management action along with prominently displayed supporting information to accelerate decision making.

7.3.8 LPD: Adoption and Resistance to Change

For those who wish to adopt LPD, it's important to start slow. Lean thinking teaches that most organizations resist change. The common pattern, shown in Figure 7.21, is that change initially meets with resistance—people who don't understand ideas often discount them or may even feel threatened by them. During this period, the agents of change must press on in an organization that may not be fully supportive. But, given time and evidence of improvement, most people start to welcome change.

The initial period of change must be managed well if the adoption is to be successful. A few things you can do are:

- **Start with low-risk ideas**

  Resistance to change decreases as people see improvement, so start your conversion to lean with low-risk ideas like:
  - Organizing confused task lists with a Kanban board

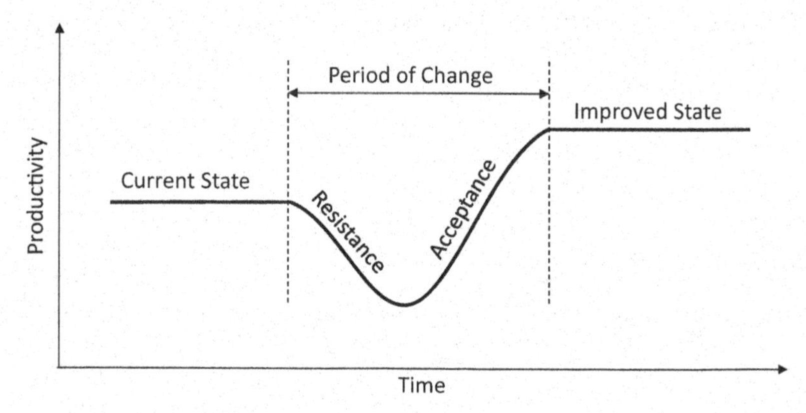

FIGURE 7.21 The resistance to change.

   - Reducing WIP in a work queue
   - Creating a score card with a few simple metrics to define what "winning" looks like

- **Choose something important to change**
  Apply change to an area where there is a broad consensus for a need to improve. Perhaps the ECO[3] process is a thorn in everyone's side. Or, maybe a certain type of lab test often causes delays. When improvements are made in areas where the team as a whole sees need, benefits will be recognized quickly, accelerating the acceptance of the change.

- **Implement nonthreatening changes**
  When people find change threatening, they are unlikely to accept it, even when there's data that shows it brings improvement. People can be threatened in many ways such as:
   - A perceived loss of prestige because they don't understand lean thinking. If a thought leader in the current organization has no exposure to lean, she may fear her approaches will become obsolete.
   - Their work will become less interesting.
     It is a common fear of developers that process will constrain their creativity, forcing them to follow a step-by-step list where before they could determine their own path. In fact, good process frees people to work on more interesting areas by making mundane tasks easier and pushing decision making down the org chart. But until people experience this, it's difficult to understand.
   - They may lose their jobs. Lean techniques are too often lauded for reducing head count. Especially in the beginning of lean adoption, avoid any change that has even the perception of putting someone's job at risk.

When introducing lean techniques, plan improvements to minimize the resistance to change. Use your transformational leadership skills—help create a vision for what the team can achieve with lean thinking and connect with those that are implementing the change so you understand their level of acceptance and their reservations. Start slow so others have time to adjust. While some people may never fully adjust to the atmosphere of change caused by lean thinking, most will become more flexible given time assuming they see the positive results.

7.4 COMPARING LPD TO OTHER PROJECT MANAGEMENT METHODS

Lean manufacturing techniques provide many powerful techniques for product development. Both domains benefit from lean thinking: focus on understanding customer value and removing waste. However, as much as lean production can provide concepts useful in LPD, there are many differences between the

3. Engineering Change Order.

two domains. The factory floor is much more homogeneous than product development. Tasks are better known on the factory floor when they start; product development, by its nature of doing things never done before, has many unknowns when starting out, from customer needs to technology readiness to needs for new manufacturing capability.

There are many corollaries between CCPM and LPD. For example, CCPM (via TOC) teaches that as the primary system constraint is relaxed (normally scheduled in project management), all facets of the project improve (see Section 6.2); this is similar to LPD's view of waste. Both teach visual management—make the critical items visible to the whole team to achieve focus (for examples of CCPM consider the fever chart and the use of a physical baton). While the similarities between the two systems are evident, there are important differences. Chief among them is mindset. CCPM is a system of project management with many key components that must be implemented at the time of adoption. It is revolutionary. LPD is evolutionary—it is first a change in mindset and second a set of techniques that come from that mindset.

In the final analysis, LPD is not a project management method in the way CPM and CCPM are. It provides a way of thinking and a set of tools to get you started. While in itself LPD is not a project management method, its clear-headed thinking can be applied to other project management types. An Obeya room can be used in CPM. MVP works in CCPM.

There is a fair amount of writing about the power of LPD and often that writing doesn't explain the need for base process. Sometimes authors criticize formal design reviews and Phase–Gate requirements for incremental funding approval as wasteful—these process steps do add work and if all requirements are met from the beginning, they are wasteful; of course, that can be said about almost any approval step. True, small companies can depend on the capability of one or two extraordinary PMs so that a full project management system can be overly constraining. But that solution is difficult to scale—relying on extraordinary personalities when the project portfolio includes 25 or 50 projects spread across 15 PMs working out of three sites usually produces disappointing results.

Perhaps a good analogy of LPD without base process is something like driving according to race-track rules—no speed limits, no minimal following distances, etc.; by comparison CPM, CCPM, and other project management systems run on something more like interstate highway rules. A single race car will go much faster than a commuter, but race-track rules cannot scale to deliver a city's full complement of commuters to their workplaces. The comparatively severe constraints of the interstate system are an important part of what enables that system to scale. In a similar way, the process of Phase–Gate and CPM/CCPM constrain the outstanding PM who can apply lean thinking with great agility and reliably uses sound judgment. But those same constraints allow a product development system that can scale to many PMs working on many projects across numerous sites within a company.

7.5 LPD KEY MEASURES OF EFFECTIVENESS

This section will provide an evaluation of the LPD method assuming it is additive to another method like CPM or CCPM. The focus here is on high-cost-of-iteration projects. (Agile, the subject of the next chapter, applies many principles of LPD to low-cost-of-iteration projects.) The results are shown in Table 7.4.

Good with high- and low-iteration-cost projects (+, +)
LPD techniques can be used with projects that have extensive planning cycles (high cost of iteration) and those that don't.

Process to coordinate varied disciplines (++)
Lean thinking with its holistic approach to understanding value and eliminating waste is ideal for pulling many disciplines in a single direction.

TABLE 7.4 Key Measurement of Effectiveness for Four Project Management Methods

Measure	CPM	CCPM	Phase-Gate	LPD
Good with high-iteration-cost projects	+	+		+
Good with low-iteration-cost projects	–	–		+
Process to coordinate varied disciplines	+	+	++	++
Mitigates risk before large investments			++	
Provides standard work for planning	–	–	++	–
Clearly defined methodology	+	+	+	–
Tools to maintain schedule	– –	++		+
Intuitive, has few adoption barriers	+	– –	+	+
Well-defined metrics/ visualization	+	++		++
Plans shared resources well	–	+		–
Availability of software tools	++	–		+
Sustainable over time	+	+	++	++
Low overhead over time	+	+	–	++

Mitigates risk before large investment
The ability to methodically mitigate risk requires LPD be built onto a process like Phase–Gate.

Provides standard work for planning; Clearly defined methodology (–), (–)
Because LPD is tailored to the needs at hand, it doesn't by itself provide standard work or a comprehensive project management method.

Tools to maintain schedule (+)
LPD with its unique ability to bring team focus to the critical issues provides tools to help maintain schedule such as the Obeya room and visual workflow management.

Intuitive, has few adoption barriers (+)
Much of LPD is highly intuitive starting with the focus on more value and less waste. There are some counterintuitive principles such as favoring the reduction of WIP above engaging resources 100% of the time. However, those principles have become so ingrained in modern industry thanks to the Toyota Production System they have reached a sort of intuition for people even modestly familiar with modern manufacturing techniques.

Well-defined metrics/visualization (++)
Visualization is foundational to LPD—it creates the demands for it and then teaches how to fulfill those demands.

Plans shared resources well (–)
LPD provides few tools to better share resources.

Availability of software tools (+)
The main reason software tools are available is that most lean techniques are simple enough that spreadsheets can support them. In fact, often poster board and sticky notes are sufficient to support many lean methods.

Sustainable over time (++)
Because LPD is based on continuous improvement it provides all the tools needed to sustain and improve the system over time.

Low overhead over time (++)
Because LPD focuses on removing waste it not only adds new features to process as requires, but also provides a framework for removing or trimming process that no longer delivers value.

7.6 SUMMARY

The concept of applying lean thinking to product development is widely used with success. The approach of applying the simple principles of maximizing value and minimizing waste to product development is almost self-evident. But LPD by itself does not create a project management method; instead it can be used to improve CPM, CCPM (actually, a substantial portion of CCPM is adding lean thinking to CPM), and, as will become clear in the next chapter, Agile Scrum through the lean thinking of extreme programming. The overriding advantage of LPD is it takes an organization where it is and helps it improve at a sustained rate.

7.7 RECOMMENDED READING

Coletta, Allan R. The lean 3P advantage: a practitioner's guide to the production preparation process. Productivity Press; 2012.

Martin Karen, Mike Osterling. Value stream mapping: how to visualize work and align leadership for organizational transformation hardcover. McGraw-Hill; 2013.

Mauer Robert. One small step can change your life: the Kaizen way. Workman Publishing Company; 2014.

Reinertsen, Donald G. The principles of product development flow: second generation lean product development. Celeritas Publishing; 2009.

Ward Allen, Durward Sobek. Lean product and process development. 2nd ed. Lean Enterprise Institute, Inc. 2014.

7.8 REFERENCES

[1] Womack, JP, Jones, DT. Lean Thinking: Banish Waste and Create Wealth in your Corporation, Productivity Press, 2003, p. 16f.

[2] Womack James P, Jones D, Roos D. The machine that changed the world: the story of lean production—Toyota's secret weapon in the global car wars that is now revolutionizing world industry. Free Press; 2007. Reprinted.

[3] www.GoLeanSixSigma.com.

[4] Tapping D, Tapping S. The lean pocket handbook for Kaizen events. 3rd ed. MSC Media, Inc., Kindle Edition; 2014. 21%. See also. www.dropbox.com/s/v1250notuci3hur/WasteWalk.xlsx.

[5] Mauer R. One small step can change your life: the Kaizen way. Workman Publishing Company; 2014. p. 5.

[6] Tapping D, Tapping S. The lean pocket handbook for Kaizen events. 3rd ed. MSC Media, Inc., Kindle Edition; 2014. 63%. See also. www.dropbox.com/s/v1250notuci3hur/WasteWalk.xlsx.

[7] Mauer R. One small step can change your life: the Kaizen way. Workman Publishing Company; 2014. p. 113.

[8] United States Environmental Protection Agency. Lean thinking and methods, http://www.epa.gov/lean/environment/methods/threep.htm.

[9] Taylor R. The Kaizen facilitator: how to manage a Kaizen. Amazon Digital Services; 2012.

[10] Tapping D, Tapping S. The lean pocket handbook for Kaizen events. 3rd ed. MSC Media, Inc., Kindle Edition; 2014. 4%. See also. www.dropbox.com/s/v1250notuci3hur/WasteWalk.xlsx.

[11] Bresko M. Production preparation process (3P): lean concepts for project planning. GP Allied; 2011. Available at: http://gpallied.com/wp-content/blogs.dir/27/files/2011/06/015.-3P-Lean-Production-Preparation.pdf.

[12] Ibid, p. 7.

[13] Coletta AR. The lean 3P advantage: a practitioner's guide to the production preparation process. Productivity Press; 2012. p. 185.

[14] Martin K, Osterling M. Value stream mapping: how to visualize work and align leadership for organizational transformation hardcover. McGraw-Hill; 2013.

[15] Womack J. Take the value-stream walk. Industry Week; April 1, 2011. Available at: http://m.youtube.com/watch?v=M2lp1QDbXWE.
[16] Poppendieck M, Poppendieck T. Lean software development: an agile toolkit. Addison-Wesely; 2003.
[17] McMahon T. The eight wastes of new product development. A Lean Journey; December 3, 2012. Available at: http://www.aleanjourney.com/2012/12/the-eight-wastes-of-new-product.html.
[18] Emmons P. Get lean: avoid 8 wastes in development. Adage Technologies; March 29, 2013. Available at: http://www.adagetechnologies.com/en/Get-Lean-Avoid-8-Wastes-in-Development/.
[19] Ward A, Sobek D. Lean product and process development. 2nd ed. Lean Enterprise Institute, Inc., Kindle Edition; 2014. Locations 5605ff.
[20] Reinertsen DG. The science of WIP constraints. Lean Kanban Central Europe; 2012. Available at: https://www.youtube.com/watch?v=gr9I1LRhI_E.
[21] Reinertsen DG. The principles of product development flow: second generation lean product development. Celeritas Publishing; 2009. p. 18.
[22] Reinertsen DG. The principles of product development flow: second generation lean product development. Celeritas Publishing; 2009. p. 56.
[23] Reinertsen DG. The principles of product development flow: second generation lean product development. Celeritas Publishing; 2009. p. 79f.
[24] Gilb T. Decomposition of projects: how to design small incremental steps. Available from: http://www.gilb.com/dl41.
[25] Ellis G. Control system design guide. 4th ed. Butterworth-Heinemann; 2012. p. 279ff.
[26] Virginia tech uses NI products to develop HIL system for competition. NI News in Real Time; March 29, 2010. Available at: https://decibel.ni.com/content/groups/ni-news-in-real-time/blog/2010/03/29/virginia-tech-uses-ni-products-to-develop-hil-system-for-competition.
[27] Ellis G. Control system design guide. 4th ed. Butterworth-Heinemann; 2012. p. 431ff.
[28] Anderson D. Deep Kanban, worth the investment? London Lean Kanban Day; 2013. https://www.youtube.com/watch?v=JgMOhitbD7M.
[29] Agan T. The secret to lean innovation is making learning a priority. Harvard Business Review; January 23, 2014.
[30] Liker JK. The toyota way: 14 management principles from the world's greatest manufacturer. McGraw-Hill; 2003.
[31] Lindlöf L, Söderberg B. Pros and cons of lean visual planning: experiences from four product development organisations. Int J Tech Intell Plann 2011;7:269–79.
[32] Jennie A, Bellgran M. Spatial design and communication for improved production performance. In: Proceedings of the international 3rd Swedish production symposium. Göteborg, Sweden. 2009.
[33] Söderberg B, Alfredson L. Building on knowledge, an analysis of knowledge transfer in product development. Chalmers University of Technology; 2009.
[34] Javiadi S. Supporting production system development through Obeya concept. [Master's thesis]. Sweden: Mälardalen University; 2012.
[35] Olausson D, Berggren C. Managing uncertain, complex product development in high-tech firms: in search of controlled flexibility. In: R&D Management, 40. 2010. p. 383–99.
[36] Andersson, Bellgran, 2009.
[37] http://www.iobeya.com.

Chapter 8

Agile Project Management: Scrum, eXtreme Programming, and Scrumban

This chapter will introduce Agile project management to those who (1) coordinate with an Agile team, (2) may be leading an Agile team in the future, or (3) want to consider the adoption of Agile in their organization. Agile is targeted at software development projects, though it has been applied successfully to hardware projects. Three of the most common components of Agile are Scrum, Scrumban, and eXtreme Programming (XP).

Scrum is the set of processes that most strongly distinguish Agile from Phase–Gate methods. Where Phase–Gate methods define a project as a series of tasks finished over many months, Agile Scrum breaks a project into short *sprints*, which are iterations that sum to the full project. Each sprint is almost a mini-project, lasting just a few weeks and ending with a new product that could be released. The team executes one sprint, and then defines the next; they loosely follow a plan to the project end, refining the project direction at the start of every sprint. Every project management method accepts the possibility of the endpoint changing as the project proceeds, though it's normally an undesired event. Scrum expects—even welcomes—the redirection and builds its processes to ensure it happens on a regular basis.

Scrumban is a leaner version of Scrum and is based on the Kanban project management of Section 7.3.5. It is a good fit for simpler product development projects where, by contrast, Scrum is built to take on projects of any complexity. Finally, XP is a set of processes for software development. Where Scrum defines interactions of the team (when they meet, what roles people take on), XP defines how the team does their work (how they code, how they test, how they build the code set). Together, Scrum and XP create a complete product development process.

8.1 INTRODUCTION TO AGILE

Before the introduction of Agile methods, software projects were managed more or less like hardware projects. The use of Phase–Gate was common and software projects started with long specification phases before code was written. A common experience was to spend several months writing lengthy documents defining

Project Management in Product Development. http://dx.doi.org/10.1016/B978-0-12-802322-8.00008-5

software functions and interfaces and only then could coding begin. Coding proceeded more or less on schedule with all teams writing large blocks of code, testing ad hoc, and then storing code away until the "integration" phase of the project. When integration started, an enormous amount of effort had been expended. Only then did the team discover that the interfaces had left out critical elements, that the software performance was too slow, or that the user interface was unintelligible. Then began a long and painful rework cycle as the project schedule slipped and budgets were consumed well before the work was done.

This was the Phase–Gate model applied to software development: first define the entire product in detail, then design the entire architecture, then code, then integrate, then test. Today, it's generally agreed this model doesn't work well for software projects. By contrast, Agile methods take an iterative approach: break off a small portion of the code function: define, design, code, integrate, and test. Break off another small portion and repeat. This is probably the single identifying practice for Agile. And it works. Agile has received accolades from the Government Accounting Office (GAO) [1], Spotify [2], Microsoft [3], and many, many more organizations.

8.1.1 What Is Agile?

Agile has several characteristics, but if there is a single defining one, it's probably the sprint cycle from Scrum, as shown in Figure 8.1. In Scrum, the project is divided into fixed-time iterations called *sprints*; sprints are typically 2–4 weeks long. During a sprint, a new iteration of software is planned, designed, coded, and tested creating a *potential release*.[1] It's as if the entire Phase–Gate process occurs every 2 weeks, once for each small increment of the product.

The length of the sprint is set typically at 2–4 weeks. Most developers find that sprints shorter than 2 weeks have too much overhead. As we'll discuss in Section 8.2.3, each sprint has several meetings that create a fixed overhead that grows in significance as the sprint time decreases. Also, when sprints are too short, it's difficult to move features through the design–code–test cycle and get them into the release set in one sprint.

When sprints are too long, the benefit of short iterations is reduced. Another factor that limits the maximum sprint length is the constraint that the work committed to at the start of the sprint should not change during the course of the sprint. Changing the work inside a sprint causes several problems including the inefficiency of having to rework the sprint plan and the loss of ownership from the team, who were (presumably) committed to the original plan. Mid-sprint changes also reduce the effectiveness of planning and make opportunities for improvement more difficult to see.

To be sure, holding every sprint unchanged is an ideal constraint that will be violated from time to time as customer needs dictate. However, the longer the

---

1. In theory, code from every sprint could be released to the market, but in practice often only a portion of sprint releases actually release to the market/customer.

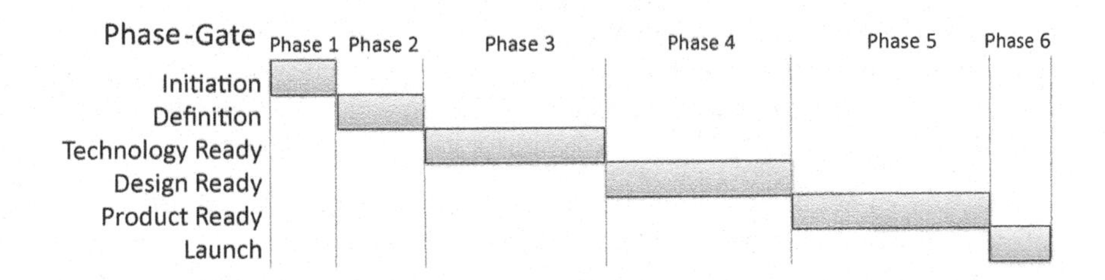

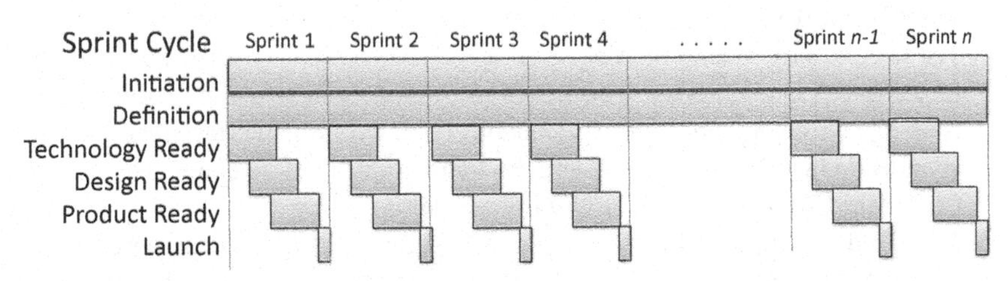

FIGURE 8.1 Traditional software development using Phase–Gate (above) versus the sprint cycle from Scrum.

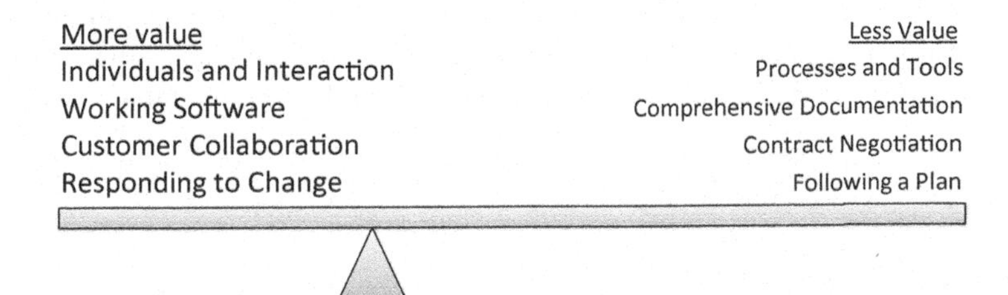

FIGURE 8.2 The defining principles of Agile from the Agile Manifesto.

sprint, the more likely it is that the constraint must be broken. The problem is that when observing the constraint, if an urgent need arises early in a sprint, the team can require almost two full sprints to resolve the issue—the remainder of the current sprint to even start (which could be nearly the entire sprint depending on the timing of the interruption) and the succeeding sprint to deliver the change. With a 4-week sprint, the minimum reaction time could be nearly 2 months. Even a 2-week sprint has up to almost a 4-week delay. With a 4-week or longer sprint, teams often find that nearly every sprint changes, significantly reducing the team's efficiency and their ability to improve. Taking these factors together, most teams have sprints between 2 and 4 weeks with the average probably being closer to 2.

8.1.2 Agile Manifesto

Agile is much more than the sprint cycle from Scrum. The first widely accepted definition is a short document resulting from a meeting of leading software developers in 2001: the Agile Manifesto [4]. As shown in Figure 8.2, the document presents four principles of relative value, each favoring agility over rigidity, although the more rigid items still retain significant value. Together, these principles form the most nimble of the major project management methods.

1. Individuals/Interaction over Processes/Tools
   The first principle of individuals over processes shifts the project management toward transformational leadership. In light of traditional project management's common overemphasis of transactional leadership, readers will likely find this logical.
2. Working Software over Comprehensive Documentation
   The second principle is the only one of the four unique to software. It results from how poorly internal documentation has worked to define software. Of course, internal documentation is valuable, but a working interface reveals more about code than a thick manual defining that interface before coding has started.
3. Customer Collaboration over Contract Negotiation
   The third principle extends the first from internal focus to customers. The gold standard here is having a customer participating in the development process, reviewing every code release and giving input for future work.
4. Responding to Change over Following a Plan
   The fourth principle is in harmony with concepts such as project innovation (Section 5.3.7) and small batch sizes (Section 7.3.4), but carries the concept farther than the other project management methods presented in this text.

These four principles are applied to many facets of project management to create the primary practices of Agile.

8.1.3 Why Use Agile?

One of the primary motivations for Agile is the need to avoid the problems created by long planning cycles; it works well for those projects that have a low cost of iteration. Software development and IT (which is essentially software development for internal use) make up the overwhelming majority of projects reported to use Agile. Software can have a very low cost of iteration—new revisions can be released with small effort assuming there are no large fixed costs for release such as certification for medical and aerospace applications. Every couple of weeks, a small amount of code can be added, tested, documented, and released on a website. Such thinking is impractical with a die for injection molding or the foundation for a building. But for those projects that have a low cost of iteration, the use of Agile can create a learning-based project management system, avoiding the major pitfall of Phase–Gate: making decisions without sufficient knowledge.

Delaying Decisions to Allow for Learning

The fundamental weakness of the Phase–Gate approach being applied to software is that it causes almost all important decisions on architecture, interfaces, and hardware to be made early, when the team knows the least about the system.

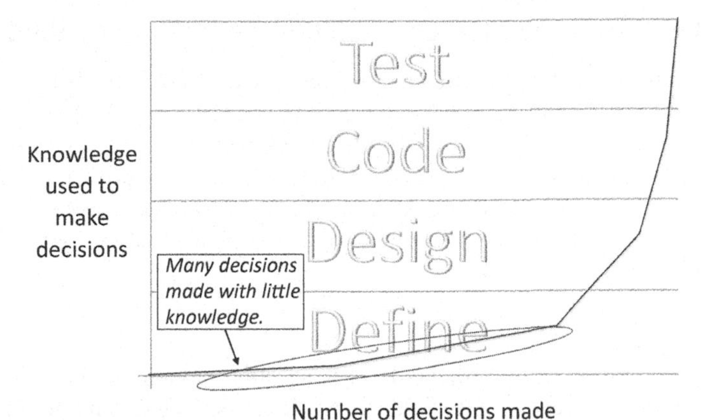

FIGURE 8.3 The traditional approach for software made most decisions when the team knew the least about the system.

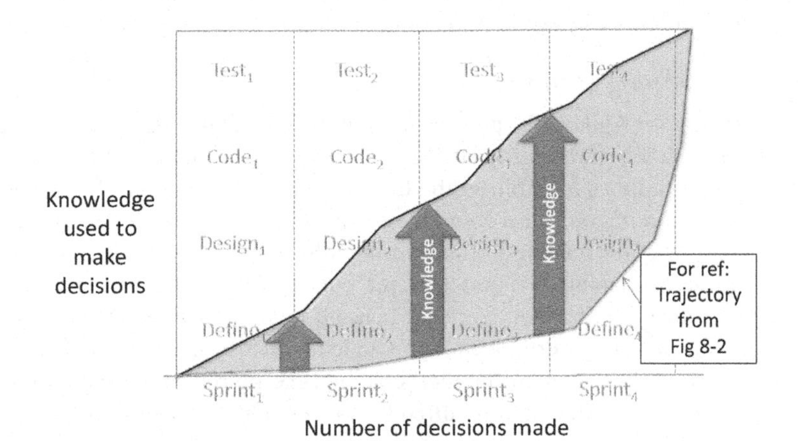

FIGURE 8.4 The iterative approach of Agile maximizes learning to improve decisions.

As shown in Figure 8.3, in Phase–Gate projects the system is completely defined and designed before any part of it is coded and tested as a releasable code set.[2] This means the most important decisions of the project are made before gaining much experience in the application.

Agile methods with their iterative approach solve this problem by delaying design decisions as long as possible. As can be seen in Figure 8.4, by coding and testing the first increment, the team is better informed to make decisions for the second increment. This trend continues throughout the project allowing many decisions to be delayed until the maximum amount of learning has been gained.

2. Releasable code sets should not be confused with proof-of-concept software, which is generated early in a project to validate technical and/or commercial feasibility. Those steps are appropriate for both Agile and traditional project management.

8.1.4 How Are Agile Methods Different from Other Methods?

In this section, the key differences between Agile and non-Agile methods will be discussed. Bear in mind that those differences can be challenging to define because both Agile and non-Agile have broad meanings. Agile includes Scrum, XP, and Scrumban. Non-Agile includes critical path, critical chain, lean product development and others, each of which can be used with and without Phase–Gate. So, any comments comparing Agile against other methods must be general.

More Iterations

As discussed above, Agile projects have many more iterations than traditional projects. With a 2-week sprint cycle, an Agile project could generate up to 26 revisions per year—the typical traditional project release cycle is perhaps one or two per year. Of course, for a project to be appropriate for Agile, the cost of iteration must be low. There must be minimal tooling, capital equipment, and certification. This is why Agile is generally used for software projects.

Minimal Planning at Initiation

Most planning for Agile is for an individual iteration and it is delayed until just before that iteration starts. Agile evolves products—get something working and deployed as rapidly as possible so the team can learn–learn about the technology the product relies on, learn about the needs of the end users, and learn about the capabilities of the team. All this learning allows for better decisions. Some level of project planning is required at the outset of the project to ensure the project will meet its financial targets when the many iterations are complete. The areas that need long-term planning include foundational technical decisions and long-term product feature targets. Even so, the upfront planning required for Agile is much less than that required for most competing methods.

Permanent Teams

Agile forms cross-functional teams like traditional projects, but they are generally permanent. In most Agile organizations, an Agile team is formed and remains together moving from one project to the next. Agile teams are cross-functional, but that cross-functionality is generally related to disciplines within software development: for example, database analysts, system programmers, user interface programmers, and testers. It's common to have more overlap within Agile teams than traditional project teams. Because of the overlap, Agile teams are more flexible than traditional project teams, so the same team can take on many projects.

By contrast, a traditional project team will generally have more specialization. A traditional team might be made up of a chemical engineer, an electrical engineer, two mechanical engineers, a manufacturing engineer, a sourcing engineer, and a quality engineer. The next project might need a team reformed with two more electrical engineers, one fewer mechanical engineer, and no chemical engineers. So traditional project teams often form around the needs of a single project and disperse when that project is complete.

The following sections will discuss three of the most popular types of Agile: Scrum, XP, and Scrumban. Over time, the differences between these types is blurring since most practitioners of Scrum and Scrumban often have a large dose of XP. So, these methods with be covered separately, understanding they are generally mixed by the teams that use them.

8.2 AGILE SCRUM

Agile Scrum defines a project team and how they interact with each other. Compared to critical path management (CPM) or critical chain project management (CCPM), there is less attention given to prescribing standard work such as the processes, practices, and tools the team uses to complete daily tasks.

8.2.1 The Product Backlog and Grooming

The product backlog is the container for all the work the team will do on a product. Rather than specify a full product at the outset of the project as in traditional management, features are entered into the product backlog in a form called a *story*. When a story is entered into the backlog only a coarse description is required. The backlog can be thought of as an evolving specification where only the stories about to be worked on are defined in detail. Most stories in the backlog are unready for the team to work on either because the description is too coarse, the story will take longer than one sprint to complete, or both.

The transition from a coarsely defined feature to a sharply defined set of tasks is accomplished through the process of *grooming*. Grooming refines the initial description step by step until its stories are crisply defined and have a granularity small enough that they can be completed in one sprint.

Years ago, software was typically defined from an implementation viewpoint, listing the interface and what methods need to be coded. In Scrum, the focus turns to understanding the customer's viewpoint above all else. To this end, each feature is defined in a form that explains who will use it, what they expect it to do, and why it's valuable to that user with the following template [5]:

> *As a* ***<define user>*** *I need to* ***<statement of function>*** *so that I can* ***<statement of value>***.

For example:

> *As a* ***mobile ap user*** *I need to* access the ***secure data base remotely*** *so that I can* ***pull reports when I'm traveling***.

Of course, this story is too coarse to implement. We don't know which reports, which mobile devices, or what performance is needed. Over succeeding sprints, this story can be broken down step by step until one or more of the stories is fully groomed, which is to say it can be implemented by the team in a single sprint.

Grooming is a major task for the team, taking about 10% of total resources [6]. It can be led by one person or be split up among many team members. Poor grooming is often quoted as the most common reason Scrum can yield poor performance. If a story is improperly prepared, it causes turnbacks during the sprint when the team recognizes there's not enough detail to complete or when the story isn't divided into small enough chunks to complete in one sprint. Worse, it may create turnbacks later in the project when features are found to be improperly implemented. Either way, poor grooming compromises the efficiency of the team.

8.2.2 Scrum Roles

Scrum defines two primary leadership roles: Product Owner (PO) and Scrum Master (SM). The responsibilities of these two leaders total those of the project manager and marketing lead in traditional project management. However, as shown in Figure 8.5, these roles divide up leadership responsibility differently. Leadership functions can be separated into three main groups according to whom they face: external (to customers and end users), organizational (to stakeholders outside the project team), and the project team.

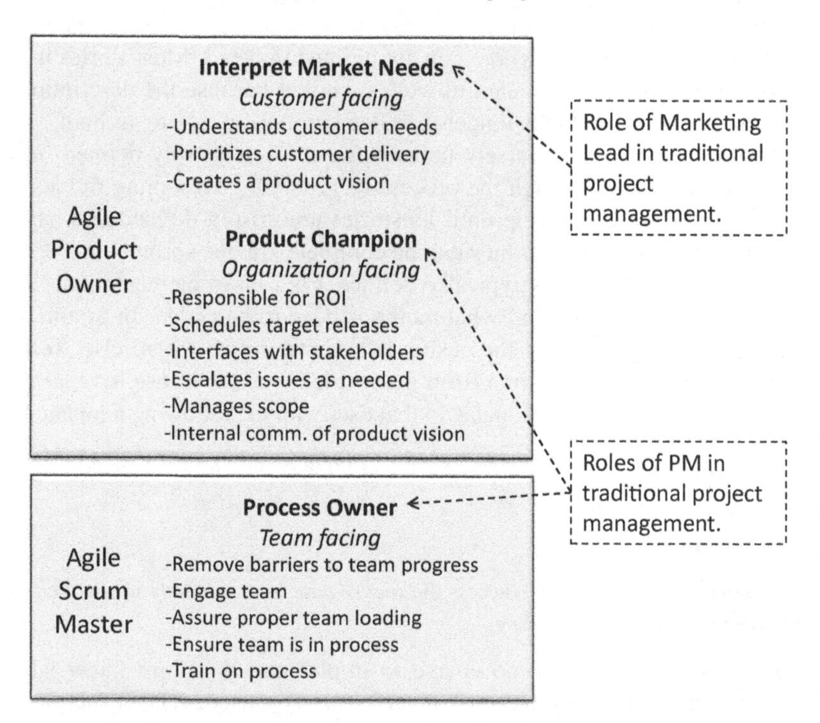

FIGURE 8.5 Leadership roles in Agile compared to traditional projects.

- Interpreting market needs (external facing) includes:
  - Understanding what customers need and which of those needs are not currently met.
  - Prioritizing the stories and performance the customer wants, and the order in which they want them.
  - Creating a long-term vision for the product capability that will guide the technical team and the organization in the development and launch of the product.

  In Agile, *interpreting market needs* falls to the PO while in traditional projects it falls to the marketing leader.

- Product champion (organization facing) includes:
  - Estimating and fulfilling return on investment.
  - Scheduling product releases to best meet customer needs.
  - Interfacing with the management team formally (e.g., at project reviews) and informally, as well as escalating issues that need management attention.
  - Presenting the vision of the product to the organization over the life of the project and managing the scope of the project as perceived customer needs change.

  In Agile, *product champion* also falls to the PO, but in traditional projects it typically falls to the PM.

- Process owner (team facing) includes:
  - Removing barriers from the team.
  - Ensuring the team is engaged and properly loaded.
  - Ensuring the team follows the right processes and follows them well.
  - Providing training and mentoring as needed.

  In Agile, *process ownership* also falls to the SM, but in traditional projects the PM is normally expected to own the processes.

Product Owner

The PO owns the financial goals for the project and manages communication outside the team, both external to the organization and internal. The PO owns the vision: what the product will do and why customers need it. But the PO does not direct team members as to what they will do. Nor does he dictate which stories will be completed during a given sprint. The PO does set the priorities for stories and the team chooses the order of implementation. The team gives great weight to the PO's prioritization but has latitude to modify the order of implementation if technical issues dictate. For example, consider a case where the PO has prioritized 5 tasks from highest (1) to lowest (5). If the Priority 2 story needs code that will be written for the Priority 5 story, the team may elect to do the Priority 5 story first. If a story is not sufficiently groomed, the team may delay implementation until grooming is complete. As a third example, if stories with Priority 2, 3, and 4 all require database analytical skills, but the team has resources only for 2 and 3, the Priority 4 story may be delayed while the Priority 5 story is accepted. So, all things being equal, the team will follow the PO's initial prioritization, but the PO and team must negotiate to arrive at the decision of what stories are planned to deliver the maximum value for every sprint.

Scrum Master

The SM is responsible for process. When a team member needs help, the SM should be there to remove barriers and review current process in order to drive improvement. The SM protects the team, reducing incoming workload when the team is stressed. The SM also pushes the team when she sees they are able to take on more.

The SM owns the process, but has no direct authority over the team for specific tasks. It's entirely appropriate for the SM to point out that the team is not following the unit test procedure or that defects are increasing so that peer review practices must be improved. However, an SM is not empowered to direct a specific team member to take on a task or rework code. Agile teams are self-directing—the SM points out the shortcomings, facilitates discussions on how to improve, but leaves it to the team members to decide what action will be taken, when it will be taken, and who will lead it. The team makes commitments that the team should meet. They succeed or fail as a team; as with CCPM, individual success is deemphasized.

SM is a part-time role, taking perhaps 25–50% of one person's time depending on how mature the team is. More mature teams that are working in stabilized projects require less time from the SM than do new teams working on new projects (see Figure 8.6). The SM effort reduces as the team becomes more self-directing, as more and more of the persistent barriers are removed, and as the team learns the processes. These factors all increase team velocity as they simultaneously reduce the time required for the SM role.

In small organizations—those with perhaps a single Agile team— the SM is likely to use her remaining time as a working team member. In larger organizations, there is an option to have one person be SM for two or three project teams. There are at least two ways to look at this issue. On the one hand, the benefit in

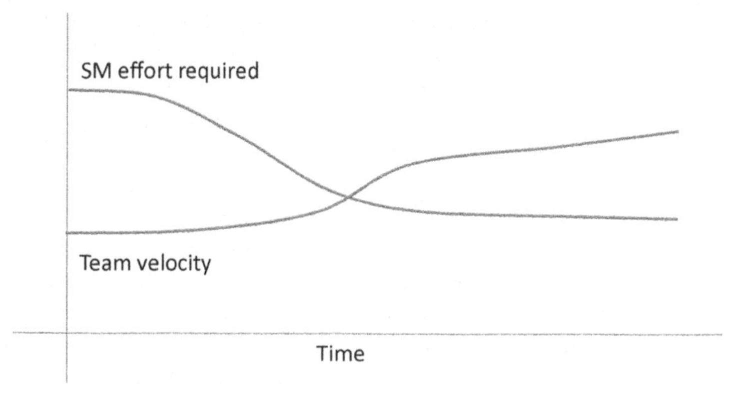

FIGURE 8.6 SM effort reduces as team maturity increases.

having her on the project full time (part time as SM and part time as team member) is that she will become more familiar with the project and the team, better informing her judgment. On the other hand, having one person be both SM and team member can cause conflict as the SM may become (or just appear to be) biased on issues she is contributing work toward. For example, if the SM were also the leader in a software test, the team might sense she managed process with bias towards testing.

Team Members

Team members form the remainder of the Scrum team. Team members commonly represent a range of software disciplines: user interface developers, client-side and server-side developers, database analysts, test engineers, and many more. Scrum teams typically have about eight members, giving rise to the famous 2-*pizza rule*, defining the team size by the amount of food required for lunch.

Team size varies from 2 to 20, but Cohen quotes research where about eight people per team is optimal. In one study, smaller teams provided team members more satisfaction because their contribution was more apparent [7]. Another study of more than 100 teams showed that small teams (four to nine members) were more effective than large teams (14–18) in commitment, goal awareness, and communication [8]. A third study quoted by Cohen shows efficiency is essentially flat up to seven people; after that it starts to fall, losing about 20% by the time team size is in the high teens [9]. Cohen continues presenting a convincing case for teams of about eight members [10].

Teams in Agile are self-organizing, meaning the team decides how much they will commit to in a sprint, who will lead each task, how people falling behind will be helped, and so on. The team decisions are generally facilitated by the SM whose goal it is to lead the team to consensus on such matters. In a well-led team, those members with the most experience and skill will carry the most weight; but they must earn the right from their colleagues to lead because authority is not handed down from management. Agile leadership roles rely on total leadership as discussed in Chapter 4, with a strong focus on *Connection*, the transformational leadership of people (see Figure 4.4).

Sometimes self-organizing is seen as a euphemism for anarchy. This is certainly not the case. Self-organized teams should be following leaders (hence the PO and SM roles) and they are responsible as a team to meet performance requirements. Ultimately, if the team doesn't deliver to management expectations over time, they would expect changes intended to improve that condition. But normally, well-led scrum teams take ownership, make and keep commitments to their teammates, and are passionate about delivering.

The differences between Agile and non-Agile here may not be as large as they seem. Well-led traditional product development teams also have a large component of self-direction. Traditional PMs often have no direct reports and leading by fiat is broadly recognized as ineffective. It is true that Agile teams generally have more latitude for task assignment, but this may be because there is more overlap in team technical capability—if a traditional project has one electrical and one mechanical engineer, the ownership for most tasks will be dictated by the domain of the task. By comparison, software developers often work across boundaries since the specialization among software disciplines is often easier to cross.

Chief Architect

Some Scrum teams rely on a chief architect to lead the design of foundational issues. These are the sorts of issues that cannot be properly resolved in the space of one sprint. Here a team member with more experience may have to create a design that will support functions that will be added in the months to come. It might be that the PO has prioritized stories so that those that place the largest demands on the database structure won't be developed for months into the project. In such a case, the role of an architect is to ensure the structure will support the demands likely to come later. Scrum delays decisions as long as it creates advantage, so it is important the architect does not take on the entire design; however, enough must be done to ensure the foundations of the code set will meet the needs of the market when the full product is released. Consistent with a self-organizing team, the architect receives most of her empowerment from the team; she cannot dictate designs to the team but must instead influence the team with a balance of evidence and a track record that creates confidence. But the architect should be a member of the team, saddled with the consequences of her decisions; Agile discourages the use of noncoding architects.

Testers

It is natural for Agile teams to specialize into developers and testers because the skill sets for the two functions are different. If so, testers are like any other team member—they don't test to a specification but to customer need as they understand it. Of course, a test specification is a good start on understanding customer need, but nothing replaces the tester's ability to know what the end user wants and to act on that knowledge. In Scrum, testers are full team members and should be allowed to and even be held accountable to influence the design. After all, designs that are hard to test are usually poor designs and no one can pick that flaw out faster than an experienced tester. And Agile discourages "black box" testing where the tester doesn't need to know how something works. "White box" testing is preferred. Knowing the software structure allows the tester to spend the most energy on the weakest areas maximizing his value to the team.

Many testers are uncomfortable in such an egalitarian team. They may have years of experience doing "what they are told": working line by line through a specification feeling their job is complete when the spec is tested rather than when the customer is satisfied. Scrum encourages teamwork—no one wins unless the team wins and everyone is expected to apply themselves with passion to satisfying the customer.

8.2.3 Team Interaction

Scrum specifies both formal and informal team interactions. The informal interactions should be constant and rich, with the SM mentoring and reviewing process and performance on a regular basis. The PO should work closely with team members to ensure customer needs are understood. Finally, team members should work together closely to encourage each other as well as helping and getting help from one another.

The formal interactions of Scrum are specified as four team meetings, each repeating every sprint cycle as shown in Figure 8.7.

Sprint Planning

Sprint planning occurs once per sprint cycle, typically 1–3 days before the sprint starts. Here stories are selected to move from the product backlog into the next sprint. The team pulls stories from the product backlog according to the priority set by the PO. The team can modify that priority if technical need dictates. The team decides how many stories they can complete in the sprint. The team also determines if stories are groomed well enough to move into the sprint; those that are not must be refined to prepare them for implementation in future sprints.

Daily Scrum Meeting

Daily Scrum is a short meeting where each member conveys progress and plans to the team. The SM facilitates the daily Scrum and has the initial responsibility to remove barriers. A common prescription is to ask each member to spend 2–3 min answering these questions:

1. What did I do yesterday?
2. What do I plan to do today?
3. What are the barriers that might block my progress?

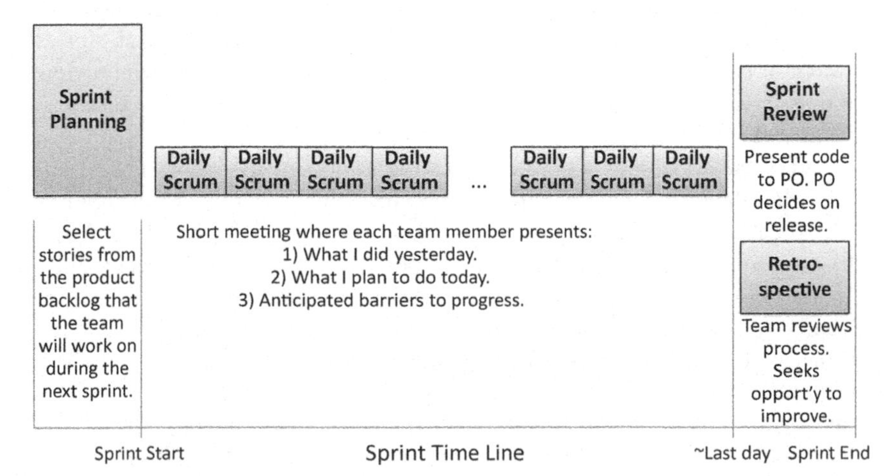

FIGURE 8.7 Four common Scrum meetings, which repeat each sprint.

Daily work is usually managed on a Kanban board, which is similar to the lean product development (LPD) process presented in Section 7.3.4 except there is a time component: the board is loaded at the start of the sprint and there is an expectation it will be cleared by the end. At the far left, the Kanban board stores the *sprint backlog*, which will be discussed below. The columns of the Kanban board are steps required to complete each story such as: design, code, unit test, document, and so on. The Kanban board in Agile provides the same benefits as it did in LPD: visualization of work in progress (WIP), management of flow, and minimization of multitasking.

Sprint Review

The sprint review is held once near the end of each sprint. Here the team presents the PO and any other interested stakeholders with the working code set that is the output of the sprint. The PO evaluates the code according to whether it meets customer needs. The PO decides whether the code will release as presented and, if not, which parts are ready for release. Team *velocity*, which will be discussed below, is measured at the sprint review.

Sprint Retrospective

At the end of each sprint, the team meets to discuss what they did well and what they can learn for the future. This is the opportunity to discuss quality of planning estimates, team efficiency, and process that needs to be improved, among other topics. The sprint retrospective is the primary vehicle for continuous improvement.

8.2.4 Story Flow: From Product Backlog to Sprint Backlog to Release

At the start of a project, the product backlog holds all known stories for the full product. At each sprint planning meeting, the team selects a set of stories to flow from the product backlog into the sprint backlog. During the sprint, stories flow out of the sprint backlog into the working code set. After the sprint review meeting, those stories that are accepted by the PO flow into the release code set. This flow is shown in Figure 8.8.

Stories can come into the product backlog in many ways. Most come from the PO, but they can be entered from any team member and other stakeholders in the organization. There are few limits on what comes in, but the PO prioritizes stories; those prioritized near the bottom of the list will receive little attention. Those near the top will be groomed so the team can implement them.

Let's review the flow shown in Figure 8.8 in detail. On the left, we can see that of the original 15 stories in the product backlog, some have flowed toward release (1–7) and others remain in the backlog for future sprints (8–15). Four stories (1–4) were pulled into earlier sprints and have already flowed into release. Stories 5–7 were pulled into the sprint backlog during this sprint. We

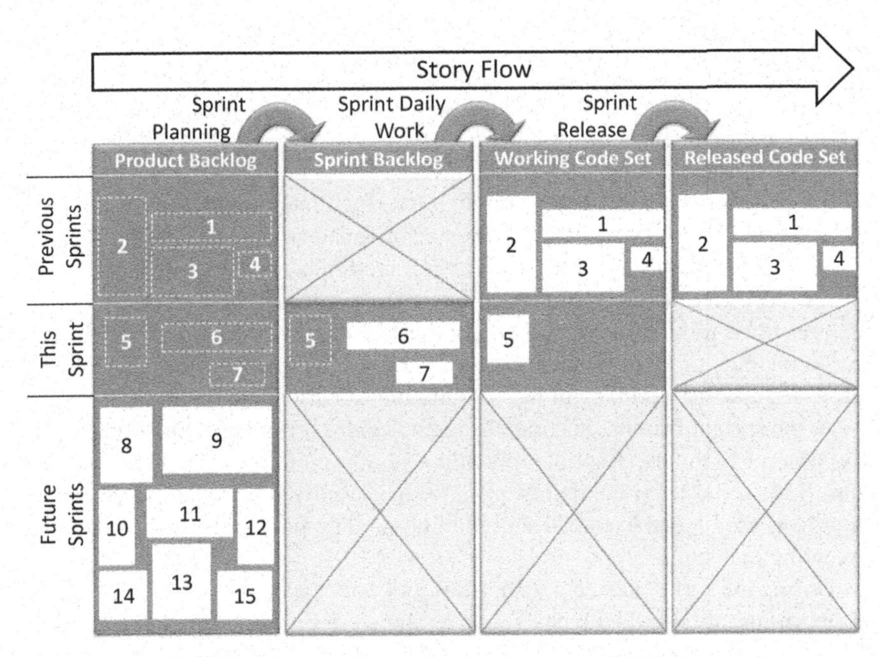

FIGURE 8.8 Story flow during a project managed with scrum.

can see Figure 8.8 must have been drawn mid-sprint because the third column shows that Story 5 has already been implemented in the working code set, but the second column shows Stories 6 and 7 are still in progress. In the final column we see the released code set contains Stories 1–4, so the PO must have released them in prior sprint reviews.

This example is simplified to focus on flow; real projects are more complex. First, a product backlog for a full product will contain many more than 15 stories. Further, stories come in coarsely defined and then are groomed where they are usually broken into multiple smaller stories. So in addition to the flow from left to right, we would expect to see grooming break stories into smaller chunks before they exit the product backlog.

8.2.5 Measuring Velocity

One of the most important measurements of a Scrum team is *velocity*, defined as the number of story points completed per sprint. Each time a story comes into the product backlog, there is an estimate of effort to complete it in units called *story points*. Story points are typically defined as one point equals the smallest story the team will take on.[3] Estimates for stories are then in proportion to that smallest

3. As an alternative some teams estimate stories in ideal days or ideal hours—the time it would take if someone were uninterrupted.

story: if a story is estimated to take double the time of the shortest story, it would be assigned two story points. To ensure appropriate granularity, the estimated points are often limited to the Fibonacci series: 1, 2, 3, 5, 8, 13, 22, and so on. Trying to make finer estimates is seen as wasteful—you can't reliably differentiate between 15- and 17-point stories, so choose either 13 or 22, and move on.

When a story is entered in the product backlog, the description is coarse and the estimated story points are for reference, intended to inform team of the scale of the feature. However, as the story is groomed, the estimates should become increasingly accurate. By the time stories are pulled into the sprint backlog, they are presumed to be as accurate as is possible because the team is committing to complete them based in large part on the estimated story points. Estimation errors degrade the accuracy of the velocity measurement.

At the sprint planning meeting, the team decides how many story points they will accept. In the early sprints this might be more of a guess than an estimation. But, as the team builds history and their ability to estimate story points improves, they are able to better predict how many points they are capable of executing in a sprint.

During the sprint, the team tracks the story points completed on a daily basis in a burn-up (or burn-down) chart such as shown in Figure 8.9. In this example, the team has finished 13 story points by the end of the 8th day (Feb 3) against a plan of 17 points. In this example, they are behind, having nine story points to finish in just 2 days. At end of the sprint the total story points completed is recorded and graphed against prior sprints to provide a history of team performance; an example is shown in Figure 8.10. You should expect the velocity to be volatile, especially with new Scrum teams.

The SM will normally lead the team to improve velocity and this is commonly a point of discussion at the retrospective. The discussion will often fall into two categories: issues to improve the average velocity and issues that were

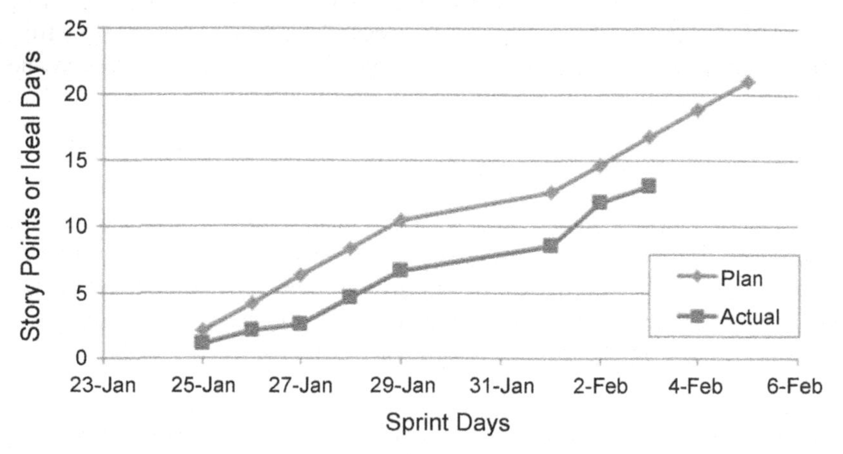

FIGURE 8.9 Sprint burn-up chart.

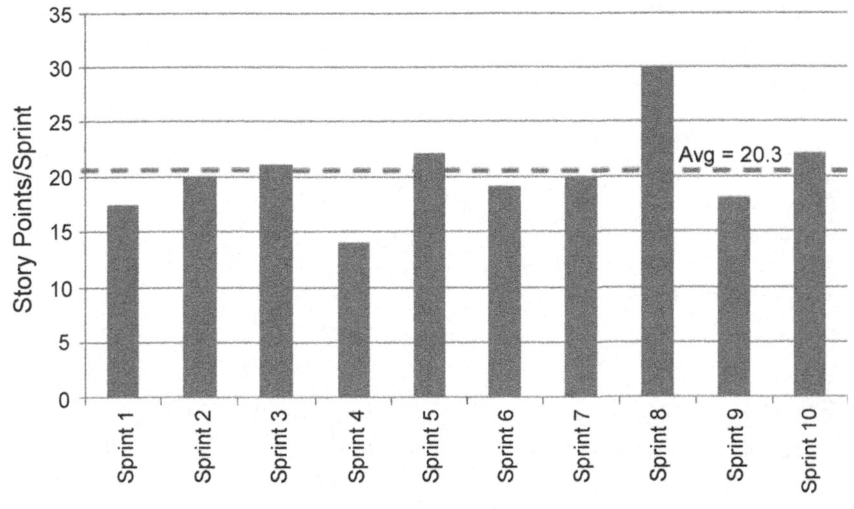

FIGURE 8.10 Sprint velocity over 10 sprints.

peculiar to the last sprint, differentiating between *normal variation* and *special variation*, to borrow terms from quality assurance.

Normal variation is the variation that occurs because of process capability. In this case, there is a limit to accuracy of estimations, there are ordinary uncontrollable interruptions to the team, and there is variation in coding speed across different topics (Alexia may be competent at both client-side and server-side coding, but faster at client side). Normal variation can be improved, but only by changing the process. While it's always a goal to reduce variation, some level of normal variation is acceptable. In these cases, the team's focus will be less on the variation and more on increasing the average. Is there an opportunity to adopt standard work around unit tests that could speed up validation? Would more diligent review increase coding performance? In the case of normal variation, the questions are general, not particular to the previous sprint.

Special variation is the condition where something changed that was essentially unpredictable. On the factory floor, an example of special variation is that damage to a tool went undetected for several days. On the sprint team, special variation might be that Greg worked on database queries but learned he is not competent in this area (Greg should avoid this area until he receives the proper training and demonstrates capability). Another example is both Jenny and Sanjay were out of the office for most of the sprint due to illness. Looking at the example in Figure 8.10, there does seem to be special variation between Sprints 8 and 9, where velocity dropped 12 points; perhaps that's because in Sprint 8, where the velocity was very high, those points were not completed properly and so Sprint 9 required extra rework. Or it could be due to many other reasons. Here the SM would probably facilitate a team discussion to establish root cause and determine what action should be taken.

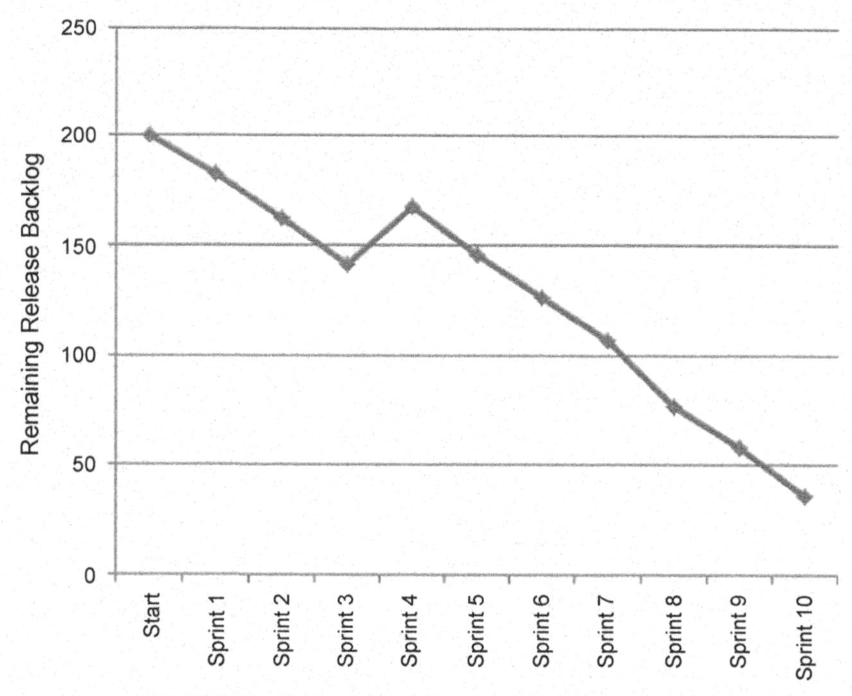

FIGURE 8.11 Release backlog burn-down chart through sprint 10.

Measuring Backlog for a Release

The product backlog is the complete list of remaining stories for the product. The backlog goes on as long as work is being done on the product. However, an upcoming major release can be defined and the stories to support it can be segregated into a release backlog. The release backlog can then be tracked as shown in Figure 8.11, which is a burn-down chart for the release.

8.2.6 Progress Reporting

So far, the reporting that has been presented is specific to Scrum. Charts for sprint burn-down, product backlog, and sprint velocity are unique to Agile. This type of reporting is highly useful for the project team. However, when reporting outside the team, the usefulness of this information will be limited. For managers or coworkers in other departments who have limited exposure to Scrum, these charts will have little meaning. This is especially true for organizations that are running Agile and Phase–Gate projects in parallel; this can happen even on a single project, for example, where the software for a new frequency generator is managed in Scrum but the hardware remains in Phase–Gate.

This issue can present a barrier to adoption of Scrum. The GAO pointed out that there are numerous problems identified with Scrum progress reporting such as the lack of milestones or an estimate of total budget at the project outset. While the Agile iterative approach is sensible, obtaining a large investment to start a project to create a new Digital Video Recorder (DVR) without an estimate of total investment will be unacceptable for many organizations. So many times those adopting Agile must make estimates on release dates in order to get organizational support for larger projects.

It should be possible to create portfolio-level reporting—information needed primarily to justify ongoing investment in projects and to coordinate activities with other projects and other parts of the organization (sales, ops, etc.). There are many examples of requiring disparate activities to conform to a single format for high-level reporting. Consider how differently progress is measured in high school physics versus a French class; yet they both conform to an A-F grade for a report card system. Similarly, among different project management methods, there may be a great deal of variation—for example, burn-up charts versus Gantt charts. Still, at the portfolio level, the company will benefit if reporting conforms to a single standard. This issue is shown in Figure 8.12. Unfortunately, much of Scrum writing downplays the importance of being able to report to management in a standard system. In companies where the management is not familiar with Scrum, it will be difficult for managers to understand the details

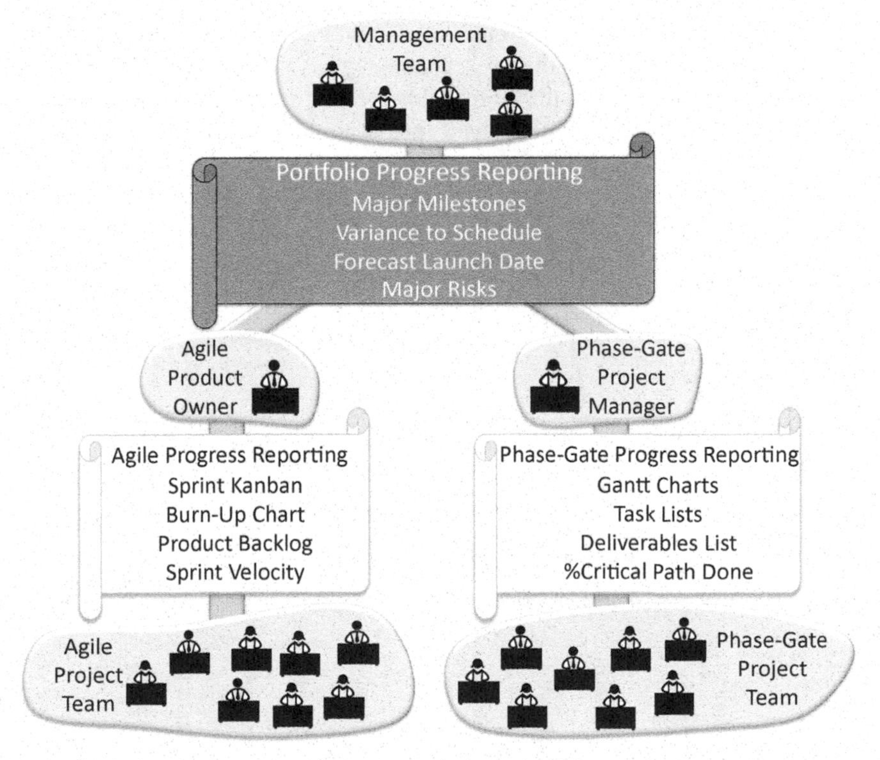

FIGURE 8.12 Progress reporting within the team versus to management.

like story points and velocity. And in most cases, that is probably unnecessary. The management team is primarily interested in data that drives portfolio-level decisions: Which projects need more resources and which could manage with less? Which projects should be placed on hold and which should be started? Which projects are experiencing serious delay or risk? It should be possible to communicate at this level to those with a minimal understanding of Scrum.

One area to consider for portfolio reporting is creating a forecast release date. If the release backlog is tracked in Figure 8.11 and average velocity is tracked as in Figure 8.10, then the release date can be forecasted and tracked with the following steps:

1. Segregate the stories in the backlog necessary for the release.
2. Sum the remaining story points for release stories.
3. Divide the story points by the average velocity to determine the number of remaining sprints. (Use average velocity from Figure 8.10.)
4. Forecast the release date as No of sprints × sprint length + current date. This is shown in Figure 8.13, which is the release backlog from Figure 8.11 with the forecast date added (note the forecast date uses the vertical scale on the right).

This can be done at the start of each sprint, but it can also be done daily to include the progress of the existing sprint.

Now that we have a forecast release date that can be tracked daily, we can use a run chart to track variation to schedule as was done in Figure 5.16. The only missing information is the target release date, which can be estimated at the outset of the project; alternatively, if the Scrum team is not able to estimate

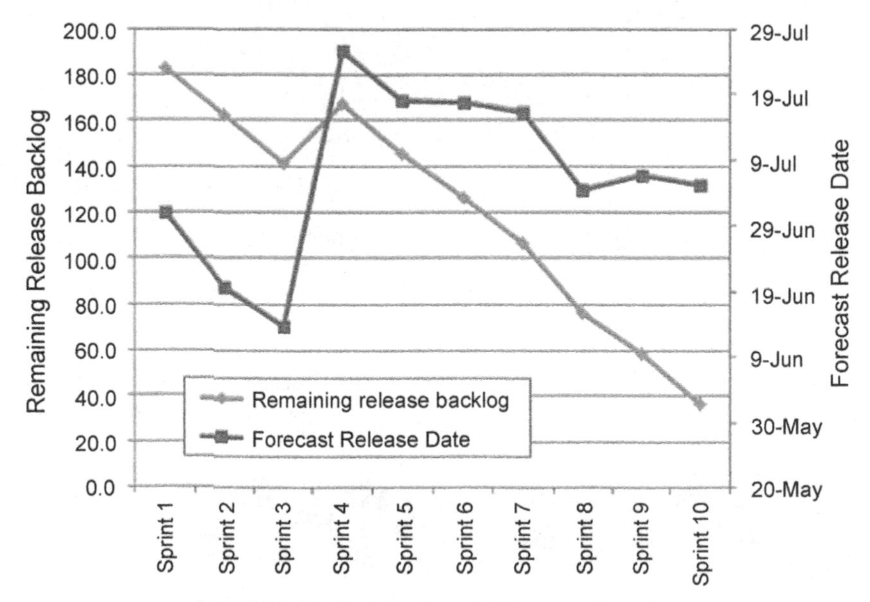

FIGURE 8.13 Backlog chart with forecast release date.

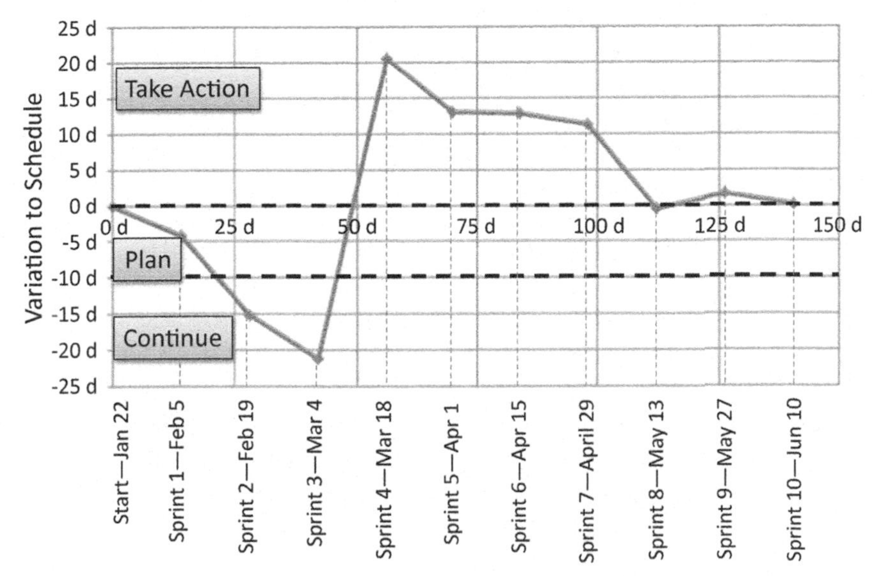

FIGURE 8.14 Run chart of variation to schedule plan.

a completion date early in the project, a target or "need" date can be used. The variation to schedule is then the target date minus the forecast date. This is plotted in Figure 8.14.

Another option is to create a fever chart similar to Figure 5.17. The vertical axis is the same as Figure 8.14 except shown as a percentage of the release plan. The horizontal axis is then the percentage of story points finished. This is shown in Figure 8.15. This has the benefit of showing turnbacks clearly. For example, notice that between March 4th and 18th (Sprint 4) the team regressed. From the backlog chart of Figure 8.13, we can see that the number of story points increased about 25 points. That may be because the release workload was increased (the PO saw that critical features were missing) or it may be from a turnback (the team recognized that story points they thought were finished required unexpected rework). Again, the SM would facilitate a discussion to arrive at the root cause and countermeasures. The point here is that even a person unfamiliar with Scrum can easily see something went wrong between March 4th and 18th. Moreover, it's clear the team recovered the schedule over the succeeding sprints.

8.2.7 Coordinating Agile and Phase–Gate Projects

One of the challenges of product development is that Agile and Phase–Gate projects are often run together. For example, if a team is developing automated lab equipment, it's common there will be Agile subprojects (such as for the process software) and Phase–Gate projects (such as for the hardware platform). Because the two projects run on such different systems, it can be difficult to

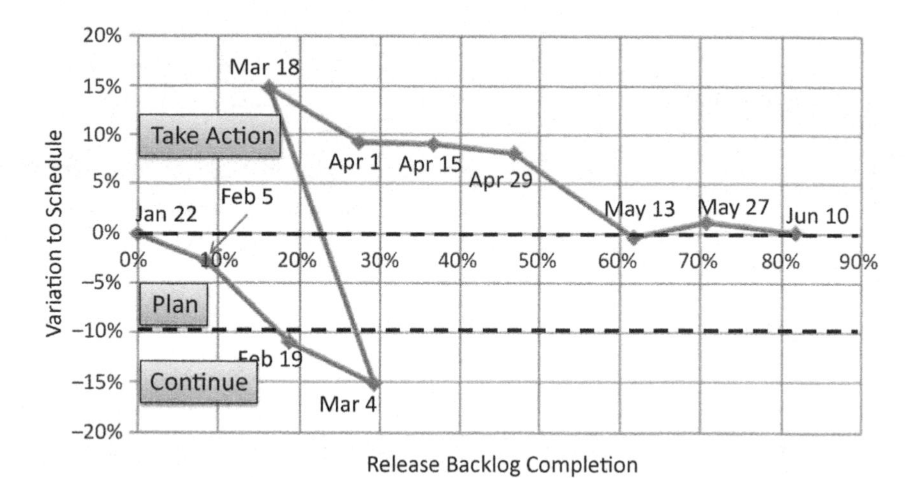

FIGURE 8.15 Fever chart of variation to schedule plan.

coordinate these subprojects. One issue is Agile projects are not fully planned at the outset and so it's common to shift requirements out to future sprints as the team gains knowledge. On the other hand, the hardware project is typically on a firm schedule—there are deadlines with manufacturing and test labs that are typically set months in advance. It's important that the two projects stay linked. Two ways to facilitate this coordination are with the snapshot fever chart and shared milestones.

With the fever chart, we can plot progress of multiple projects on the same graph similar to what was described in Figure 5.18. This concept can be used to coordinate Phase–Gate and Agile projects. For example, let's suppose the development for an automated piece of laboratory equipment requires six subprojects:

1. Process software (Agile)
2. User interface (Agile)
3. Hardware platform (Phase–Gate)
4. Automatic test equipment or ATE (Phase–Gate)
5. Regression test hardware (Phase–Gate)
6. Database development (Agile)

Since the format of the fever chart is identical for Agile and Phase–Gate projects, a single graph can show progress for all subprojects. This is true even when the subprojects start and end at different times. The result is an intuitive display of progress of disparate subprojects. An example is show in Figure 8.16.

A second technique used to coordinate Agile and Phase–Gate projects is creating shared milestones: integration points used by multiple subprojects. For example, when the first hardware sets are available the hardware team will need

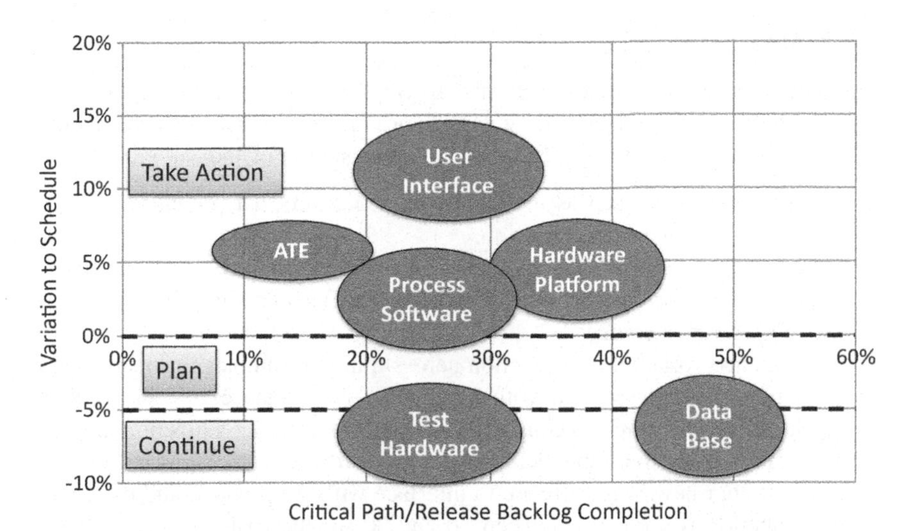

FIGURE 8.16 Progress of a project composed of six subprojects in one fever chart.

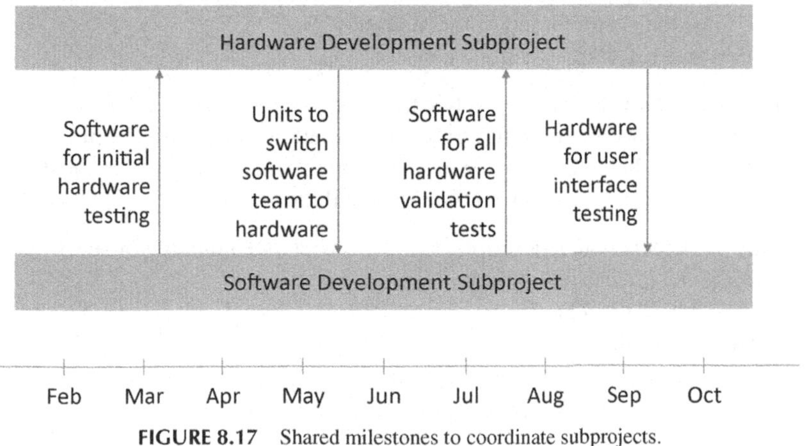

FIGURE 8.17 Shared milestones to coordinate subprojects.

enough software delivered to support initial testing such as ensuring the processor can read memory and interface to other circuitry. Later, when hardware is validated, the software team can switch from simulation to target hardware. These are two examples of shared milestones as shown in Figure 8.17. The two teams can stay coordinated for the length of the project as each delivers necessary hardware/software functionality to the other.

These measures of progress are given as examples; currently, there are no universally accepted reporting metrics that accommodate Agile and Phase–Gate projects. There are other options: defining major milestones, tracking potential revenue of each release against a target, and tracking completed customer requirements.

8.2.8 Measuring Quality

One other measurement that is part of most Scrum teams is tracking defects. A *defect* or *bug* is any condition that causes the product to produce incorrect results or behave in unexpected ways. Defect is an expansive term and can include:

- Algorithmic errors such as a square-root procedure that produces a result unequal to the square-root function.
- Timing defects such as when a process takes longer to execute than expected. This can occur for several reasons including inefficient coding and inefficient use of hardware resources.
- Performance variation of a function across multiple samples. In many cases customers may be satisfied with a range of performance so long as it is repeatable. For example, a consistent delay in carrying out a function may be acceptable where that same delay might not if it was seen only occasionally. If a customer is writing software to interface with the product, and the delay of a function is normally between 20 and 30 ms, the customer may write a timeout at, say, 40 or 50 ms. If the delay is occasionally 75 ms, the customer's timeout would generate an error. It could be that if the delay were consistently 65–75 ms, the customer might be satisfied.
- Defects that result in crashing the system.

Agile teams are recommended to capture system defects in a defect tracking tool [11]. There are many defect trackers (including many free applications), some that can be hosted by the development team and others that are hosted on a website. In general, a defect tracking tool will allow:

- Entry of defects with a description of the defect and a ranking of severity.
- How the defect was found, for example,
  - By a developer ad hoc
  - By a formal review
  - By a tester during test development
  - By a test within a battery of tests executed as a standard process
  - By a customer
- Storing files that clarify the defect such as a screen capture depicting a defective result.
- Status of the defect such as *entered*, *in progress*, *corrected*, and *test added*. ("Test added" records whether a test has been created that will detect this defect in the future.)

Typically the team will track some combination of the total backlog of active defects, the rate at which defects are being found, and the rate at which they are being resolved. For example, Figure 8.18 tracks the total active defects and the rate at which they are found (that is, the rate they are added to the backlog). This data can be used to measure team performance over time. In this example, the total defects are growing over Sprints 1 through 6. This is normally undesirable. Here is an example of where the SM might highlight this condition as

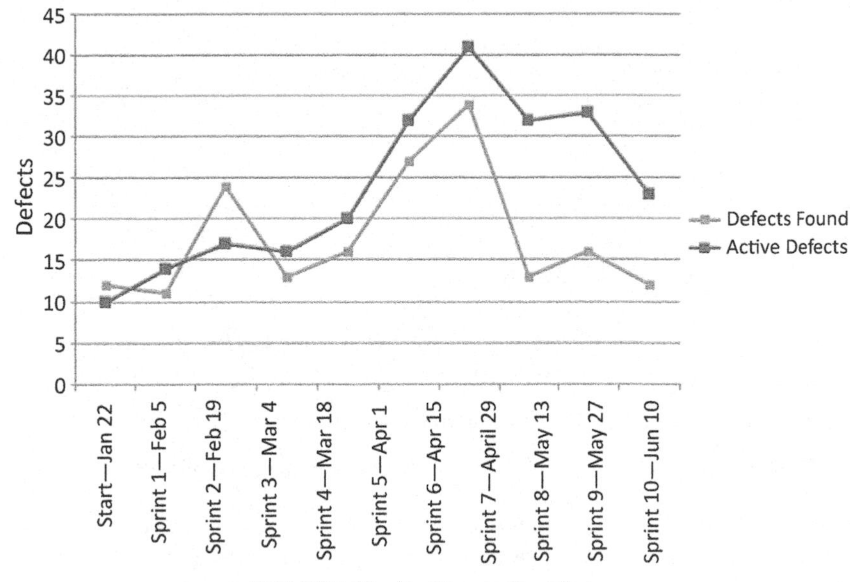

FIGURE 8.18 Tracking quality defects.

retrospective and then facilitate a discussion with the team of how to improve. This discussion might trigger actions such as:

- Increased review processes such as *pair programming*, which will be discussed in the next section, eXtreme Programming.
- Changes to test processes to detect defects faster, ideally before they enter the defect backlog.
- PO assigning more story points in upcoming sprints to reduce the defect backlog.

Using defect tracking tools allows the team to track and improve defects, which are probably the single most important indication of work quality.

8.3 EXTREME PROGRAMMING

eXtreme Programming (XP) specifies practices such as how to test, integrate, and review code. Scrum specifies how the team members interact among themselves and how the team interacts with the organization such as the many ways the team meets, how new work comes into the team, and how to measure progress. For example, Scrum may tell the team that code needs to be well tested, but it stops short of specifying how that testing takes place. By contrast XP would specify test driven development (TDD), a prescription of how the team develops code using tests; TDD, one of many XP techniques, is discussed below.

eXtreme Programming gets its name from the process of taking a best practice and then pushing it to its extreme. For example, it's well known that having team members review each other's code improves quality. XP then specifies

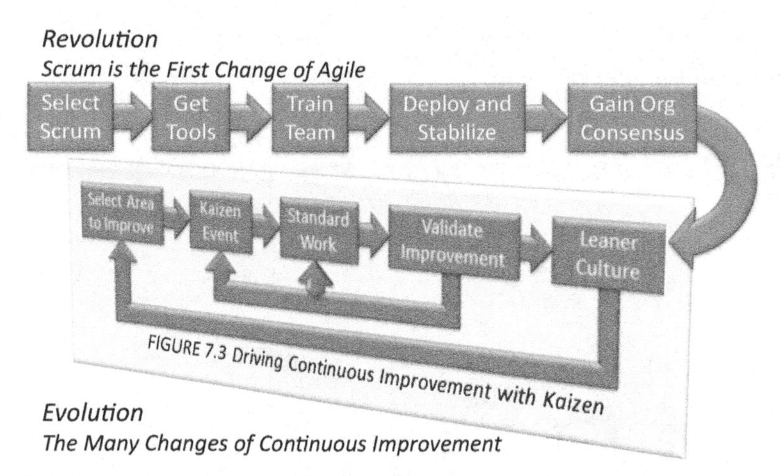

FIGURE 8.19 The revolution of scrum starts the evolution of continuous improvement.

pair programming, a technique where cross-team review is continuously performed as code is written. The approach of pushing good practices to their extreme is similar to the lean approach as seen in *Error Proofing* (Section 7.1.3) and *Single-Piece Flow* (Section 7.3.4).

8.3.1 Scrum, XP, and Continuous Improvement

The adoption of Scrum is *revolutionary*—many things must change at one time: new roles are defined, new measurement systems are put in place, and the team delivers code in a fundamentally different way. But, as much improvement as Scrum brings, it is only a first step. After Scrum is adopted and stabilized, the team must then look to *evolutionary* changes to improve over time. This is illustrated in Figure 8.19 using the continuous improvement cycle from LPD (Figure 7.3).[4] The approach to continuous improvement in Agile follows the pattern in LPD: team-driven action to bring continuous improvement generally supported by metrics.

In the continuous improvement cycle, XP offers a set of validated means to improve. For example, if during a team event (like a Kaizen), the team is in consensus that their testing practices are wanting, XP offers TDD as a process that is known to improve test coverage in Agile teams. So, XP is not a single process a team can adopt in the way it adopts Scrum. Instead, it is a list of processes that can be adopted one or two at a time to address an improvement need.

The leadership roles of Scrum are essentially unchanged by XP. In fact, the people in leadership roles defined by Scrum can use those roles to guide the

4. The terminology from LPD may not carry over to Agile in full; for example, Agile teams may meet in events that are not called Kaizens and they may not always use the term *standard work* for process change.

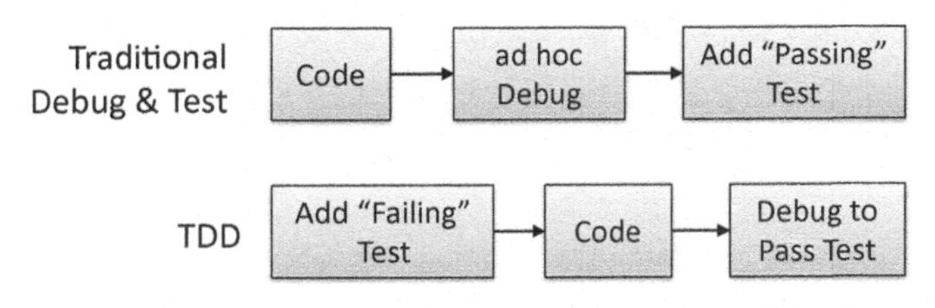

FIGURE 8.20 Traditional debug and test versus TDD.

team through the adoption of XP. The SM will normally facilitate discussions on how to make simpler changes. For larger changes, the team may engage in Kaizens or Kaizen-like events or they may attend a training class together. The PO will balance how much change the team can afford over a given time period against the need to execute backlog stories.

There is no universal definition for the practices of XP. Certainly there are core XP practices such as TDD, pair programming, and refactoring. However, most lists of XP practices number over a dozen and there is considerable variation in those lists. The remainder of this section will introduce some of the most commonly accepted practices of XP.

Test Driven Development

The intuitive coding practice for most programmers is to write code and use ad hoc testing until it seems the new code works. Then, the coder or a tester writes tests. The result is often that the tests don't fully exercise the code. Even the most diligent tester will inadvertently fail to test some functions. In the worst case, the test may exercise no important functions.[5]

TDD reverses the cycle (Figure 8.20). First, the coder writes a test for a function that does not exist. Of course, the test fails. The tester then codes so the test passes. A feature can be broken down to many tests that together define the majority of behaviors of that feature. The coding then proceeds in a cycle: pull out a testable function, write a failing test, code until the test passes, and repeat. Compared to writing tests after debug, TDD provides much more confidence that tests are effective and comprehensive.

Continuous Integration

In traditional software development, coders kept a copy of the code set on their machines and then developed functions on it for months, merging in code from other developers ad hoc. At an appointed date, everyone merged their code into a single code set in a painful integration step. Integration might take days or

5. Once a new tester on my team wrote a test that was intended to exercise a complex function. It ran successfully for weeks and I was impressed with how rapidly a new team member wrote such a difficult test. When I reviewed the test code, I found that it simply returned "test passed"! TDD certainly would have caught that.

even weeks as dozens of defects were discovered by having the new code from many developers interact for the first time. In lean thinking, the integration was a "large batch" (Section 7.3.4).

Over the years, the time for the integration cycle was reduced and by the 1990s, daily integration was a widely accepted best practice [12]. XP takes this concept to its limit: continuous integration, a process where every time new code is checked into the repository, the application is built and tested. Continuous integration requires a high level of automation to build each revision[6] and execute a set of tests to at least partially validate the code.

Pair Programming

In traditional development, coders worked alone most of the time. In some cases, code might be reviewed by a colleague—perhaps at a customer-mandated design review or when a piece of code has a difficult bug. Very often, the process of review brings unexpected benefit—developers can share best practices so that both the writer and the reviewer learn. Peer review is well known to benefit other areas of product development; for example, mechanical engineers by nature often review each other's drawings. The nature of code development, where developers write thousands of lines of text, makes peer review less natural.

XP teaches that if review is good, do it continuously using pair programming: two people developing code sitting at one machine. Pair programming can be done 100%—all code written by two people. There are known problems with pair programming, especially that many people don't enjoy working side by side with another programmer all day. It also is probably more expensive, though the expense may be offset by reduction in coding defects. For those organizations where 100% pair programming does not fit well, it can be used partially. Mike Cohen recommends using it at least for the riskiest code development [13].

Refactoring

Refactoring refers to changing the way code implements a function without changing the behavior of the function. Refactoring is done to make improve code structure and make it more readable; ultimately refactoring makes code easier to maintain and extend. Examples of refactoring include (1) removing duplicate code, (2) changing from complex logic to a software state machine, (3) breaking a large method into several smaller methods, and (4) making a data structure more efficient. In XP, refactoring is done continuously; it's a part of every story that touches existing code. Without refactoring, changes to code accumulate, slowly making the code more complex, a phenomenon sometimes called "code rot." With refactoring, each time the code is changed, small improvements are made so that the structure of the code slowly improves over time and code rot is adverted.

---

6. Cruise Control was one of the first tools that did this and remains a viable alternative.

Coding Standards or Coding Conventions

Coding standards are rules the team writes to ensure the team members reliably follow identified best practices. Coding standards include concrete rules like naming conventions, source-code organization, and comment requirements and templates. They can also specify programming principles such as code reuse, polymorphism, and standard interface design. With coding standards, there is less variation in code between developers making it easier for one developer to debug, expand, or refactor another developer's code. It also improves the quality of the code set in general by prescribing best practices to each team member, something especially useful for less experienced team members.

The On-Site Customer

The on-site customer is a team member that is always available to resolve questions about how the application is used. The on-site customer has broad domain knowledge of the area the application supports; they do not develop code. As an example, if the team is developing billing software, an on-site customer could be a bookkeeper. Of course, the on-site customer need not sit with the team full time, but they need to be available on short notice.

These are examples of XP practices—there are many others. The value of XP is providing best practices and process for code development, something Scrum does not address. For more information on XP, see *eXtreme Programming Explained* [14].

8.4 SCRUMBAN

Scrumban is Kanban project management applied to software development. It is a simplified version of Scrum, keeping the daily Scrum meeting and the Kanban board (hence the name), but eliminating the planning activities and velocity measurement. Scrumban focuses on smooth flow and minimizing WIP. Scrumban is oriented toward simpler projects—the Kanban board presents each task in isolation so there's no easy way to represent complex interactions among tasks. Also, there is no velocity planning or measurement system and thus no means of predicting when a product will launch the way Scrum can (see Figures 8.13–8.15). As a result, Scrumban is not usually a good match for large product development projects. It can be a good fit to sustaining activities and smaller development projects.

Scrumban does present advantages for those projects where it does fit. The barriers to adoption are much smaller than Scrum—there is no requirement for the large changes Scrum requires (see the top half of Figure 8.19) [15]. There are no special tools or reporting conventions; there is no need for large training programs. There is also no need to change the organization and no new roles like SM and PO. And there's no need to change an entire organization to Scrumban—a single group within a company can use Scrumban indefinitely. Finally, adopting Scrumban is simple: a team can decide they will move to

TABLE 8.1 Agile Scrum versus Scrumban

Scrum	Scrumban
Tasks are planned in 2–4 week sprints.	Tasks are single-piece flow.
Tasks in backlog have effort estimated in story points.	Task effort is not estimated.
Focus on story point velocity.	Focus on task flow and minimal WIP.
Requires a large effort to start (see Figure 8.19).	Requires little effort to start.
Targeted at teams with deep collaboration.	Targeted at teams with lots of interrupts and simple projects.
Team size 5–9.	Any team size.
Easy to integrate XP practices.	Easy to integrate XP practices.

Scrumban in the morning and be up and running with sticky notes and poster boards that afternoon.

Scrumban is evolutionary; in that sense, it may be closer to LPD than Scrum. Table 8.1 gives a brief comparison between Scrum and Scrumban.

8.4.1 Six Key Scrumban Practices

David Anderson suggested six key practices of Scrumban [15], focusing on the Kanban process almost to the exclusion of the Scrum meetings. In fact, most of what he presents would be valuable for Kanban project management in general (see Figure 8.21).

Visualization

Make sure the Kanban board is visible at all times. Avoid an electronic board on someone's PC. If the team does elect to use an electronic version (something often necessary for distributed teams), display it continuously with a dedicated monitor or projector.

Limit WIP

Limiting WIP is a basic function of Kanban project management. The Kanban board reveals WIP and provides a simple way to limit it: the team cannot advance a task into a Kanban column when the column is at its limit.

Manage Flow

Again, this is a function of Kanban project management. Flow is visual—if a task is stuck, it's easy to see. As in LPD, smooth flow is desirable; uneven flow

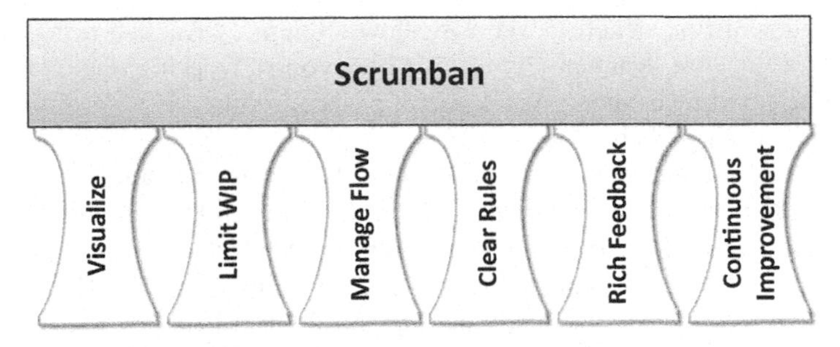

FIGURE 8.21 The foundational pillars of Scrumban.

(often called "mura") is a sign of waste. So teams should strive to flow items at a consistent pace.

Create Explicit Rules

The team should write down the Kanban rules. What are the WIP constraints? How will tasks be added to the backlog and how will they be pulled out of the backlog to be processed? If an inexperienced team member has advanced a task improperly, how will this be reversed? Does someone on the team have authority? Is there a small group that can override team member decisions? Scrumban doesn't have defined roles like PO or SM, so the team must decide how to manage interactions.

Implement Feedback Throughout the Organization

Make the Kanban board visible across the company and invite feedback from every corner. Focus on how to increase value and reduce waste.

Continuous Improvement

Typical of lean thinking, Scrumban starts simple and brings little benefit in the beginning. Scrumban brings value through evolution powered by the collective ideas of the team. The team can choose a handful of key performance indicators to measure their performance: defect backlogs, on-time delivery to internal customers, customer satisfaction, turnbacks (task flowing to the left), or task time in a column. When performance metrics are stabilized, the team can take action to improve.

8.4.2 Adopting Scrumban

If your projects are a good fit for Scrumban and you're thinking of adopting, the right answer is probably to try it. Anderson recommends you start with a personal Kanban board, a recommendation I followed and can validate (see Figure 7.19). By spending a few months managing your daily work with Kanban, you will become familiar with the workings of Kanban project management in a near-painless process. There's virtually no downside of using Kanban—the investment is small and if it's not beneficial, simply stop. But it's likely you will find benefit. Even a shallow attempt of Kanban brings value because, at its core, Kanban lets you see how you're working; that alone can justify the small investment. As a leader it has a secondary benefit of showing your team how you're working.

In one of Anderson's talks on Kanban, he spans the range of Kanban from simple boards that visualize work process to deep Kanban that changes the organization. He talks about scale with an example of 75 people working from one board or interlinking multiple Scrumban boards. He gives an example of a company, Corbis, that changed their culture for the better. If you have interest in considering Scrumban, spend a few minutes watching his talk [15].

Scrumban is as open as Scrum is to XP: TTD, pair programming, and continuous integration fit just as comfortably in Scrumban as they do in Scrum. Defects can be tracked and improved just as well in Scrumban. The development processes are not changed by the removal of sprint planning and the specialized roles of Scrum.

8.5 BARRIERS TO ADOPTION OF SCRUM

As powerful as Scrum is, there are barriers to adoption as shown in Figure 8.22. To be clear, one organization after another has overcome these barriers making Agile probably the most popular technique for managing software projects today. Even so, those considering Agile for their organization should be aware of the barriers so they will understand adoption issues if they do occur.

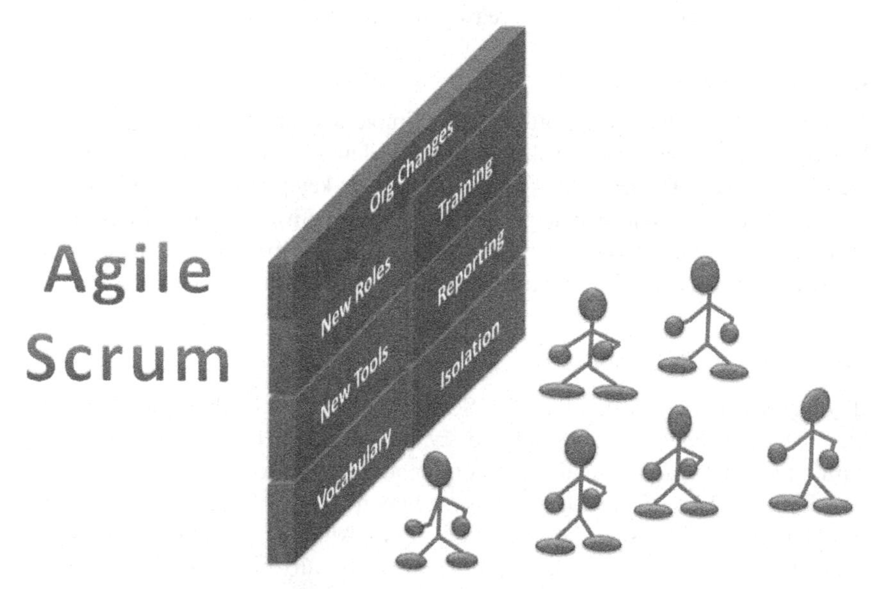

FIGURE 8.22 Barriers to adoption of Scrum.

8.5.1 Changing the Organization

The introduction of the PO and SM along with other Agile changes can create org chart winners and losers. The new authority given to the PO, the SM, and other Scrum leaders may be taken from someone else who is likely to be resistant.

8.5.2 New Roles

Scrum brings new roles to an organization, the most important being the PO and the SM. These roles can cause confusion because there are no exact equivalents in other project management systems.

8.5.3 New and Complex Tools

At its most basic implementation, Scrum requires a Kanban project management board and a few other simple tools for planning. But a full set of Scrum tools must track burn-down points, measure velocity across sprints, and manage the product, release, and sprint backlogs. The full set can be difficult to use and they are often expensive when deployed to a large team.

8.5.4 Training

Converting an organization to Scrum often requires extensive training. The Agile teams may have to travel to a class for a week or two, or a consultant may come on site. There is often follow-up training as the adoption stabilizes.

8.5.5 New Reporting Methods

Agile brings a host of new reporting tools: burn-up charts, backlog sprint points, and average team velocity are a few. The performance measures and formats are new for Phase–Gate-based development organizations. This raises issues at two levels. First, at the project level, it can be difficult to coordinate Agile and Phase–Gate subprojects creating a single product. Here, teams need to synchronize, at least for major milestones. Second, at the management level, the reporting of Agile and non-Agile projects should be unified as is shown in Figure 8.12, for example using run charts or fever charts in similar formats. Unfortunately, Agile and Phase–Gate are often reported in uncoordinated ways placing a larger burden on the management team because they must adapt to different measurement systems.

8.5.6 Vocabulary

The vocabulary of Scrum can make adoption in an organization more difficult. Of course, new vocabulary is common in all project management methods: CCPM brings a host of new terms: fever chart, relay-racer mentality, feeder

[3] Bjork A. Scaling Agile Across the Enterprise. Microsoft Visual Studios Engineering Stories. Available at: http://stories.visualstudio.com/scaling-agile-across-the-enterprise/.
[4] Beck K, Beedle M, and about 15 others. Manifesto for Agile Software Development 2001. Available at: http://agilemanifesto.org/.
[5] Cohen M. Succeeding with Agile. Software development using Scrum. Addison-Wesley; 2010. p. 238.
[6] Cohen M. Succeeding with Agile. Software development using Scrum. Addison-Wesley; 2010. p. 244.
[7] Steiner ID. Group process and productivity. Academic Press; 1972. p. 220.
[8] Bradner E, Mark G, Hetzel TD. Effects of team size on participation, awareness, and technology choice in geographically distributed teams. In: Proceedings of the 3th Annual Hawaii International Conference on System Sciences, vol. 271a. IEEE Computer Society; 2003. p. 7.
[9] Putnam D. Team size can be the key to a successful project. http://www.qsm.com/process_01.html.
[10] Cohen M. Succeeding with Agile. Software development using Scrum. Addison-Wesley; 2010. p. 179 ff.
[11] US Government Accounting Office (GAO). Software development: effective practices and federal challenges in applying Agile methods. July 27, 2012. p. 16. Available at: http://www.gao.gov/products/GAO-12-681.
[12] Cohen M. Succeeding with Agile. Software development using Scrum. Addison-Wesley; 2010. p. 162.
[13] Cohen M. Succeeding with Agile. Software development using Scrum. Addison-Wesley; 2010. p. 164.
[14] Beck K, Andres C. Extreme programming explained. 2nd ed. Addison-Wesley; 2004.
[15] Anderson D. Deep Kanban, worth the investment?. London Lean Kanban Day; 2013. Available at: https://www.youtube.com/watch?v=JgMOhitbD7M.

Part III

Advanced Topics

Part III will present three advanced topics. You can read those you find most interesting and they can be read in any order.

CHAPTER 9 RISKS AND ISSUES: PREPARING FOR AND RESPONDING TO THE UNEXPECTED

Describes risk management using two lines of defense. First, prepare diligently—adequate planning, choosing a strong team, and having the right processes in place. Then use leadership skills to resolve the issues that survive the preparation.

CHAPTER 10 PATENTS FOR PROJECT MANAGERS

Discusses patents and patent law as they relate to project management for product development. It is likely that as project manager (PM), you will at some point come into contact with patent law. Unfortunately, patents are often misunderstood by development teams. PMs are in a unique position to help the company accomplish its goals related to patents. So, you will want to know the basics.

CHAPTER 11 REPORTING

Provides detailed discussion on reporting with a focus on reporting up—to the sponsor or steering committee. The first section focuses on oral presentations given for project reviews and/or approvals. The remainder discusses the use of quantification in project management, beginning with metrics and then developing key performance indicators (KPIs), and finally creating a project dashboard.

Chapter 9

Risks and Issues: Preparing for and Responding to the Unexpected

This chapter will discuss how to deal with the effects of risks and issues identified during the execution of a project. Product development is nondeterministic and projects that create innovative products will at some point almost always face risks and issues that are not fully understood at the outset. Unfortunately, there is a common expectation of project managers (PMs) that a well-managed project will be orderly and, further, that deviations from orderly execution are due in large part to poor planning. However, even well-planned product development projects will have unexpected risks and issues.

This chapter will approach unexpected events with two lines of defense. The first line is diligent preparation—adequate planning, choosing a strong team, and having the right processes in place. But even after the most diligent preparation, some number of risks will present themselves. So, the second line is strong leadership in resolving those issues that survive the preparation; this includes identifying, tracking, reacting to, and reporting on the risks that affect the project.

9.1 RISK IN PRODUCT DEVELOPMENT PROJECTS

Most of this text has focused on actions PMs take to bring order to their projects: selecting the right processes, planning thoroughly, and exhibiting solid leadership skills. It is certainly true that a great deal of good project management proceeds from the order that is created by these kinds of actions. In fact, there is a general expectation in many organizations that projects should go smoothly, that customers will be satisfied, and that objectives will be met. Figure 9.1 represents the extreme case: an implicit belief that a perfect, laser-focused plan will create a perfect project. The unfortunate corollary to this is: the main reason projects don't deliver as predicted is because PMs don't plan as well as they should.

In fact, product development projects normally follow a less orderly path. Like winning a football game, raising a child, or any of a myriad of complex human endeavors, competent planning is required but not sufficient for success.

Project Management in Product Development. http://dx.doi.org/10.1016/B978-0-12-802322-8.00009-7

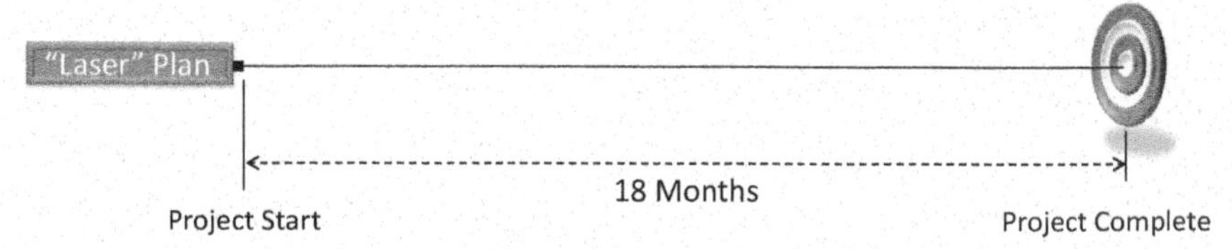

FIGURE 9.1 A common expectation: the perfect plan results in the perfect project.

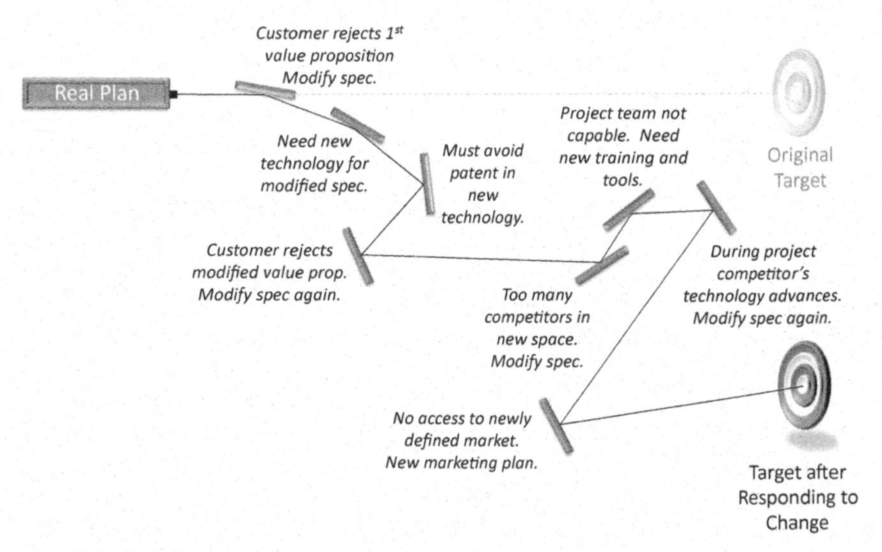

FIGURE 9.2 A common experience: unexpected events interfere with initial plans.

In a product development project of ordinary complexity, there are simply too many unknowns at the project start. And while there is no doubt that good preparation is critical in creating a well-run project, success will also depend on the PM's ability to respond to the issues and risks that do occur. As shown in Figure 9.2, sometimes those events can cascade, changing the product, the team, and even the target markets substantially from what the original plan predicted.

The world of product development is full of stories where the initial plan encountered a challenging issue and was abandoned for something better. Perhaps one of the most famous comes from the development of 3M Post-It® Notes. Spencer Silver was searching for an adhesive stronger than anything known. Instead he discovered the opposite. Says Spencer,

> *"It was part of my job…to develop new adhesives, and, at that time we wanted to develop bigger, stronger, tougher adhesives. This was none of those" [1].*

It took years of persistence for Silver to sell the idea internally at 3M and then to successfully market Post It® Notes. Eventually, the product went on to be recognized as one of the top consumer products in the 1980s. Not every product development effort undergoes the level of change that Silver's did; but many face significant changes due to facts that cannot be (or, at least, are not) known at the outset.

A person might wonder why these types of risks cannot be planned for. For example, referencing the top left of Figure 9.2, someone might wonder, "Why couldn't the team predict the customer would reject the first value proposition before the project started?" One reason is that at the project start, the team is usually limited to traditional marketing—asking hypothetical questions to target customers through surveys, interviews, and focus groups. But until a customer has to make a real decision (for example, to purchase the product), you don't know how they will act. And in most projects, you cannot offer a customer a real decision until the project is running for some time. So, there's a possibility the first prototype will uncover unwelcome surprises well after the project has started. And there are other events that can occur during a project that are nearly impossible to predict. For example, a new patent is issued to a competitor for technology the project relies on or a competitor releases a new product that is superior to what is being developed by the project team.

Beyond the unknowable issues and risks, there are risks that are too expensive to identify. At some point, planning becomes sufficiently unlikely to identify important issues that the likely returns on further planning don't justify the costs. Think of the issues that might arise just driving to the grocery store: there might be an accident that stops traffic for an hour or your car might break down. These risks are so unlikely that it's not worth planning for their occurrence. Instead, we strap on the seat belt, check the gas, and go. Product development projects follow a similar pattern. Thorough planning is necessary for success; but when the planning is sufficient, the project begins with the team knowing some unexpected events are likely to occur. At that point, the team's ability to react becomes critical to success.

9.1.1 Traditional View of Risk in Project Management

Traditional writing on project management defines risk management in ways that don't always help those of us in product development. For example, one text defines it as "a formal process whereby risk factors are systematically identified, assessed, and provided for" [2]. Further, risk identification "must be seen as preparation for possible events in advance, rather than simply reacting to them as they happen." But many risks in product development don't fit this description well. If you know your competitor is regularly applying for patents, one risk is they may be granted one for technology used in your project. But preparing for this threat is quite difficult until the patent is published. Similarly, that competitor may have a history of releasing new products at your industry's largest trade show each year. That's another identified risk because it's possible that this year's new product will affect your project; but, how can you prepare for this risk before the show? This approach to risk management is appropriate for project types that have fewer unknowns at the outset (see Section 1.3.3); for product development, there are many risks that are not sufficiently known at the beginning of a project to allow them to be "identified, assessed, and provided for."

Phillips gives a list of the top 15 reasons projects fail including "lack of a clear vision," "conflicting priorities," "poor planning," and "poor communication" [3]. The reasons are valid, but the list is essentially a set of mistakes made by the PM, the project team, or the organization. This approach treats project management as if it were a deterministic undertaking: do the right things and the project will turn out well. In fact, product development is stochastic—there are important events that occur that are out of the organization's control. What if a supplier of a new technology, who was customizing parts for your project, goes out of business? What if your main contact at a key customer leaves and is replaced by someone who has a strong relationship with your main competitor? What if a technology fails for reasons the team could not contemplate such as unexpected sensitivity to ultraviolet rays? These are all examples of what Loch calls "unknown unknowns" [4]—things the team didn't know they didn't know. Product development projects of medium-to-large complexity will have unknown unknowns; the more novel the product, the more likely these factors will affect the outcome. Standard project management tools are not able to deterministically deal with these types of issues [5].

A risk management method more appropriate for product development is made of two lines of defense:

- **Preparation.**
  Avoiding as many risks as practical through:
  - Selecting the right team
  - Using the right processes
  - Planning the project thoroughly
- **Responding.**
  Dealing with the risks and issues that do occur after the project starts by:
  - Identifying unknown risks and issues
  - Tracking those risks and issues throughout the project
  - Reacting to risks that affect the project goals
  - Reporting the exposure of risks to the sponsor and management team

This approach will not guarantee success. The goal is rather to deliver the highest likelihood of success and simultaneously minimize the investment for those cases where projects do fail.

9.1.2 Total Leadership and Unexpected Events

As with all important areas of project management, the PM will need both transactional and transformational leadership skills to deal effectively with unexpected events. Transactional leadership skills here include following process, following up with the team on areas of concern, and ensuring mitigation plans are executed. These skills are necessary to manage the large number of activities

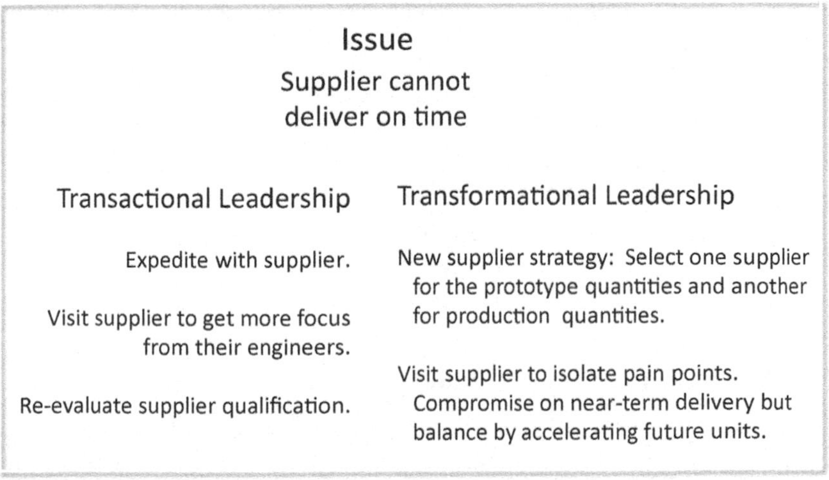

Issue
Supplier cannot deliver on time

Transactional Leadership	Transformational Leadership
Expedite with supplier.	New supplier strategy: Select one supplier for the prototype quantities and another for production quantities.
Visit supplier to get more focus from their engineers.	Visit supplier to isolate pain points. Compromise on near-term delivery but balance by accelerating future units.
Re-evaluate supplier qualification.	

FIGURE 9.3 Transactional and transformational leadership in dealing with issues.

that may accumulate related to identifying and mitigating risks and issues. Transformational skills include using project innovation, staying connected with the team, and maintaining the product vision in difficult times. These skills are necessary to deal with those events that are not resolved with process and best practices.

A simple example is presented in Figure 9.3. Suppose an issue has arisen with a supplier delivering a component on time. There are actions that are transactional in nature—the standard things a PM might do in such a situation, specified either by formal process or by the organization's best practices. At the same time there are steps that are transformational in nature—new ways of doing things, at least for this class of issue. In dealing with more complex issues PMs will usually apply a mix of transactional and transformational thinking to devise the best solutions.

9.1.3 Terminology: Risks and Issues

It's common in writing on project management to substitute the term "risk" for any potential or actual unexpected event. That blurring of terms can cause some confusion, so here we'll use two terms to discuss unexpected events: risks and issues.

- *Risks* here are events that might or might not occur. The likelihood of occurrence of a risk is not sufficiently high for the team to take action beyond investigating and monitoring.
- *Issues* in this context are risks that are either certain to occur or have a high enough likelihood of occurrence that the team reacts to them, for example, by modifying the design of the product.

For example: suppose the team is engaged in activities to secure medical certification for a new product.

Risk: An internal test indicates there is a possibility the product may not pass the certifying body's test set, but that possibility is too small for the team to modify the design at this point in time. However, further testing is ordered to better understand (investigate and monitor) the risk.

Issue: After the second set of tests, it seems all but certain the product will fail the final test from the certifying body, so the team modifies the design. At that point, the risk has matured into an issue.

9.2 TYPES OF ISSUES AND RISKS

Project management risks are commonly broken down into many categories, often 50 or 75 types loosely connected into larger groups. For example, risks can be broken down by technical, management, commercial, and external [6]. Further, technical can be divided into scope definition, technology readiness, reliability, and many more. They can also be broken down by who presents the risk: *sponsor* (slow decisions, poor decisions), *project management* (PM skill deficiency, incorrect information, poor communication, etc.), *team* (insufficient technical capability), and *external* (economic, legal, etc.).

Matta and Ashkenas break down risk in a manner interesting for product development: execution risk, white-space risk, and integration risk [7]. Further, they point out that traditional risk management is focused on execution risk—the risk that planned activities won't be completed properly. Ways to reduce execution risk include training PMs, selecting the best project management method, and defining processes; these are, of course, important actions for a project to be managed well. However, traditional risk management has less emphasis on white-space risk (necessary activities that are overlooked at planning time) and integration risk (the many pieces won't come together as predicted). As we'll see later, one effective tool for treating the last two types is simplifying the project—breaking it into smaller pieces and delivering smaller increments of value to the market more quickly.

For the remainder of this section, the most common risk types in product development will be discussed in detail. Here they will be broken down into 10 categories, mostly focused on who will generate the issue with the exceptions of technology and specification issues, which are broken out separately. These lists are common in writing on risk and certainly bring value because they help the team think about issues that they have not experienced. However, they can be overwhelming; like a person with health anxiety reading a medical book, a little imagination can make it seem that your project is assailed by every risk on the list. So, skilled facilitation for team events relying on such a list is important: the review can bring value, but it also can be a heavy process that generates an unwieldy result.

9.2.1 Project Management

Project management issues focus around the duties of the PM and the consequences of those duties being performed improperly. Examples include:

- **Delay due to poor execution**
  - Unclear project planning so resources are working on the wrong tasks.
  - Not properly coordinating with supporting functions such as labs.
  - Not properly coordinating with related projects (for example, an electronic hardware project and its companion software project).
  - Not managing suppliers to expectations.
- **Quality of project work**
  - Accepting poor quality work product from team members, suppliers, or support groups.
  - Not ensuring the team follows company mandated process.
- **Team management**
  - Poor leadership causing team members to be disgruntled.
  - Failure to track resources applied to project properly.
  - Not effectively communicating product vision to the team.
- **Reporting issues**
  - Incorrectly tracked budget or progress, leading to a too-late discovery of schedule slip or spending overruns.
  - Poorly tracked product cost estimates leading to a too-late discovery of lower-than-acceptable financial margins.
- **Escalation issues**
  - Not escalating important issues rapidly enough to allow early resolution.
  - Escalating poorly, not communicating appropriate urgency or recommending action.
  - Overescalation (escalating issues that could be dealt with inside the team).

9.2.2 Team Members

Team member issues include:

- **Retaining team members for the life of the project**
  - Team members changing projects.
  - Team members multitasking and giving this project inadequate attention.
  - Team members leaving the organization.
- **Quality of work**
  - Poor workmanship due to lack of capability.
  - Poor workmanship due to inadequate effort.
  - Features that don't meet customer needs due to team having poor understanding of those needs.

- **Relationships**
  - Chemistry problems between difficult individuals.
- **Tools**
  - Poor tools.
  - Tools unable to adapt to changing project scope.
  - Inadequate training on tools.

9.2.3 Specification Issues

Specification issues include:

- Product specification vague or incomplete.
- Specification incorrect (defining a product that customers won't value).
- Specification changing.
- Value proposition unclear or unrealistic, or product not aligned with value proposition.

9.2.4 Organizational Issues

Organizational issues include:

- Poorly defined business objectives for project.
- Management not fully bought into the project or losing management support during the course of the project.
- Resources taken from the project during execution phase.
- Slow or opaque approval processes.
- Poorly defined or undefined development processes.
- Unreasonable expectations—overworking team members or PM, or never satisfied with results.
- Unwilling to invest in necessary infrastructure, tools, or training for PMs or project teams.
- Project not aligned with organizational goals.

9.2.5 Manufacturing Issues

Some examples of potential manufacturing issues are:

- Lack manufacturing technology to support new products.
- Schedule delays in building or expanding factory equipment necessary to support the project.
- Processes/equipment unable to meet market quality requirements.
- Factory associates incapable of producing product or not properly trained.

9.2.6 Technology Issues

Several technology issues are:

- Technology proves to be ineffective at meeting customer needs.
- Technology is too difficult to use.

- Lack of core competence prevents appropriate use of technology.
- Technology improvements during project allowing competitors to "leapfrog" this project.
- Discovering patents that protect technology the team thought it was free to use.
- Missed opportunity to patent key technologies.

9.2.7 Customer/Market Issues

Some issues related to customers and markets are:

- Customer does not value product above competitor's product.
- Customer cancels or postpones the program that uses your product.
- Customer delays qualification testing.
- Customer's organization has high-level relationship with competitor that drives their acceptance criteria more than expected.
- Customer is evaluating your product primarily to leverage price at one of your competitors.
- Customer values product but doesn't have the financial resources to purchase it.
- Customer inadvertently gives requirements that do not meet their needs (customers often don't fully understand their needs).
- Key contact at customer changes roles.
- Discovering key contact at customer has little influence over buying decisions.

9.2.8 Competitor Issues

Issues related to competitors are:

- Competitor releases a product superior to the one you are developing.
- Competitor is able to copy your product quickly and sell it at a lower price.
- Competitor attracts key team members to join their company.
- Competitor has relationship with supplier that causes supplier to deprioritize your project.

9.2.9 Supplier Issues

Examples of supplier issues are:

- Supplier late on delivering key technology or components.
- Supplier not capable of producing components to advertised/required specifications.
- You are a low-priority customer to your supplier; your projects get few resources.
- Supplier has a patent issue with their competitor.

9.2.10 Regulatory Agency Issues

Some issues related to regulatory agencies are:

- Delays in completing certifications.
- Testing method misunderstood so your product unexpectedly fails tests.

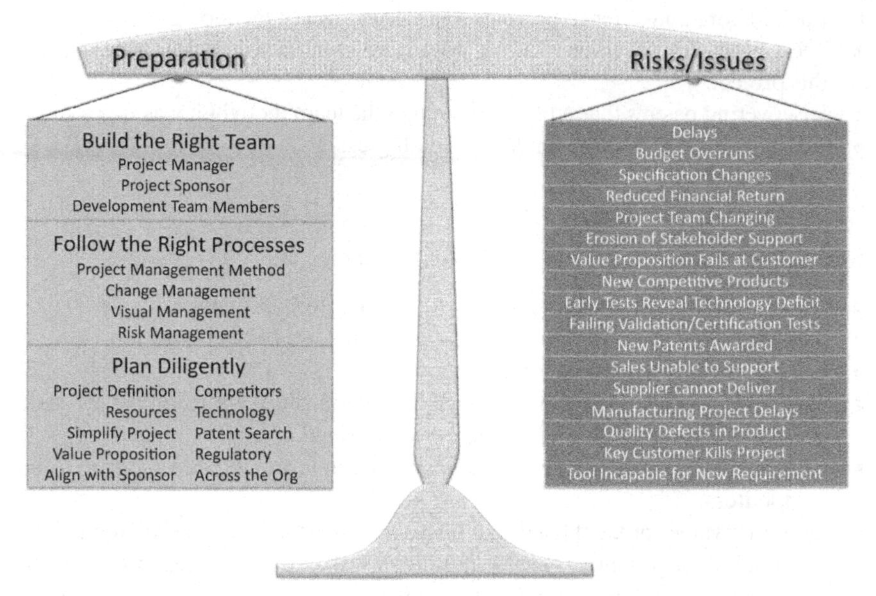

FIGURE 9.4 Preparation to avoid issues and to allow rapid response to those that occur.

- Agency changes certification requirements during project.
- Market requires certification that was not expected at the project start.

9.3 PREPARATION: AVOIDING AND REDUCING ISSUES

The best way to deal with many issues is to avoid them before they happen. Three ways you can prepare for issues are shown on the left side of Figure 9.4:

- Build the right team
- Follow the right processes
- Plan diligently.

Build the Right Team
Project Manager
Project Sponsor
Development Team Members

9.3.1 Build the Right Team

A strong team is probably the most important component of a successful project. The Standish Group lists the main reasons IT projects succeed and three of the top five are related to the team: a skilled PM, skilled developers, and support from the sponsor [8].

A strong team begins with a strong PM, so execute your job diligently—use transactional and transformational leadership with your team. Follow process, create and sustain vision, and stay connected with your team members. Ensure your team is composed of people with the talent and experience the project needs. If you're trying to develop a high-efficiency positive displacement pump that is lower cost than all your competitors, you'll need technical experts in pump design. There is no process or project management skill that will create innovation from people that lack the expertise to innovate. This doesn't mean every person on the team needs to be star; there is room on most teams for beginners and journeymen—but be sure the team contains the technical ability to create the innovation your product will need.

The sponsor has been called the most important member of the team [9]. A good sponsor is your advocate to the company management, a coach when your skills need improvement, and an encouragement in difficult times. If you're a PM, you probably won't choose your sponsor; but you can chose to develop a strong relationship with her: stay in contact, be transparent when things go wrong, and seek out her advice.

9.3.2 Follow the Right Processes

Follow the Right Processes
Project Management Method
Change Management
Visual Management
Risk Management

When process is well designed, up to date, and complied with throughout the team, it defines what needs to be done in normal circumstances. It's a sort of autopilot that can guide most of the decisions the project team faces. Without process, a product development project of any size will overwhelm a PM because the response to every situation must be reinvented. Just managing the project through ordinary circumstances is challenging. When a difficult issue does occur—and they almost always do—the PM can't give it full attention because daily work is absorbing all the available mindshare. It's like a road system without standards: no lines in the streets, every car built with a unique set of signal lights, and every sign and signal varying town by town. There would be more accidents, more traffic jams, and driving would be exhausting. Of course, you cannot rely on traffic rules in every situation; good drivers augment the rules of the road with good judgment.

Process in product development is similar to rules in traffic—it is not meant to cover every situation. But for the situations process does cover, it simultaneously guides and empowers individuals to make the right decisions. Without proper process, the PM and sponsor must needlessly invest time and energy in decisions team members could make, wasting their time and the time of the team member who either must ask for guidance or rework due to making the wrong decision. And every hour the PM is engaged in issues that could be resolved with process is one less hour that can be used for scouting risks and resolving complex issues.

Project Management Method

Project management processes such as CPM (Chapter 5), CCPM (Chapter 6), and Agile (Chapter 8) guide the planning and execution of ordinary project tasks. They create a common language throughout the organization for what *done* is. They standardize reporting and set standards to help ensure quality work product in every phase of the project. They are the primary processes that guide team members in normal circumstances, freeing the PM and sponsor to deal with issues that occur.

Moreover, strong project management process helps avoid issues before they happen. It defines standard work, which a company can continually augment, helping teams avoid mistakes that have been made in the past. Let's say a company has had several projects that provided products to the market that flopped—customers didn't find the products appealing. In such cases, the senior leadership could invest a lot of energy determining the root cause: Was the perceived quality of the product poor? Was the price too high? Was the product unable to meet the customer's needs? Questions like this take time and energy to answer, but let's suppose a thorough review revealed the price was too high. Further, let's say the development group had a history of projects that started with low cost estimates that proved overly optimistic over time. You might expect the company to install a process that tightens up the work around accurate prediction of cost at the project outset and diligent tracking as the project executes. Such processes, when well developed and followed, create team behavior that reduces risks and issues across the portfolio.

"Project management process" is not a single process. At a high level, it specifies the many steps the company expects to be completed as part of the project. But each step within the larger process is a process itself: how to conduct design reviews, how to qualify suppliers, how to estimate market size, and how to test prototypes. Well-designed and well-followed product development process can remove risks and issues by the hundreds.

Risk Management Process

A risk management process is a defined method to deal with risks and issues throughout the life of the project. There are usually four key parts of risk management:

1. Method(s) to identify risks and issues.
2. A method to track risks that have been identified and determine when they mature into issues.
3. A method to react to issues.
4. A method to report on the risks and issues.

We will discuss methods for each in Section 9.4.

Change Management Process

A change management process is used to reduce issues that result from changes to product definition or project scope. Change management processes specify who has authority to approve changes and under what conditions. For example, a simple process is that any change that negatively affects project goals must be approved by the senior management.

Change management processes are often targeted at preventing "scope creep," the phenomenon where product marketing or the customer adds features throughout the project. The phenomenon often results because the longer the project lasts, the more time there is to learn about the product and its applications. As marketing visits more customers and trade shows and as they spend more time investigating competitors, they are more likely to conceive of new features that will benefit the customer. The problem with scope creep is it causes rework and delays. Those changes often appear at a slow but steady pace (hence "creep") and so the project scope can increase so slowly that changes may not be noticed. Scope creep also occurs in long-running projects because the market solutions are changing as the project proceeds. Customers unaware of a feature at the start of a long project won't request it initially; but that can change if a competitor introduces them to the previously unknown feature.

The simplest method of change management is "freezing the spec," the practice of ignoring all change requests until the project is completed. This, of course, protects the team from all changes; unfortunately, that includes changes that would ultimately benefit the project. A more balanced method of evaluating changes is to compare project financial performance with and without the change, for example, by calculating the NPV (net present value, see Chapter 5) in both conditions. If the cost of the change is more than compensated by increased revenue, the change may be regarded positively.

A pure financial review ignores some problems that change brings to the project. All of the effects of a change are difficult to predict, so often the full consequences of changing a specification are not understood until later. Also, the team can feel a loss of ownership of the initial commitments. The team will be most committed to the project when they are part of the planning. When changes are forced on them, their ownership may diminish, a condition well known to reduce the enthusiasm of the team, which can, in turn, increase the likelihood of schedule delay and reduce work quality. This effect can be reduced by making the team part of decisions that bring significant specification changes.

So, taken together, a decision to change the specification significantly should be financially justified by a wide enough margin to deal with unforeseen effects of the change. The team should be involved when possible, and always informed immediately. These are things a change management process can help to ensure.

Visual Workflow Management

Visual workflow management was discussed in Section 7.3.1. Strong visual management brings the most important information concerning a complex topic together to create an intuitive view of the current situation. It is one of the best ways to bring consensus in the team concerning a risk or issue by providing a common understanding of (1) the technical and commercial factors, (2) the need to take action, and (3) how the issue is being led.

Consider this example: a project team is working on providing cable harnesses rapidly. A cable harness is an assembly of wires and cables shaped to fit precisely in an electrical assembly. They are highly customized with special connectors, insulation, and wire size. They are often time consuming to produce, but our project team is building a system that will deliver prototypes in 24 hours at a low cost. This project is planned to have several subprojects:

- A web interface that brings in designs from multiple CAD systems.
- A robotic assembly station that can assemble the wires automatically using the CAD files as input.
- A specialized rapid-prototype assembly line where workers can fix a wide range of connectors to the conductors.
- A test station that is configured automatically.

Suppose the team has determined one of the major issues revolves around the number of connector types the assembly line can support. Each connector type supported makes the system capable of satisfying more customer needs and if the system doesn't support enough types, few customers will use the service. But each connector type also brings substantial cost for stocking many parts, providing a means of making electrical connections, and supporting the electrical interfaces to the test machine. It also brings quality concerns because workers must be trained on different attachment methods (soldering, crimping, etc.) and tools.

This issue requires balancing many factors: cost to support each type, processes each type requires, complexity added to the automatic test equipment, and how much the customer values each type. This could be a dynamic issue, with the team learning about connectors the customer will want as the project continues, then having to learn the technical issues around those types. Keeping track of these issues and keeping the team aligned is needed for early identification of issues such as overrunning the budget for the assembly line, exceeding the capacity of the test station, and excessive risk of product defects. Here, the key information could be displayed visually such as shown in Figure 9.5.

A chart like Figure 9.5 brings together a great deal of information. If it's accurate and updated often, many issues can be identified early. Consider the graph in the upper left as an example: if the total budget for supporting connectors is \$45k and 10 connectors are expected, the average is \$4500 per connector. The graph makes it easy to see which connectors are likely to bust the budget. Tracking customer requests over time (top right) shows which connectors are likely to be popular and lets everyone see how often that data is being collected.

Beyond specific examples, visual management brings an intangible advantage: a common view for the most important decisions that can evolve as the project proceeds. Let's say some team members see another issue they consider large enough to be critical to the decision of which connectors to support—for example, the details of some connectors might be more difficult than others to import through CAD interface tools. If the team picks too many of

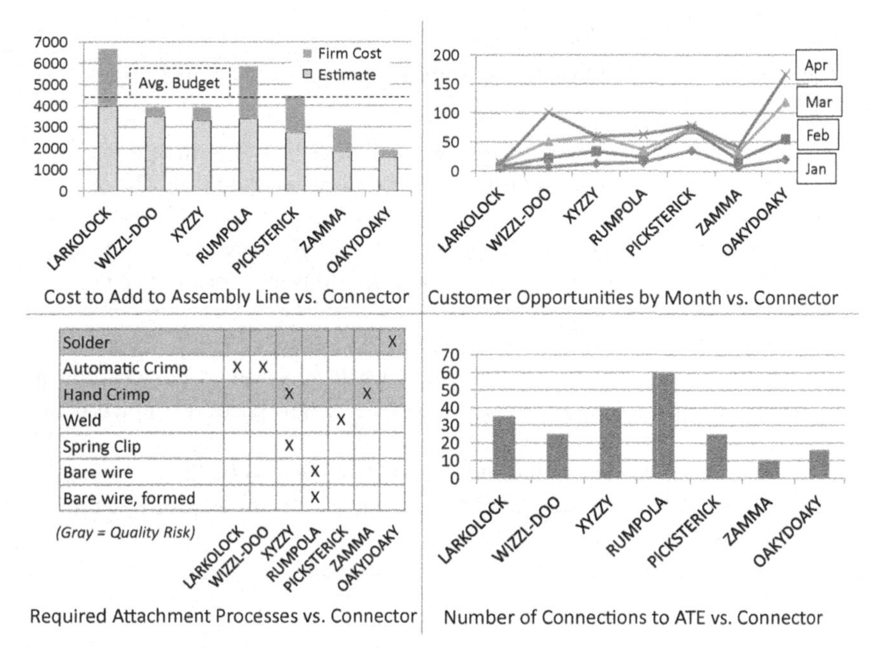

FIGURE 9.5 Visual management: example for connector types.

the "difficult" connectors, software resources might be overloaded. By making the primary factors of the most important decisions clear, visual management provides a platform for this type of discussion. And, if a convincing case is made, the chart would be modified accordingly. Visual management provides a shared view of the critical decisions and the key data that drives them so every team member understands the project's direction. This enables everyone to participate in the identification of risks and issues, increasing the likelihood that problems will be found early.

9.3.3 Plan Diligently

Plan Diligently

Project Definition	Competitors
Resources	Technology
Simplify Project	Patent Search
Value Proposition	Across the Org
Align with Sponsor	Regulatory

Diligent planning is one of the most recognizable characteristics of a skilled project manager. It is a combination of transactional and transformational leadership: requiring that processes are followed and detailed plans are documented, but also that the PM is well connected with a team that understands the vision of the product. Planning is a broad topic, but this section will address 10 areas where planning will help the team avoid risks and issues.

Clear Project Definition

The PM should ensure the project is well defined. Here are some measures of clarity:

- The goals and vision of the product should be clear and written down. The expectations for financial return should be clear as should be the path to achieve it.
- There should be an explicit value proposition. Everyone involved should know the key features of the product, why someone will want to buy it, and how the company will be able to deliver it.
- The specification, target markets, and key customers should be clear at the outset. All critical features and performance factors should be specified in measurable terms—beware of terms like "very high speed" or "much better than…"
- The schedule estimation should be clear and understood by the team and the stakeholders. The commitment of financial and human resources to the project should be written down.
- The critical risks and issues should be stated and understood by all stakeholders.

Regarding quantifying features and performance: those factors that are necessary for success are worth taking the time to quantify [10]. Even subjective performance qualities can be quantified; for example, ease-of-use can be defined by what a novice can accomplish after a specified amount of training. Tom Gilb, a well-known expert in software development, makes a compelling case that every quality can be quantified and describes a process to do it [11].

Every area of the project that is poorly defined is an opportunity for misalignment within team and between the team and the rest of the organization. At the outset of a project, the team forms a sort of contract with the sponsor and the rest of the organization; as with any contract, chances of success increase when everyone understands what they are committing to.

Resource Planning

Resource issues are among the most prevalent in projects. The team may lose a resource to another project or they may discover they are unable to complete required work due to capacity, capability, or both. When resources committed to the project change, it can create chaos. It can also be discouraging for the team, who can feel the loss of a key resource has them striving against impossible odds. When these things happen, morale will fall, and this can cause the quality and quantity of work to decline.

The RACI Chart

The project manager can protect the team and the project by working out clear agreements with the sponsor and management team as to who works on the project and with what portion of their time. The first step is to clearly define roles,

		Roles					
	Activities/Tasks	PM	Sponsor	Marketing Lead	Tech. Lead	Tech. Writer	Test Engr
RACI Chart	Functional Spec.	C	C	R A	C	I	I
	Set Price Targets	C	C	R A	I		
	Set Cost Targets	A	C	I	R		
	Estimate Financial Returns	R	A	C	C		
	Complete Risk Register	A	I	C	R		C
Allocation Chart	Person	Greg	Kim	Ethan	Lisa	Kenneth	Betty
	Request for project	30%	10%	100%	100%	25%	100%
	Allocation to project	30%	10%	50%	100%	25%	100%

FIGURE 9.6 Sample RACI and allocation chart.

for example, using a RACI chart [12,13], which is a matrix of tasks/activities versus roles. Each entry in the matrix shows how the role relates to the activity using five options:

1. **Responsible**—the role(s) performing the tasks or activities.
2. **Accountable**—the role making decisions regarding the tasks.
3. **Consulted**—the role(s) that must be part of decisions regarding the tasks.
4. **Informed**—the role(s) that must be informed about decisions regarding the tasks.
5. **Blank**—roles not substantially involved in the tasks.

A sample RACI chart is shown at the top of Figure 9.6 for a handful of tasks. Notice that only one role can be accountable. In most cases, only one role is responsible, but sometimes responsibility can be divided among two or three roles. Responsibility can be delegated, but accountability cannot. One role can be simultaneously accountable and responsible for an activity. By contrast, many roles may need to be consulted and many others may need to be informed. The RACI chart can be the same for all projects in a portfolio.

While the RACI chart makes clear how roles interact with a project, it does not formalize the allocation of people to those roles for a given project. For that, the RACI chart can be augmented with an allocation chart such as is shown on the bottom of Figure 9.6. The allocation chart then clarifies who is on the project, what their responsibilities are, and what portion of their time is committed. If gaps exist, call them out explicitly as with the dark gray for Ethan in Figure 9.6. Ensure the sponsor and the team members are in consensus and then manage people to their commitments. Ensure there is a defined process when changes are made to the project team including who is empowered to make the change, how the PM will participate in the decision, and how the effects on project goals will be reviewed.

Resource changes will come to all projects; the longer the duration of the project, the more changes the PM can expect. However, effects can be minimized by clarifying commitments at the start and managing the commitments throughout the project. It's a good idea to make the RACI/allocation chart part of the project approval.

Simplify the Project

Ensure the project is as simple as possible. Where possible, divide a 2-year project into two 1-year projects, with the first project delivering a subset of products to the market. According to the Chaos Manifesto, a report on IT projects published annually, large projects (>$10M in effort) are much more likely to fail than small projects (<$1M in labor):

> *A large project is more than 10 times more likely to fail outright, meaning it will be cancelled or will not be used because it outlived its useful life prior to implementation [8].*

That report rates complexity as the single largest factor in project success. As shown in Figure 9.7, there are many factors that increase the complexity of a project including:

- The number of markets and regions being served, especially when those markets or regions are new to the organization.
- A multisite development team, with complexity increasing with increasing number of sites, larger distances and more time zones between the sites, and the greater cultural and language differences among the sites.
- The complexity of the product being introduced including the number of features and the number of model number variations.

New Target Markets | Long Project Execution
New Technologies | Many Features
Large Manufacturing Changes | Large Number of Variants
Geographical Sales Expansion | Multi-site Development Team

FIGURE 9.7 Factors that increase project complexity.

- The amount of new technology including technology that may be known elsewhere but is new to the development team.
- The amount of operational infrastructure work to support the new products including the complexity of factory equipment and supplier tooling, and the expansion of the supply chain to new suppliers, especially those located far from the project team.

The combination of large projects and unanticipated events can force a project from well managed to chaotic (see Figure 9.8). A 12-month project carried out on one site is going to be better able to handle a series of specification changes than a 36-month project executed on three continents. And if a tool the team is using needs to be replaced with something more capable, organizing that change is easier if the team is located together. More complicated projects need more time to respond to change, which gives rise to more *work in progress*. More complicated projects also have more opportunities for unanticipated effects due to issues and risks because of factors like larger teams, larger scope, and more coordination with other groups in the organization.

Steps the PM can take to reduce the project complexity will increase the likelihood of success. This is directly related to the discussion from Section 7.3.4 recommending reducing the batch size. Examples of reducing project complexity include:

- Reducing the feature set and number of variants at launch to the minimum number that brings value to the customer. Add other variants and features in future projects. By focusing on the minimum feature set, you reduce the

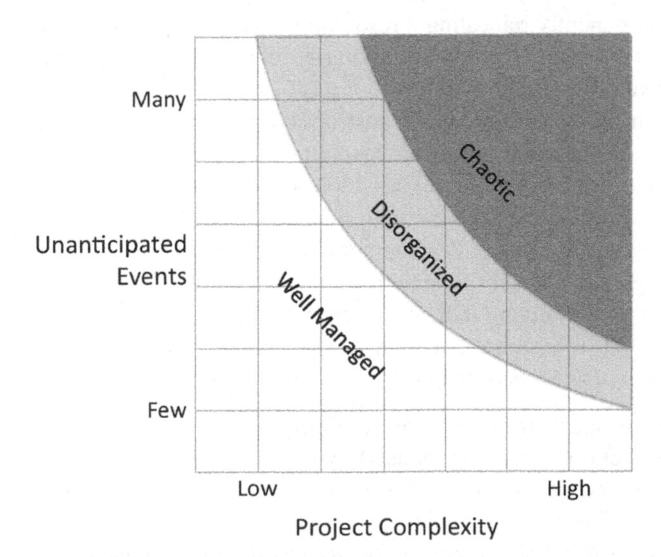

FIGURE 9.8 Unanticipated events combine with high complexity to reduce order.

"batch" size—smaller batches move through the development system faster so quality defects are discovered more quickly. The terms *quality* and *defects* are expansive, applying to any step in the project where value is added including assembly, design, coding, testing, writing the specification, commissioning, certifying, and documenting.

- Minimize technology development to what is needed for the introduction. For example, let's suppose you have a product with many new features including a technological advance that reduces assembly time by half. As is true of many steps taken to reduce cost, consider implementing this later, after volumes are significant. Normally products take time to ramp up so production costs are typically less important at launch when volumes are likely to be lower. Focus on the technology critical for launch, adding other technologies later when that's possible.
- Delay the fabrication of tooling and factory lines as long as possible. Use less automated means of production while you can. The reason is turnbacks on tooling and factory line creation can generate some of the most expensive rework cycles in a project. The longer the team can wait before committing to a design, the more time they will have to learn.
- Limit the regions of the world where the first products will be launched, staying as close to the development team as possible in the beginning. When products are first launched, they can contain defects that escape the most diligent validation processes. If the unit can be launched locally or in the same region, it simplifies the activities that can be necessary to stabilize the product. Having fewer products to support reduces the workload as does limiting issues that do occur to nearby places.
- Locate the team in as few sites as possible, especially early in the project.
  - If permanently relocating a resource is impractical, consider having team members from other sites temporarily relocate to the main project site for an extended period of time.
  - When multiple sites are the best option, minimize the interdependencies between sites such as having the entire software team at a single location.
- Minimize use of overly complex tools for the project. For example, avoid complex financial analysis methods that require prediction of revenue streams for many years. Product development is by its nature difficult to predict and investing a great deal of time in data that is unreliable is wasteful. Also, these tools are often lagging indicators—by the time they indicate a problem, it's often too late to respond to it. Focus on low-burden tools that provide a leading indication of success or failure [14].

Of course, there are limits on the ability of a PM to reduce project complexity. Project complexity is reduced by dividing a project only when the first project delivers a product that customers will value; when the smaller projects don't each deliver value, complexity is not reduced. The goal then is to reduce the complexity of a project to what the organization can manage well while still delivering value to the market.

Validate the Value Proposition with an MVP

Rapid validation of the value proposition requires the iterative release of value to representative customers. The goal is to work in fast cycles, learning the customer needs and adjusting direction accordingly. The process runs from the project start until the full product family is competitive in the market place. It's especially important to release value to some set of customers early in the project, when the ability to react is so much higher. This is lean innovation from Section 7.3.6.

Consider this example for rapid-results initiatives: RME Widgets has been producing a series of widgets that have a lead time of 4 weeks and this is recognized within RME as the longest lead time in the industry. RME Sales complains that they lose orders every week because of this disadvantage. In response, RME's development team is proposing a more flexible design together with advanced manufacturing technology that promise to reduce the lead time to 2 days. The new design will require a $750k investment: $350k in design tasks and $400k in licensing, tooling, and development to deploy the new manufacturing technology. Based on extensive interviews, surveys, and focus groups, RME Marketing is able to predict an increase in volume for their Flagship Widget of 40%, which will pay back the $750k of development costs in less than 10 months. In traditional project management, this project might be given a green flag, but lean innovation might guide the team to validate the untested assumptions in the value proposition. How could this value proposition be tested?

The first step in testing the value proposition is to state it: "our competitor's customers will value a reduction of lead time from 4 weeks to 2 days on our Flagship Widget family enough to displace our competitor at key customers." What is the weakest assumption in this value proposition? Perhaps the direct connection of reduction in lead time and a large sales increase. There is no doubt that customers value shorter lead times, but often customers have multiple reasons for turning down a product and lead time might just be the most obvious one. So, let's create an experiment that helps validate that value proposition.

First, let's choose a key customer. Suppose Global Machines is a key customer identified for this project: Marketing has listed them as one of the most likely customers to switch to RME because of the benefits of the project; sales in that territory agrees. Further, while the full Flagship Widget line comprises thousands of part numbers, suppose Global Machines uses only 75 different widgets; this makes it practical for RME to create an inventory buffer so Global Machines could experience a much shorter lead time (say, 5 days) using the current manufacturing process. Simply putting in an inventory buffer may be unacceptable for the market in general, but it might work well for this experiment because of the small number of types Global uses. So, the company can offer Global Machines a contract for 12 months with a 5-day lead time and see if that converts Global to a new customer. There are several possible outcomes, including:

1. Solid validation: RME wins the order. This part of the value proposition is validated with a key customer. Judgment is required to decide if this is sufficient or if the same experiment should be run with more customers.

2. Inconclusive: Global might say something more difficult to interpret. For example, "We will be reevaluating the design of our machines in 18 months and we will consider you at that point. However, we see the shorter lead time as a benefit and we don't know of anything else that would exclude you." In this case, RME has some positive feedback, but without a firm decision; more experimentation may be the wisest path.
3. Solid invalidation: Global's response is negative: "the lead time of 5-days is acceptable, but to win our business you will need a 15% price reduction, MED111 certification, and a manufacturing base in South America to support our factories in Brazil and Argentina. That's what your competitor offers us." So, the project as scoped will not satisfy Global Machines. From here, RME can build similar experiments and try again with other customers with the thinking that Global Machines doesn't represent the market as well as first thought. However, multiple responses like this may lead RME to rescope the project.

Experiments are likely to generate results that are somewhere along the continuum of full validation and full invalidation. As shown in Table 9.1, results must be interpreted along two dimensions: how positive the results are and how complete the list of customers is that have participated in the experiment.

What makes the RME example an experiment versus a new product project is that it's not a sustainable solution. The inventory buffer will increase production costs and so bite into margins. It may consume too much headcount to allow it being offered to the market in general. And the 5-day lead time achievable with the experiment is less valuable than the 2-day lead time promised by the full project. So, the experiment is not a substitute for the project, but rather an opportunity to learn what customers really need before making a full

TABLE 9.1 Interpreting Results of Value Proposition Experiments

	Data **Validates** Value Proposition	Data **Invalidates** Value Proposition
Larger number of customers, many key customers.	**Positive** Move to the next increment of value and continue validating.	**Negative** Value proposition probably needs to be revised.
Smaller number of customers, few key customers.	**Inconclusive** Results are optimistic, but consider more conclusive validation before proceeding.	**Inconclusive** Results cause concern, but you may experiment with more customers before deciding.

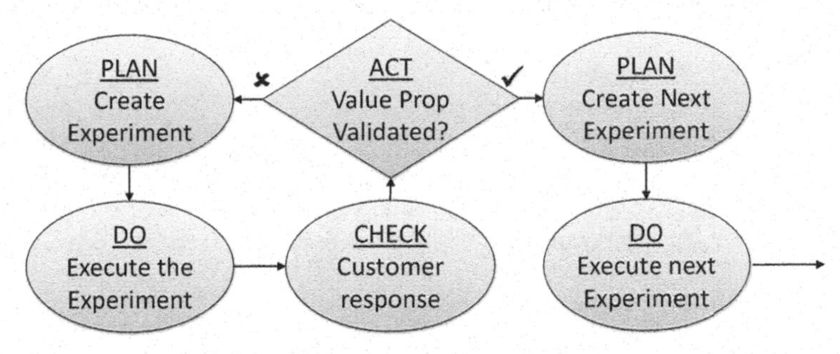

FIGURE 9.9 The Deming (PDCA) cycle applied to value experiments.

investment in the project. The goal of the experiment is to create a temporary set of conditions rapidly and at modest cost that allow a limited number of customers to make decisions that validate or invalidate the value proposition.

This approach to experimentation can continue though the life of the project. Cisco called it "rapid iterative prototyping" [15], where the increments of value "are viewed as probes—as learning experiences for subsequent steps." Iterating from experiment to experiment, each time able to deliver more value to the customer, allows experiments to more precisely validate the value proposition. The closer the launch date, the more the value proposition is validated. By launch, the untested assumptions should be nearly removed, making a successful launch likely. Compare this to the traditional project where the team works 12 or 18 months to release the product; only at launch is the value proposition tested with real decisions from customers. If the team then learns the customers do not value the product, a great deal of time and resources are likely waste.

Eric Ries calls this method of obtaining customer input "Scientific" [16]. He states "we must learn what customers really want, not what they say they want…" [17]. As shown in Figure 9.9, this approach is like the Deming cycle (refer to Figure 7.1) applied to experimenting on the value proposition.

By comparison traditional marketing asks customers hypothetical questions: if this product were available for that price, how many would you buy? The customer must imagine her state of mind at some point in the future and then imagine her decision. Experimentation, sometimes called "validated learning," places some portion of the value proposition in front of the customer and asks for a decision: will you purchase this today? According to Ries:

> *Validated learning is the process of demonstrating empirically that a team has discovered valuable truths about [an organization's][1] present and future business prospects. It is more concrete, more accurate, and faster than market forecasting or classical business planning [17].*

---

1. Ries uses the term "Startup," but means it to apply to organizations of any size: "A startup is a human institution designed to create a new product or service under conditions of extreme uncertainty." From The Lean Startup, p. 27.

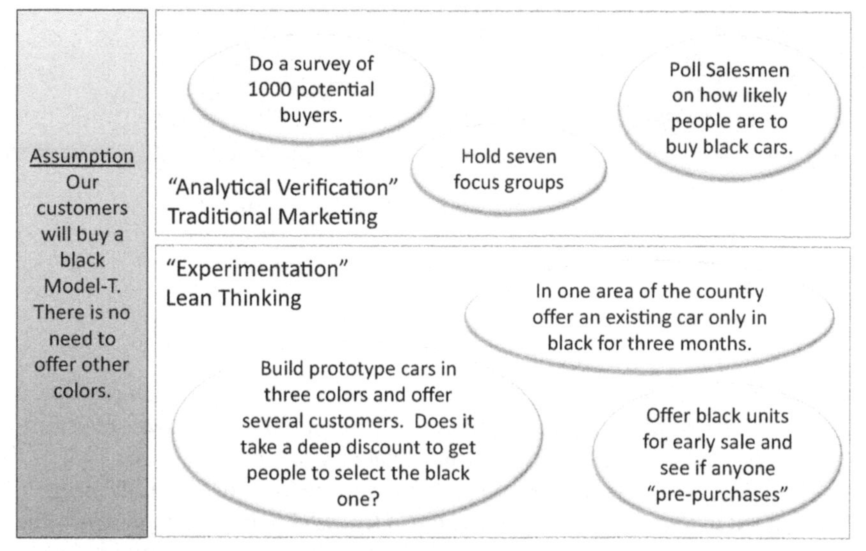

FIGURE 9.10 Comparing analytical marketing with experimentation for the Model T.

An experiment on just a portion of the value proposition reveals things that hypothetical questions about the full value proposition cannot review. The two approaches to marketing are demonstrated in Figure 9.10 with the Ford Model T, which was famously offered to customers with a limited color choice, often paraphrased as: "You can have it in any color you want, as long as it's black."

An example of the power of experimentation is how Nick Swinmurn, founder of Zappos, used an experimental approach to build the world's largest online shoe store. He saw a need for an online shoe store with a wide selection. But, rather than relying in traditional marketing and then making large investments in an infrastructure of warehouses and distributors, he used experiments to validate what customers would value in an online shoe store. He started by photographing the inventory of a local shoe store and agreeing to pay them the full price for any shoes he sold online. His experiment had him interacting with customers: transacting payments, providing support, and dealing with returns. By building a tiny business rather than asking customers what they wanted (or, more accurately, what they thought they wanted) he gathered reliable data about how customers behave—what they really want. For example, he could test how discounts affected buying behavior of online shoppers. The approach started slowly, but was ultimately successful. In 2009 Zappos was acquired by Amazon.com for $1.2 billion [18].

Minimum Viable Product

The concept of experimentation leads to a definition of the minimum viable product or MVP, the first variant of product that delivers real value to the customer. Though there are a range of definitions for MVP, it's generally thought of as the earliest product that can be used in real-world conditions. The MVP is a fast way to get reliable feedback based on customer experiences: a low-feature version of a planned website, a new temperature controller that works in the one mode the customer needs, or a remote video streaming device that works only for the one file type that a key customer needs. An MVP is more than a mock-up or wireframe model; those embodiments are meant to communicate a concept. While a mock-up is valuable earlier in the project, it is not an MVP.

An MVP is normally less than the first product to be commercially launched. The MVP is created to solve a real problem, something more than a mock-up can do. At the same time, it need only solve problems for a subset of customers that will provide the needed learning. If that subset is too small, it will not generate enough revenue to make a full commercial launch desirable. According to Ries:

> *The minimum viable product is that version of a new product which allows a team to collect the maximum amount of validated learning about customers with the least effort [19].*

So the "viable" in MVP refers to the product's ability to solve a problem, not to commercial viability. MVPs are not half-baked ideas with buggy code and unreliable hardware—they are minimal in the sense that they solve problems in a limited subspace. However, in that subspace, they solve those problems well.

Consider an example: suppose a company wants to offer a new range of Internet-connected kitchen appliances: an oven with remote temperature control and monitoring, and with integrated video webcam that includes thermal imaging of the food being cooked. In addition, the line will include a matching cooktop and refrigerator. What's the ideal MVP?

To answer this question, it's necessary to return to the concept of experimenting from above: what's the weakest part of the value proposition? It may be: customers will value the features of an oven webcam enough to pay the added price. If so, the MVP need not include the full line of appliances: the oven is probably sufficient. And since the goal is to produce the MVP rapidly, it may be that only the features to support the webcam are needed. If so, the MVP may be constructed from an existing oven, adding a webcam that may be too expensive for volume production. But this product could be built quickly and offered for sale in limited quantities. Most importantly, it can be used in real-world conditions—in a kitchen cooking meals.

Another concept that goes hand in hand with the MVP and experimentation is the capability for fast prototyping as discussed in Section 7.3.4. These include building up a supply chain of rapid prototype houses for printed circuit board assemblies, molded parts, 3D printed parts, and CNC (computer numerical control) machined parts. For software, it includes the ability to automatically build and test software quickly, and to manage the various revisions. Because the MVP may not satisfy customers on the first iteration, it's wise to be prepared to generate several variations until you find the ones your customers prefer.

The MVP then occupies an important position in the iterative sequence of releasing customer units. As shown in Figure 9.11, there is a path of iteration: first provide mock-ups, then early prototypes that demonstrate the concept, then

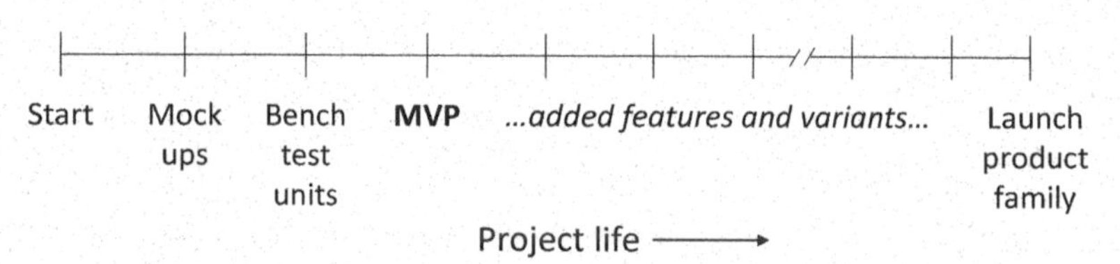

FIGURE 9.11 A series of experiments iterating from project start to launch.

the MVP, and then continue experimenting with more and more mature versions until the product family is launched. At each step, the value proposition is validated more completely so the team is continuously learning what customers really need. This learning approach protects the project from one of the greatest risks of innovative projects: the possibility the team will create a product that few customers are willing to pay for.

Align with Sponsor

The PM and project team can build a relationship with the sponsor to prepare for risks that can originate with the management team: slow decisions, reassigning team resources (or inadequate actions to replace those that are lost), and reduced funding. It's normal for a project to go through low points from time to time. These low points may result from mistakes made by team members such as designs that fail critical tests, delays that come from poor coordination, or disappointing customer response due to technical issues. Low points may also result from factors out of the team's influence such as competitors releasing new products, key customers selecting competitors for nonproduct-based reasons (for example, historical quality or delivery issues), or new technology emerging that obsoletes the product under development. Whatever the cause of that low point, the project will need support from senior management to weather it; for these cases, the relationship with the sponsor will be important.

A few steps the PM can take to create alignment with the sponsor are:

- Enthusiastically integrate the sponsor's input into the project. If the sponsor thinks a feature is important, diligently evaluate that feature. If the feedback is positive, include it. Going the sponsor's way on a close call can make sense partly because it helps maintains a strong relationship with the sponsor and partly because the sponsor's experience will often give him insight that can "break a tie vote." Of course, if the data clearly speaks against the sponsor's inputs, adoption is not called for; in such cases, be sure you've fully evaluated the data and present your findings to the sponsor. You probably want to pull him into the decision process: "That concept for using the less expensive hardware looked like a good cost reduction, but our analysis shows it will reduce response time in fault condition by 40%. Marketing thinks that's unacceptable based on XYZ Machining, Inc. input. What do you think?"

- Presell the sponsor on changes to the project. Let him see the ideas early. This has several benefits. First, you build trust by reducing exposure of your sponsor to unwanted surprises. You also benefit from his input, which can improve your presentation and allow him to prepare others in the management ahead of team.
- Make sure the sponsor is aware of bad news early. A new found source of project delay, an expensive rework cycle due to a design error, and a potential customer deciding to reject the new product are all examples of information you want to funnel to the sponsor rapidly. Again, you're building trust and you'll get the benefit of the sponsor's feedback on how to react. Giving bad news to your sponsor can be painful, but not as painful as him being surprised by the news in a large meeting or finding out through someone else.

Competitive Review

Knowing your competitors' products will help you avoid risks. First, you'll be better prepared to make daily decisions. If a part of the marketing specification turns out to be unclear, knowing how the competitor's product works will inform the situation. This doesn't suggest you should follow the competitor's lead in all cases, but knowing the competitor's product will give more data, which normally drives better decisions. Also, you'll be capable of better interpreting customer feedback. If a customer says they want "an easier user interface—we spend too much time training our people to use equipment" you'll be able to probe more deeply if you know the strong and weak points of the product they are using. And having a working knowledge of the competition will gain you more respect with the customers, your project team, and your organization as a whole. So, spend a few hours on key competitive websites, visit competitor booths at trade shows, and reverse engineer leading products in your markets. Remember, your competitor probably looked at many of the same factors you're looking at, and their interpretation of those factors drove their product features. You need not defer to your competitor at every turn; just recognize that successful companies make a lot of good decisions—it is prudent you understand how they navigated as your company charts its course through similar waters.

Technology Evaluation

Another important step to prepare for risks and issues is to evaluate the new technology your team is using. Discovering in the middle of the project that technology is not ready for use in a new product can cause long and expensive turnbacks. A few questions to ask about any technology you are using are:

- Has it been thoroughly validated in similar conditions to what is planned for this product? Think about the many factors that make up the environment: physical environment (temperature, vibration, contaminants, pressure), product environment (hardware, software architecture), and user environment (technical competence, culture).

- Is the team competent in the technology? If not, can you obtain training or are they going to "learn as they go"?
- Is any new manufacturing technology needed to support the product? If so, is the responsible team competent in this area?
- Are others using this or similar technology successfully in your market?

If these or similar factors bring concern, it's probably time to accelerate testing on the many facets of a new technology. You'll also want to ensure an appropriately diverse group reviews these issues—new technology brings risks and issues from many quarters, so you'll need a strong cross-functional team reviewing the results.

Patent Search

Patents can bring large issues after the product is released, so ensure the team is diligently searching the literature, especially issued patents and published patent applications. We'll talk about the benefits and risks patents bring in Chapter 10.

Regulatory Issues

Thoroughly evaluate regulatory requirements at the start of the project. Understand what regulatory requirements the competitors are meeting; if your project isn't planning to meet all of them, be sure there's clear reasoning. Also be certain to understand any other regulatory requirements customers might value. Finding out about new regulatory requirements late in a project can generate a range of new issues that can be painful to deal with. It's usually wise to meet with regulatory inspectors early, especially concerning new designs and components.

Across the Organization

Finally, be prepared for issues that come from other parts of the organization. The project team will depend on many parts of the organization: internal laboratories, manufacturing, sourcing, sales, finance, and senior management to name a few. The best way to ensure coordination is to build a cross-functional team with representatives from areas outside product development. However, this approach works best only when there's a substantial amount of work for these team members. If cooperation is only occasional (for example, with a test lab or finance), having someone on the team may not be practical. Whatever the case, as PM, build good relationships with all the groups you'll be working with. Establish trust by being transparent and pulling affected people into issues early—don't surprise your colleagues in manufacturing by presenting a goof-up on their part at a key meeting. Instead, pull them in early and work with them to reduce the impact before the big meeting. Building relationships throughout your organization will prepare you for risks and issues: a cooperative spirit will help you learn of looming issues earlier and if you help others when they have a problem, you're a lot more likely to get help when you have a problem.

9.4 RESPONDING TO RISK: LEADING THE TEAM THROUGH THE UNEXPECTED

Even after diligent preparation, a number of things will likely go wrong in a product development project of any complexity. This is depicted in Figure 9.4. On the left side of the scale are the preparation steps, which were discussed in Section 9.3. But no amount of planning can make the items on the right side of the scale entirely predictable at the start of a project; in fact, a few can hardly be predicted at all (for example, "New patents awarded"). That does not discount the need for thorough preparation; preparation reduces the magnitude and severity of the issues that will come into the project. But for those issues that do appear during the project, the PM will need to lead the team to resolve them. The requirement to deal with dynamic issues is a difficult role of the PM because of the versatility of skills required:

- Leading the team to devise creative solutions to newly discovered technical and commercial issues while the project proceeds at full speed. This is sometimes called "working on the plane while it's flying."
- Managing the expectations of management as the exposure to newly identified risks comes to light.
- Maintaining a team spirit when stress can lead to blaming and strife.
- Sustaining the vision of the project in the presence of doubts from some of the stakeholders.
- Keeping a sense of order in the face of change driven by factors outside the team's control.

At a low point in a project, things will usually seem worse than they are. It is here that strong PMs are the central stabilizing force. When the primary customer backs out of a purchase order or when a major certification test fails, spirits are likely to be low. Anyone who's familiar with competitive sports knows that when the team gets behind, leadership is critical in maintaining focus and morale. With hard work, focus, and a little luck, success is still probably attainable. The PM can make the difference with well-developed transactional and transformation leadership skills.

The process of dealing with issues and risks is depicted in Figure 9.12 as having four steps:

1. Identify previously unknown issues throughout the project.
2. Track issues using a risk and issues funnel.
3. React to issues when necessary.
4. Manage the issues and report out on progress and exposure.

9.4.1 Step 1: Identify Issues throughout the Project

The first step in dealing with new issues is creating a means to identify them. This typically has two facets that complement each other: formal risk identification

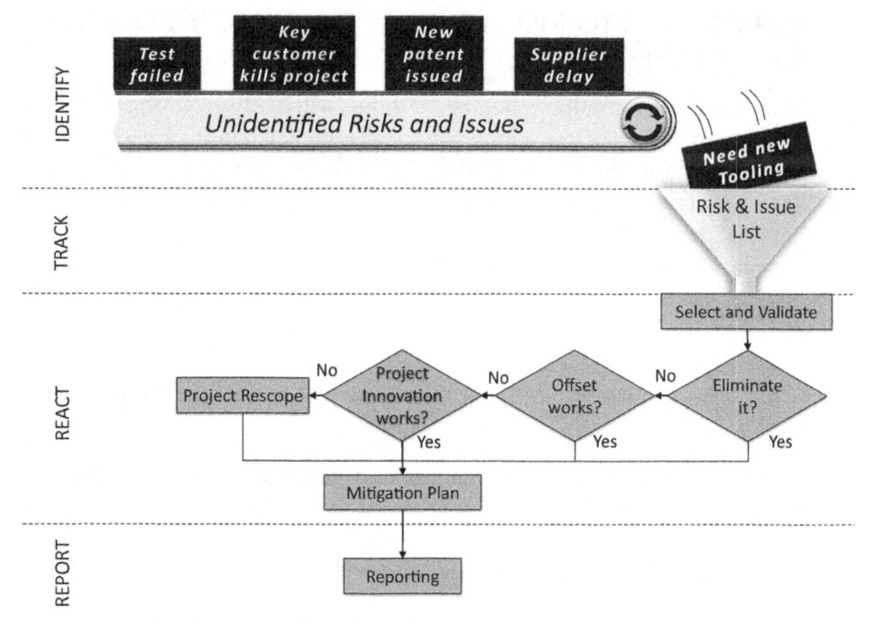

FIGURE 9.12 Identifying, tracking, and reacting to risks and issues.

events, which are typically held early in the project, and the ability to leverage other project activities to identify risks throughout the life of the project.

Risk Identification Events

Risk identification events have been a staple of project management for decades. Typically one is held early in the project with the full development team, any interested stakeholders, and other subject-matter experts. A cross-functional approach is required because different team members will have different perspectives on risk. Perhaps only one person at the table will accurately foresee the full exposure of a given risk—having the team assembled allows that person to present her reasoning and gain team consensus on the topic. Because risks come from so many quarters (internal vs external, technical vs commercial, design vs manufacturing, etc.) a varied set of skills is needed for a thorough evaluation [20].

There are several types of risk events including:

Work through a List

A facilitator walks the group through a battery of known risks and evaluates the exposure of the project to each. This is a transactional approach: if we walk through a complete list of known risks, we'll identify most risks. The battery typically looks something like the list of about 60 risk categories presented in Section 9.2. Each risk is evaluated first by having an open discussion so those most knowledgeable in the area can add information that will help the group. For example, the discussion may proceed along these lines:

Facilitator: Risk 23: Product specification vague or incomplete. Does anyone have knowledge regarding the completeness or clarity of the current specification?

Ethan (engineer): One of the most difficult technical areas in the project is creating an optical system that can stand up to the temperatures seen in the oven. The cameras that look most promising have relatively low resolution. Is that acceptable? There's nothing in the specification regarding this.

Tucker (engineer): We have the same problem with thermal imaging. The resolution of these sensors is low and it will be hard to tell the food from metal pans in small dishes.

Rachel (marketing): This is a problem. The feature is new so we don't have any history. We're going to need a prototype to validate this with customers.

And so the conversation goes until everyone has an opportunity to speak to the issue. At that point, the facilitator polls the team to determine if the risk needs to be resolved. Those issues will be added to the risk and issue funnel and tracked as discussed in the next section.

Freeform Discussion

A competing approach is to hold a freeform discussion. It's more transformational in that it relies on the vision for the product and connecting with the individuals. It might begin with a project review, discussing the project value proposition, or walking through the known risks such as new technologies and commercial issues. The conversation then proceeds with the facilitator asking people to describe the most important risks they see:

Facilitator: What do you think is the largest risk this project faces?

Rachel (marketing): We haven't clearly worked out how the oven will improve the way the food is cooked. Are the optical and thermal images of the food going to help people cook meals better?

Tucker (engineer): I agree, but to know that, we need to know what resolution the thermal image will improve cooking. It's hard to proceed without a specification for resolution—that drives everything else in the camera: cost, ruggedness, size, and sensitivity to heat.

Rachel (marketing): It seems like we have a gap in the specification: we don't know what resolution is needed until we can run tests with consumers and we can't run tests without a camera in place.

The facilitator keeps the conversation going until enough information has been gathered to define the risk. Team consensus should play an important role in selecting issues to resolve.

Combining the Two Approaches

As with other facets of project management, transactional and transformation leadership are complementary, each bringing advantages and disadvantages as shown in Table 9.2. Picking the right approach for a project depends on the level of risk in the project, the company culture, and the time that can be made available for a risk-identification event. Some companies allocate a couple of hours for just the core team; others pull a large team together for days. The right choice for a given project may be one, the other, or a combination of the two. A few examples for combining the two methods include:

- Before the freeform discussion, walk through the various categories of risk. The facilitator may also bring a list of risks similar to what's in Section 9.2, perhaps giving a copy to each person or posting them on the meeting room walls. As different risks and issues are identified, the list can be checked so the team sees which areas have received focus and which might have been overlooked. For example, a technical team is likely to focus on technical risks—this method can show the team they haven't identified any significant commercial risks.

TABLE 9.2 Comparing Risk Identification Approaches

	Advantage	Disadvantage
List based	• The methodical process covers a wide area. Less dependence on the team assembled. • Process can be built up over time to capture issues experienced in the past. • Easier to facilitate.	• Can be a heavy process, requiring many hours. Energy may be high at the start, but difficult to maintain through the process. • Requires so much effort, the project team may avoid the event all together.
Freeform	• The largest known issues are discussed allowing the team to come to consensus in these areas. • People get to discuss the issues that concern them the most, so the meeting stays more interesting. • Meetings are shorter since the team covers only the main risks.	• Tends to bring up issues that the team already knew about. • Can easily miss risks and issues the team is not already focused on. • Requires a more highly skilled facilitator.

- Start the event with a freeform discussion that is limited in time—say 1 h—and then move to the list, skipping past those areas already covered. This gets the risks that are most on the team members' minds quickly identified so they can concentrate more on areas of the list they might otherwise give less attention to.

SWOT Matrix

The strengths-weakness-opportunity-threat (SWOT) matrix is a simple analysis that can be used to identify risks and issues. This method creates a two-by-two table that allows the team to consider competitors and their products as they identify risks. The strengths and weaknesses compare your company to the competitor; the opportunities and threats compare external factors such as the product under development and competitive products. Typically a separate SWOT matrix is done for each competitor. An example of a SWOT matrix is shown in Table 9.3.

The SWOT method was created in the 1960s as a tool to analyze the case for new businesses and is often used as part of business plans. SWOT is also applicable to projects and project portfolios [21]. Aside from being simple to understand, SWOT analysis also brings the benefit of evaluating organizational and product issues at the same level; this helps overcome the tendency for product

TABLE 9.3 Example of SWOT Analysis

	Favorable Factors	Unfavorable Factors
Internal factors	**Strengths**	**Weaknesses**
	• Strong commercial presence in target market. • Infrastructure in place to identify and protect intellectual property. • Have the resources to complete a project of this scale.	• Higher cost of manufacturing than most competitors. • Minimal history of developing web applications. • Historically late on projects; fast delivery is critical here.
External factors	**Opportunities**	**Threats**
	• Validated need for people to be able to monitor and/or control cooking remotely (for example, stepping out to grocery store or visiting neighbor). • No identified competition to thermal imaging in commercial/consumer space.	• Competitor Loki and Zeus White Goods has low-cost camera mounted outside oven door. • Identified patent from Rigel-Betelgeuse Appliances close to our technology. • Question our ability to find supplier for optical system capable of functioning in high temperature.

development teams to focus too little on nonproduct issues. It should be pointed out that while SWOT is popular, some authors have expressed doubts about its effectiveness [22]. The SWOT method is probably most valuable to augment a more detailed risk identification program.

Premortem

The premortem is a modification of the freeform technique discussed above. It overcomes one of the primary problems with risk-identification meetings: the reluctance of people to express reservations for fear of being seen as overly negative. Most people feel free to bring up one or two unidentified risks, but at some point, their enthusiastic identification of risks can start to sound like they don't fully support the project. That can kill the creativity that might identify a risk that could cause harm in the future. Enter the "premortem," a risk identification meeting that begins by postulating the project fails and then asks the team to imagine why. The taboo of mentioning possible failure is eliminated with the starting assumption. Instead of feeling like naysayers, meeting participants who identify risks feel valued for their intelligence, creativity, and experience.

The premortem is based on the principle of "prospective hindsight." According to Gary Klein, "Research conducted in 1989…found that prospective hindsight—imagining that an event has already occurred—increases the ability to correctly identify reasons for future outcomes by 30%" [23]. Klein gives several examples where the premortem has been successful including a case where a team member, who had been quiet during a lengthy kick-off meeting, voiced a serious concern for the first time at the premortem: the software being developed in the project wouldn't fit on a laptop and that would cause serious problems for the application. It turned out the team had contemplated a short-cut that was pivotal in allowing the software to fit on a laptop; the PM immediately included that in the plan. Had it not been for the premortem, that problem might have been discovered much later in the project where it would have been much more difficult to correct.

A premortem, the temporally displaced cousin of the postmortem (an investigation into why a person has died), determines why a project failed before the event. The process is:

- Pull together the full project team and any technical or commercial experts that can aid in risk identification.
- The facilitator begins the meeting with: "Assume this project fails. Not just that it fails, but it fails spectacularly. Imagine what risks we can see today that might have been responsible for that failure."
- Each person silently writes down the reason(s) for the terrible outcome.
- The facilitator then starts a discussion with each person reading one reason from their list.
- The risks that cannot be resolved in the meeting are then tracked throughout the project.

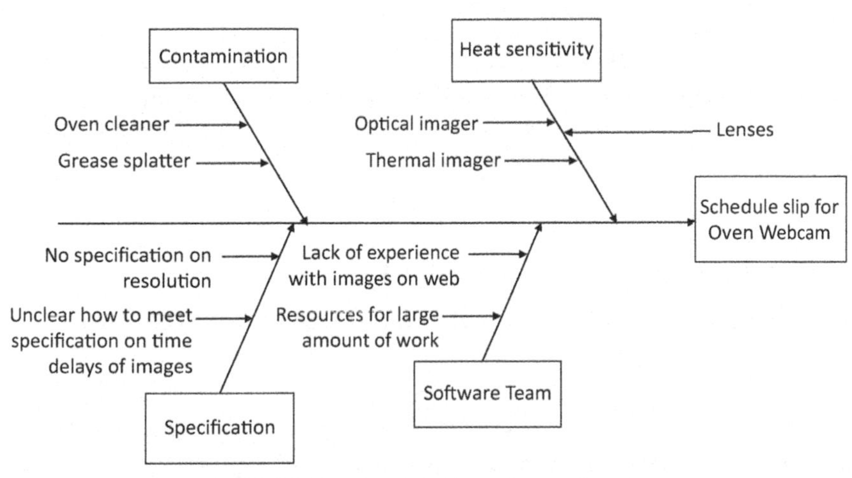

FIGURE 9.13 Sample fishbone diagram for risks.

Fishbone Diagram

The *fishbone diagram* can be used to bring order to the many items that can be discussed in a risk-identification meeting. As each risk is defined, it can be categorized so that similar risks are grouped together. The fishbone diagram of Figure 9.13 is a simple example (fishbone diagrams can become quite large). The benefit of the fishbone is the order it brings to a large set of unorganized risks. It also reveals where the team is focusing their energy—the fishbone might show, for example, that no commercial risks have been considered (as is the case with Figure 9.13).

Leveraging Other Events to Identify Risks throughout the Project

While risk-identification events are a smart way to begin a project, risks and issues come to light in many other functions of a project. Events where the design is critiqued such as critical design reviews and failure mode and effects analysis (FMEAs) often bring new risks and issues to light. Similarly, new risks may be identified at regular team meetings, customer and supplier visits, trips to trade fairs, and even hallway conversations. Risks can also come into focus when people are working alone, for example, reviewing competitive literature, studying patents or academic papers, reading component specifications, interpreting test results, or just working through a problem.

Delay sometimes reveals risks and issues. This delay can be detected with progress tracking such as the fever chart of Figure 5.17 or by missed milestones. Sometimes risks come to light through the four early warning signs of Section 5.3.6: slower-than-expected progress, low quality work product, signs of faltering engagement, or concerns from others.

In all these cases, transactional leadership is important—develop and follow process to identify and record risks and issues. Then thoroughly track the root cause of all known failures whether at a customer site or from internal testing. But transformational leadership is also important: stay connected with the team, encourage open and honest discussion, and reward those who discover serious risks.

9.4.2 Step 2: Risks and Issues Funnel

The second step is creating a method to collect identified risks and issues in a funnel. The following questions should be answered:

- What is a description of the risk/issue?
- How likely is the risk to occur?
- What is the likely severity of the risk or issue?
- What is the total exposure (often a function of likelihood and severity)?
- Who owns the management of this risk/issue?
- What is the mitigation plan?
- When will the mitigation plan be complete?

An example risk and issue funnel is shown in Table 9.4. Maintaining a risk list requires strong transactional leadership: updating, reviewing, and following up.

The likelihood and severity scales are somewhat arbitrary. The example in Table 9.4 uses a 1–10 scale so the worst exposure of a risk is 100. Sometimes the likelihood scale is defined numerically (1 = 1 chance in 100,000, 2 = 1 chance in 25,000, 3 = 1 chance in 10,000, …10 = 1 chance in 1). Other times they are defined subjectively (10 = certain, 9 = very high probability, …5 = moderate chance, …1 = almost no chance). Similarly the severity scale can be objective (10 = will cause hazard without warning, 9 = will cause hazard after warning, …1 = no effect); it can also be subjective (10 = most serious, 1 = least serious). If your company doesn't have such a process, you may choose to follow the FMEA scales for severity and likelihood, which are published widely [24]. However, the ranking of severity and likelihood of imagined risks is usually subjective, so don't spend excessive effort creating precise rating scales.

Focus can be brought to the risk tracking list by sorting the risks so those that need immediate attention rise to the top of the list. For projects of low and medium complexity, it may be suitable to simply sort by exposure. When a risk has been mitigated, its likelihood and severity should drop so the exposure will be much smaller causing that risk to fall to the bottom. Automated schemes like these work well for projects where the number of risks active at any one time is small—say 25 or 50—so the sorting mechanism need not be highly flexible.

For more complex projects where the risks may number over 100, the sorting mechanism needs to be more sophisticated, separating between importance of the item (which is represented by the *exposure*) and urgency—the need to act now versus later. For example, early in a project there may be several risks

TABLE 9.4 Sample Risk/Issues Funnel

Risk	Likelihood	Severity	Exposure	Owner	Status	Mitigation Plan	Date Due
Oven cleaner may damage lens.	8	9	72	Michele	In progress	Work with lens supplier to provide harder coatings.	14-Feb
Thermal imager maximum temperature may be exceeded.	5	10	50	Mercedes	In progress	Mock up oven with nonfunctional thermal sensor and measure max temp.	3-May
Grease spatter will cloud lens.	6	8	48	Olivia	Monitor	Develop screen to block grease spatter.	15-Oct
Team lacks experience to transfer web-based images rapidly.	7	6	42	Henry	Monitor	Select training course for software team in web-based communication.	14-Nov
Lack resources to build prototypes for environmental tests.	1	2	2	Tucker	Resolved	Fully mitigated.	–

related to production equipment that the team decides to delay addressing until risks related to the value proposition are resolved. Simply sorting by exposure may push the value proposition risks too far down the list. Good visual management requires the chart to tell the story in seconds; that isn't going to be met if the team must scroll past a dozen risks to get to the ones that need immediate attention. Other automated schemes can be investigated. For example, adding an urgency column and adding an algorithm to combine that column with exposure. Another option is to use a manual ranking column where the team picks the top five or 10 risks that need to be addressed now and then sorting them based on that rating. There are any number of schemes that will work; the PM should ensure the risks are sorted in a manner that is intuitive (that is, the team understands how the risks are sorted), easy to update, and reliably moves the risks that need attention to the top of the list.

There's no need to develop a new risk tracking tool (often called a *risk register*). There are hundreds of templates available. An Internet search for "risk register" will show a long list. A large portion of those are Excel spreadsheets or Excel add-ins and many are free or come at nominal prices. If your company doesn't have a risk tracking form, you're likely to find what you need in a short time.

9.4.3 Step 3: React to the Tracked Risks and Issues

Now that you have built a system to identify and track, it's time to decide how to react to these risks and issues in order to mitigate the effects. There are several mitigation strategies to work through (refer to the "React" section of Figure 9.12).

Selecting Risks and Issues to Address

The first step is to select which of those risks need a reaction. If a risk has a sufficiently small likelihood of occurrence or its severity is modest, the team may elect to track it for some time without taking any action. And the team may lack the resources to react to all important risks simultaneously. So, there needs to be some mechanism to decide which risks will be actively addressed and which will wait until resources are available. For example, the team may review the risk list weekly and use team consensus to decide which ones to actively address.

Reacting to selected risks and issues will follow a flow such as:

- Evaluating the risk/issue and developing alternatives.
- Selecting the best alternative(s) to use for a mitigation plan.
- Reviewing the mitigation plan for approval, especially if the plan will negatively affect any of the project goals.
- Managing the mitigation plan to completion.

A simple way to manage a risk/issue list is to use a Kanban board as shown in Figure 9.14. As discussed in Chapter 7, Kanban boards provide strong visual management for issues that are largely unrelated, which is usually the case for

Selected Issues	Developing Alternatives	Create Mitigation Plan	Review for Approval	Managing Mitigation
No sales team in Argentina	Need Quality Engineer on team	Bearing life below specification	Product standard margin below 50%	Find new supplier for battery
Enclosure Patent issued to ARealPain, Inc.	Life test failure	Sandra leaving team to lead Project Zonkdoo		Rework wiring to pass safety certification
Supplier for optical tubes	Faster widget released from Kantstandthem			Training team for newest release of Java

Flow

FIGURE 9.14 Kanban management board for active risks and issues.

risks and issues. Alternatively, the PM could merge the mitigation plans into the project plan, and manage the project out of a single task list.

Mitigation plans will generally fall into one of the following categories.

Mitigating by Elimination/Avoidance

The ideal solution is eliminating or avoiding the issue. For example, if a risk is presented that a certain component may not be able to stand up to an environmental specification like temperature or vibration, the team may be able to find an equivalent component that can. If the new component brings no new issues (for example, higher cost), that risk can be fully resolved without affecting project goals.

Mitigating by Offsetting Elsewhere

The second approach is to reduce the effect of the risk by offsetting it elsewhere in the project. For example, suppose the team is licensing a real-time operating system (RTOS) that was selected because it has the lowest latency in responding to events. An issue is identified that the supplier will increase prices 10% this year and that increase pushes the product cost over the project goal. First, the team tries to eliminate the issue—searching for a lower cost RTOS or reviewing the specification to ensure the low latency is really required. If neither alternative bears fruit, they can offset the issue by searching for cost reductions elsewhere in the product. Perhaps the product enclosure can be simplified enough to reduce the costs to offset the RTOS cost increase. An example in the commercial domain is if a certain group of countries is excluded from the market because of regulatory issues, the team can add other countries or expand the opportunity in existing countries that do not have the regulatory issue. Risks in revenue projections, product cost, project investment, and schedule delays can often be offset by looking elsewhere in the product or project. Quality and reliability issues can be difficult to offset.

Mitigating through Project Innovation

Recall the term *project innovation* from Section 3.1.2. It is a class of actions that significantly reorganizes the project in reaction to new information with no negative affect on project goals: moving team members to different tasks, adding or removing tasks, and changing the critical path. Project innovation can be an effective means of dealing with some risks, especially schedule delay. For example, let's assume the team discovers the controller for a radio controlled toy car occasionally becomes inoperable. Once every few hours, the toy stops and it must be power cycled to restart. Such a flaw would greatly reduce the attractiveness of the toy. Let's say this project has a software team of three, all of whom are working on completing various tests for the launch, which is in 12 weeks. This is a case where project innovation might be used: pull the full software team onto this issue, with the goal being to resolve the issue fast enough—hopefully within a few days—so that when the team returns to their original tasks, they can finish fast enough so that launch remains on schedule. Project innovation seeks to maintain project goals, so the team is generally free to use the technique without having to seek approval.

Mitigation through Project Rescoping

Often in a project, risks and issues are severe enough that even the best mitigation plan that the team creates will compromise some of the project goals. In such a case, the project may need to be rescoped. Rescoping includes such actions as:

- Delaying launch to allow more time for product development.
- Increasing budget, for example, to hire a consultant or rework capital equipment.
- Reducing scope such as when variants are removed, features are reduced, or lower-than-specified performance is accepted.
- Reducing revenue or margin projections and so reducing the ability of the project to meet financial goals such as payback, NPV, or internal rate of return.

As an example, consider the optical/thermal imaging cameras in our hypothetical smart oven project. Suppose the bulkhead that was planned for isolating the thermal imager from high temperature was constructed from a new material developed by a small company. While initial samples delivered the needed characteristics, the company was never able to manufacture the material in production quantities. This leaves the team needing to find an alternative way to isolate the thermal sensor. Of course, the team will start with simpler steps such as finding a substitute material or relocating the sensor to a lower-temperature area. But suppose no workable alternative is identified. In such a case, it may be necessary to remove the thermal sensor, which would require rescoping the project: modifying the specification, reviewing and probably reducing revenue projections, resetting the price point, and perhaps reducing margins.

Having to rescope the project in response to an issue is, of course, one of the least desirable outcomes. Sometimes, the team can get lucky, especially if the need to rescope is discovered early in the project. Project management literature is full of stories where Plan B eventually turned out to be stronger than Plan A—sometimes much stronger: an early discovery of a flaw in the value proposition leads the team to a stronger value proposition or discovering a technology lynchpin that won't meet customer needs leads the team to replace the technology with one that turns out to meet needs better. But the later the rescoping comes in the project, the less likely a positive outcome. If $800k is spent in a $1M project before the team finds a show-stopping issue, graceful recovery is unlikely. At the very least there will be disappointment from senior management. In the most severe cases, the project may be terminated.

When you read about projects that experience failure, there is often a sense of predictability that comes only when looking into a rear view mirror. You rarely read a story about the PM who did everything well but the project eventually failed anyway; it just doesn't make good copy. But project management is a nondeterministic endeavor. The fact is a PM can do an excellent job in every phase of a project and the project can still fail. A devastating patent can be issued to a competitor. A critical customer can have an unexpected leadership change and cancel your contract. A competitor can launch a revolutionary product that obsoletes your project overnight. We accept this in other parts of life: a person can take care of their health—eating well and exercising for decades—only to be fatally struck by a drunk driver. If you are a PM of a project that failed, you may find some people that will blame you. But don't rush to blame yourself. Of course, do learn everything you can from the event so you'll be all the better managing your next project.

9.4.4 Step 4: Reporting

The final steps in dealing with risks and issues during a project are managing the mitigation plans and reporting out. When reporting, you'll need to balance the need to be open against the need to maintain order.

Being Open

Being open implies that your reporting is clear, accurate, and available to stakeholders as appropriate. Never hide bad news once that news is known with enough certainty that stakeholders would expect to be made aware of it. If a schedule delay is virtually certain, announce it. If product cost has risen enough to injure the margins, announce that too. The same is true if a patent has come to light that the team has studied and believes will affect the project. Any issue that is known with reasonable certainty to affect project goals should be reported. But don't simply report a problem: report one or more alternatives in dealing with it. The management of your company will want to know things like:

- In laymen's terms, what happened?
- Are you sure you understand the root cause?
- What is the likely effect (or range of effects) on project goals?

- Why did this catch us by surprise?
- What are your recommendations to mitigate this?
- How are you planning to manage the mitigation plan?
- Do you need help in the form of more people or budget?
- What processes should we look at so this doesn't happen again?
- When will we know we're on the right track? What evidence are we looking for to tell us that?
- When do you estimate this will be resolved based on what you know today?

Three common mistakes PMs make when reporting bad news should be avoided:

1. Don't go into excessive technical detail. Explain the problem at a high level focusing on how project goals are affected. If managers want more technical detail, they will ask for it.
2. Don't be defensive. It's easy to look in the rear view mirror and imagine how all this could have been predicted. Report with a mindset of what was known before the issue became evident or what could have been known. Blame-free phrases can be helpful such as: "the team had no evidence to suggest we should have expected this" or "our processes are not built for this since it hasn't affected projects in the past" (assuming, of course, that these things are true). Protect the team and your own reputation: if the work was done well but something largely unpredictable occurred, represent the situation accurately.
3. Don't tell a mystery story. Get to the point, quickly giving an overview of the problem and focusing first on what most managers want to know: how will this affect the goals of the project, especially the financial goals? Avoid the temptation to start with the twists and turns the investigation has taken. If managers want to know these details, they will ask for them.

Maintaining Order

While you want to be open, the PM must also maintain order in the project. Two ways to bring chaos are reporting information before you are ready and raising an inappropriate level of alarm. Being open doesn't mean you need to report every potential troubling piece of news as it occurs. Patents from competitors are notorious at the first reading for seeming like they bring worse news than they really do. So, take time for the team to review the information so you're not raising the red flag only to learn the next day that the team chemist now thinks the problem was the test results were recording inaccurately. So be open, but do it with reliable data.

The second issue is maintaining the appropriate level of alarm. Issues usually have a range of outcomes with the worst cases usually being extremely unlikely. Report the expected outcomes with the most likely outcome getting the most emphasis. So, if the first round of certification fails, it may be possible that a major redesign will be necessary, but if the likely outcome is that moderate redesign will probably let the product squeak by, lead with that. Use your judgment when reporting the worst case and be clear if the likelihood is low.

9.5 RECOMMENDED READING

Gilb T. Adding stakeholder metrics to agile projects. Cutter IT J, 2004;17(7). Available at: www.cutter.com.

Loch CH, De Meyer A, Pich MT. Managing the unknown. A new approach to managing high uncertainty and risk in projects. John Wiley and Sons; 2006. p. 71.

Matta NF, Ashkenas RN. Why good projects fail anyway. HBR The Magazine. Available at: http://hbr.org/2003/09/why-good-projects-fail-anyway/ar/1; September 2003.

Ries E, The lean startup: how today's entrepreneurs use continuous innovation to create radically successful businesses. Doubleday Religious Publishing.

9.6 REFERENCES

[1] Post-it (R) note notes: here's how it all Began, Postit Website, http://www.post-it.com/wps/portal/3M/en_US/PostItNA/Home/Support/About/.

[2] Dinsmore PC, Cabanis-Brewin J. The AMA handbook of project management. Amacon Books; 2011. p. 9.

[3] Phillips JJ, Bothell TW, Lynne Snead G. The project management scorecard: measuring the success of project management solutions. Butterworth Heinemann; 2001. p. 6.

[4] Loch CH, De Meyer A, Pich MT. Managing the unknown: a new approach to managing high uncertainty and risk in projects. John Wiley and Sons; 2006. p. 71.

[5] Loch CH, De Meyer A, Pich MT. Managing the unknown: a new approach to managing high uncertainty and risk in projects. John Wiley and Sons; 2006. p. 30.

[6] Dinsmore PC, Cabanis-Brewin J. The AMA handbook of project management. Amacon Books; 2011. p. 196.

[7] Matta NF, Ashkenas RN. Why good projects fail anyway. HBR The Magazine; September 2003. Available at: http://hbr.org/2003/09/why-good-projects-fail-anyway/ar/1.

[8] The Chaos Manifesto. Think big, act small. The Standish Group International, Incorporated; 2013. Currently available at: http://www.versionone.com/assets/img/files/CHAOSManifesto2013.pdf.

[9] ibid.

[10] Gilb T. Adding stakeholder metrics to agile projects. Cutter IT J 2004;17(7). Available at: www.cutter.com.

[11] Gilb T. Quantify the un-quantifiable: Tom Gilb at TEDxTrondheim. TEDx Talks, Publ; November 3, 2013. Available at: https://www.youtube.com/watch?v=kOfK6rSLVTA.

[12] Raci charts—how-to guide and templates, http://racichart.org/.

[13] RACI explained it's simple, yet powerful, Raci.com. Available at: https://www.youtube.com/watch?v=1U2gngDxFkc; October 12, 2013.

[14] Managing projects large and small: the fundamental skills to deliver on budget and on time. Harvard Business School Publishing Corporation; 2004. p. 117.

[15] Harvard Business School. 2004. p. 117.

[16] Ries E. The lean startup: how Today's entrepreneurs use continuous innovation to create radically successful businesses. Doubleday Religious Publishing; 2011. p. 9.

[17] Ries E. The lean startup: how Today's entrepreneurs use continuous innovation to create radically successful businesses. Doubleday Religious Publishing; 2011. p. 38.

[18] Ries E. The lean startup: how Today's entrepreneurs use continuous innovation to create radically successful businesses. Doubleday Religious Publishing; 2011. p. 58.

[19] Ries E. The lean startup: how Today's entrepreneurs use continuous innovation to create radically successful businesses. Doubleday Religious Publishing; 2011. p. 77.

[20] Harvard Business School. 2004. p. 104.

[21] Maculley JR. A closer look. Project Management Institute; July 1, 2003. Available at: http://www.pmi.org/learning/swot-analysis-electric-power-system-3349.
[22] Hill T, Westbrook R. SWOT analysis: it's time for a product recall. Long Range Plann Int J Strategic Manage 1997;30(1):46–52. Available at: http://www.sciencedirect.com/science/journal/00246301/30/1.
[23] Klein G. Performing a project premotem. Harvard Bus. Review, The Magazine; September 2007. Available at: http://hbr.org/2007/09/performing-a-project-premortem.
[24] Quality 101: demystifying design FMEAsQuality Magazine; March 1, 2005. Available at: http://www.qualitymag.com/articles/84015-quality-101-demystifying-design-fmeas.

Chapter 10

Patents for Project Managers

Warning

Always consult with licensed legal representation before taking or foregoing actions related patent law.

This chapter introduces project managers (PMs) to some basic principles of patent law. It is based on the author's experiences with patents over the past 25 years. Readers should be aware that:

1. The author is not licensed to practice law.
2. Patent law is complex and general principles are difficult to apply to specific circumstances.
3. Patent laws vary between countries and over time.

Do not use the information contained here to make legal judgments.

This chapter will discuss patent law as it relates to project management for product development. When developing products, there is a significant likelihood that your team will at some point come into contact with patent law, either through a desire to obtain a patent for your company or through concern that your new product might infringe patents from another company. That likelihood increases for more innovative products. Unfortunately, patents are often misunderstood by development teams. That's understandable because our education system doesn't usually teach future developers much about patents. As a PM, you are in a unique position to help the company accomplish its goals related to patents, which are usually that important inventions get patented and that future patent infringement cases are avoided.

10.1 INTRODUCTION TO PATENTS

There have been some famous patent cases over recent years. Three examples were cases where an inventor sued major auto manufacturers for the intermittent windshield wiper, DEC sued Intel for infringement related to their Pentium chips, and the Apple versus Samsung cell-phone case that settled for $1B in 2012 [1]. For every well-known patent case there are countless smaller legal actions related to patents, so every PM should have knowledge of basic patent practices so they can guide the team in these areas.

Project Management in Product Development. http://dx.doi.org/10.1016/B978-0-12-802322-8.00010-3

10.1.1 Patent Attorneys and Patent Agents

When your duties as a PM require you to work with patents, you will need to consult with an expert. Patent attorneys can advise and represent you in any activity related to patents including searching patents, preparing a patent application for your invention, *prosecuting* the application (the process of obtaining a patent), and litigation. US law allows you to also be represented by a patent agent, a nonlawyer who has passed the USPTO Registration Exam and met other requirements [2]. Patent agents can represent you when applying for a patent, but not for litigation. Whoever you select, ensure they are licensed to practice in the appropriate jurisdiction(s).

10.1.2 What Is a Patent?

A patent is an agreement between an inventor and the government. When the government allows a patent, it grants the inventor a monopoly on the use of the invention for up to 20 years. In exchange, the inventor discloses the technology to the public. The inventor benefits from this relationship primarily through the economic advantage of the monopoly—either barring others from using the invention or licensing some or all of the rights to others.

The public benefits from patents in at least two ways. First, patent law requires a full disclosure of an invention so a great deal of knowledge is disseminated to the public. Second, patents encourage companies and individuals to invest in technology by creating protection when important inventions are discovered through their efforts.

For a patent to be awarded, the invention must meet three main requirements: it must be novel, nonobvious, and useful. The details of those terms are complex and legal counsel is required to understand how they may apply to your invention. From a layman's point of view, if your team has created a way to do something that has not been done before or (as is more common) a way to do something better than anyone has done it before, the possibility of getting a patent is high. A common misconception is that patents need to be groundbreaking; in fact, most patents are for relatively simple improvements over what's been done in the past. The sum of the millions of patents that have been awarded do represent impressive advances in nearly every area of technology. But, the average patent is humble, advancing knowledge a small amount in a narrow field.

The word *obviousness* has a legal meaning that is quite different from the common use of the term. Inventors often incorrectly conclude that their invention is obvious because it seems logical to them. They may have been working on the problem for 6 months and so the solution is clear. But in patent law the term *obvious* is applied to one of "ordinary skill," which is not usually a descriptor for someone who has studied a problem for months or years. A patent expert is required to determine obviousness, but, as a layman, I often use this short test when discussing the topic with an inventor:

- Does your invention do something better than what people do today?
- Is that new way better in some significant way—for example, does it bring lower cost, higher performance, more features, or higher quality?

If the answer to both questions is *yes*, it's probably not obvious—after all, if it really is better and it was obvious, then other people would probably already be doing it.

Deciding what to patent is also an economic decision. Patents can bring significant economic benefit—reducing competitive pressure can allow higher prices; also, your patents earn respect for your company in the market place. But patents also add significant cost. Aside from the time the team must invest to support a patent application, the lifetime cost of acquiring and owning a patent for 20 years can exceed $20k if filed in one country; file that same patent in five or seven countries and the figure can easily top $100k. So the decision to patent must consider technical, legal, and financial factors.

10.1.3 Why Is Patent Law Important to PMs?

Many PMs will wonder what the value is in their learning about patents. After all, they are not normally the inventors—that typically falls to the technical experts on the team. And PMs cannot apply for a patent—that requires a legal expert. An important reason for PMs to learn about patents and the processes around them is that the PM can serve as a bridge between the inventors and the legal experts. In this role the PM will carry out functions such as:

- Ensure the team has done diligent searches to reduce the risk of infringement.
- Ensure the team is thorough in identifying potentially patentable inventions.
- Manage the processes related to filing patent applications.
- Report to the sponsor on current state of issues related to patents.

In order to execute this role, the PM must be familiar with the basics of patents and the processes related to them. Fortunately, this does not normally require a legal background or deep technical knowledge of the inventions being patented.

10.1.4 How Do Patents Affect Projects?

Patents can affect product development projects in at least six ways.

Opportunity to Protect

Patent law brings the opportunity for your company to protect its technology. This can reduce competitive pressure, especially from companies that excel at reverse engineering and then produce look-alike products at lower costs. With patents, your inventions are protected from reverse engineering, at least in the countries where you file. Patents also increase the confidence your customers have in your company and give your sales force more evidence that your products are differentiated from those of your competitors. These factors can increase the profitability of your project.

Risk of Infringement

Patent law brings a risk of infringement that increases as the innovation in the project increases. If one of your team members uses a patented idea, even unintentionally, the products your team develops could infringe a patent. The owner of the patent may have a right to a portion of all sales that occur from product release until the patent expires. The portion of sales forfeited to the patent owner together with the costs of related legal actions can significantly reduce the lifetime profitability of a project. In the worst case, the patent owner may be able to prevent the future sale of the product until the patent has expired.

Patent infringement can also damage customer confidence in your product and your company. An industrial company using your product as a component in their product may see your litigation as a serious concern for them. They may be concerned that if your company loses in court, you might not be able to produce the product they need any longer. That loss of confidence can make your products less attractive and thereby reduce the profitability of your project.

Project Planning and Execution

If your company elects to pursue a patent, the PM will normally have to take on responsibility for planning and execution.

- You will need to schedule patent-related activities such as training for the team and team-member time for searching databases.
- You will need to manage patent-related processes to ensure they are completed on time and with high quality.
- You may need to budget for costs of legal counsel, search fees (when you engage a search professional), attorney time to complete patent applications, and filing fees for the countries you choose to file in.
- You will need to report on progress related to patents.

Learning the Technology

One often overlooked advantage of patent-related activities is the opportunity they bring for team members to gain expertise in a technical field. Few people search patents simply to learn about a technology. But when product developers search for patents, they invariably encounter many approaches to solve problems that they may not have seen before. If they discover a patent that is expired or lapsed due to fee nonpayment,[1] the technology may be free for use. In other cases, they might find useful technology that can be obtained for modest royalty payments. And, whether or not you find technology that is immediately

1. Payments are required every few years and a lapse on the part of the patent owner can the make technology free for use in much less than 20 years. The law regarding payment lapse varies by country and there are exceptions. Also, a company may let a patent lapse in one country and maintain a similar one in another. As with all actions you are contemplating with regard to patent law, consult an attorney before making a decision.

available, there is a more subtle advantage: a team member who studies just 10 or 20 important patents in a field is going to grow a certain savvy and open-mindedness that might not otherwise develop.

Monitoring Technology for Patent-Related Issues

As PM, you will often need to spur the team to identify technology that needs to be evaluated for patent-related issues. In general, new technology the team is using is of the most concern. As PM, you'll often be the best person in the company to monitor your project for these issues. In most cases, team members might not welcome the task of searching for prior use of something they are developing—they'd probably rather spend their time developing it. So expect to have to lead the team in reviewing development work to identify those areas most in need of attention, and then to follow up to ensure the related tasks are completed in a diligent and timely manner. As PM, you'll want to develop a mindset of filtering for potential patentability issues.

Managing Information during Development

There are many limits on how information about an invention can be shared without injuring your ability to patent. The responsibility for managing this information will often fall on the PM. This topic is covered below in Section 10.5.1.

10.2 TYPES OF INTELLECTUAL PROPERTY

There are many ways to protect intellectual property (IP) and they are sometimes confused. Let's quickly review the four most common types in product development organizations.

10.2.1 Patents

Patents protect an invention, which is to say, doing something in a new or better way. If someone else produces a product that uses the technique described in a patent, they probably violate that patent, even if they are not aware of the patent. There are two types of patents:

- The *utility patent* protects the function or method of a device or process.
- The *design* patent protects ornamental designs of functional items.

The utility patent covers what most people think of as an invention. It is the focus of this chapter.

10.2.2 Trade Secrets

Trade secrets are an important type of IP. They include a wide range of knowledge about your business: know-how to produce and sustain products, drawing sets, production equipment design, customer lists, and sales data, to name just a few. While obtaining one patent commonly costs more than $10,000, trade secrets require mostly that you take steps to prevent disclosure outside your organization—for example, avoiding publication. While patents last a maximum of 20 years, there is no time limit on trade secrets. Accordingly, companies will have orders-of-magnitude more trade secrets than patents. But trade secrets do not protect your company from reverse engineering or another company independently arriving at the same solution your company did; a patent does protect its owner in these circumstances.

10.2.3 Copyrights

Copyrights protect "works of authorship" such as writing, music compositions, photographs, and paintings. A copyright protects the expression of an idea, but not the idea itself [3]. Software and product documentation are usually protected with copyrights. They prevent someone from simply copying what you've done, but they don't protect the functions carried out in the software. If someone reverse engineers a product with copyright protection, they are usually free to create something similar. Copyrights do not protect technological innovation.

10.2.4 Trademarks

Trademarks are words or symbols (such as logos) that indicate the source of goods and services. They provide the public with a clear indication of who produced the products they might purchase. Products from other companies that are marked in ways that are "confusingly similar" to trademarked products are usually prohibited. Trademarks cannot protect a feature on a product or a new technology.

10.3 THE STRUCTURE OF A US PATENT

In this section, we'll look at the structure of a utility patent. We'll focus on US patents, though patents from outside the United States have a similar structure. There are four major sections of a patent, as shown in Figure 10.1: The *Front Page*, the *Drawings*, the *Specification*, and the *Claims*. Each section will be reviewed in detail. US patent 4,711,979 will be used to illustrate several points. You can find this patent at the US Patent Office (USPTO).[2]

10.3.1 Front Page

The front page of a US patent has a great deal of information. Refer to Figure 10.2 to find the numbered items on a sample front page. The item here is a method to view food cooking in a microwave oven.[3]

---

2. Go to http://www.uspto.gov/, click on "Search for existing patents," find "Patent Number Search" and click it, enter 4711979 as the number, then click "Images" at the top of the page. Alternatively, you can search Google Patents for "Patent 4711979" and click "View PDF" or "Download PDF."
3. We will also use this patent later as part of a patent search on the hypothetical camera/thermal imaging system for the oven discussed in Chapter 9.

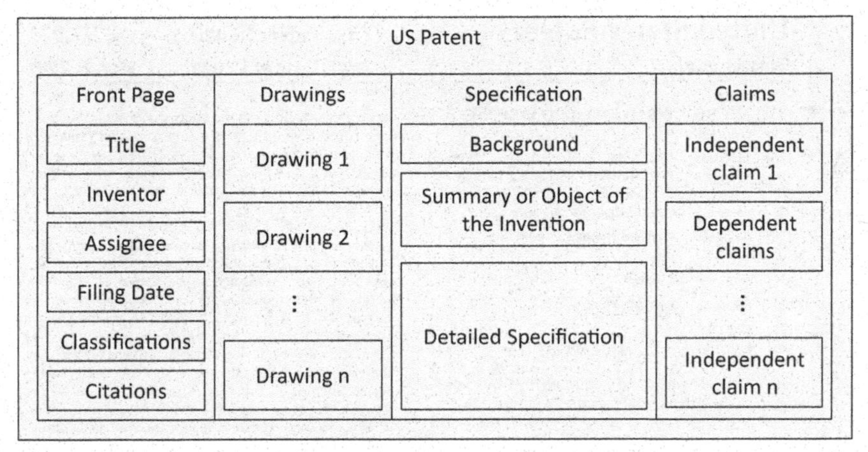

FIGURE 10.1 Structure of a US patent.

1. The patent number and date the patent was awarded. Note this is different from the filing date, the date the application was submitted to the patent office.
2. The title of the patent.
3. Inventor(s). All inventors of a patent must be listed here.
4. Assignee. The person or company the inventor has assigned the patent rights to. Independent inventors usually assign the rights to themselves; when the inventor develops the invention while in the employ of a company, the rights will often be assigned to that company.
5. Classifications. Ways this patent has been grouped with other patents. This will be discussed in detail in Section 10.4.
6. Citations. Patents and other documents this patent referred to as *prior art*, which is related technology that predates this patent. If this patent is interesting, cited patents may also be interesting.
7. Abstract. Usually a helpful overview of the patent, but sometimes just a copy of the first claim or a paragraph from the specification.
8. One figure that should aid understanding the patent quickly.

10.3.2 Drawings

Patents may contain any number of drawings. They show the components (usually called *elements*) of the invention, each numbered so they can be referred to in the text. At the bottom of Figure 10.2, there is a typical figure. An oven is shown with seven elements called out including the oven itself (10), the food being cooked (F), and the primary differentiating feature of this oven, a telescoping endoscopic tube for viewing cooking food (20). Paging through the figures can often provide a quick understanding of the patent.

Drawings sometimes show the *prior* art, what the world was like before this invention. Drawings almost always show examples of ways the invention can be used, which are called *embodiments*. A patented invention is abstract—you cannot see or touch one. However, you can see an embodiment. Our invention from

United States Patent [19]
Glasser et al.

(1) [11] **Patent Number:** **4,711,979**
[45] **Date of Patent:** **Dec. 8, 1987**

(2) [54] **MICROWAVE OVEN VIDEO VIEWING DEVICE**

(3) [75] Inventors: **George M. Glasser**, Irvington; **Robert P. DeRobertis**, Mahopac; **John H. Smithwrick**, Mt. Vernon; **William J. kelly**, Tarrytown, all of N.Y.

(4) [73] Assignee: **General Foods Corporation**, White Plains, N.Y.

[21] Appl. No.: **834,090**

[22] Filed: **Feb. 27, 1986**

(5) [51] **Int. Cl.**[4] **H05B 6/76**

[52] **U.S. Cl.** .... **219/10.55 D**; 219/10.55 R; 358/98; 358/100

[58] **Field of Search** .... 219/10.55 D, 10.55 F, 219/10.55 M; 358/98, 100, 110, 111; 128/4, 6; 174/35 MS

(6) [56] **References Cited**

U.S. PATENT DOCUMENTS

3,041,393	6/1962	Hennig	358/100
3,609,236	9/1971	Heilman	358/100
4,424,531	1/1984	Elter et al.	358/100
4,539,588	9/1985	Ariessohn	358/100

FOREIGN PATENT DOCUMENTS

1141529	12/1962	Fed. Rep. of Germany	358/100

Primary Examiner—A. D. Pellinen
Assistant Examiner—L. K. Fuller
Attorney, Agent, or Firm—Thomas A. Marcoux; Thomas R. Savoie; Daniel J. Donovan

(7) [57] **ABSTRACT**

A microwave viewing device and, more particularly, an arrangement facilitating the internal viewing of a microwave oven during the preparation of a food product or the like in a microwaving process. The viewing device incorporates a tubular support which is extendable through an opening formed in a side wall of a microwave oven, and which extends into close proximity to a food product being subjected to microwave processing in the oven. The viewing device includes a member constituted of telescopable tubular sections; in essence, which member is adapted to be varied in length by adding or removing sections, in dependence upon the size of the food product contained in the microwave oven, and which includes grid structure for supporting an optical viewing device within and in coaxial relationship with the tubular member; with the optical viewing device, such as an endoscope, fiberscope, boroscope or the like, being conducted outwardly from the microwave oven into operative interconnection with a suitable photographic or video recording apparatus. Within the tubular member, the viewing device, which is hereinafter generally referred to as an endoscope, is supported in concentric and coaxial relationship therewith through the interposition, in the annular space between the outer diameter of the shaft of the endoscope and the inner diameter of the tubular member of at least one transversely extending aperture disc component having a plurality of grid-forming through holes provided therein, and which will to a considerable degree reduce the leakage of microwave from the interior of the microwave oven.

13 Claims, 6 Drawing Figures

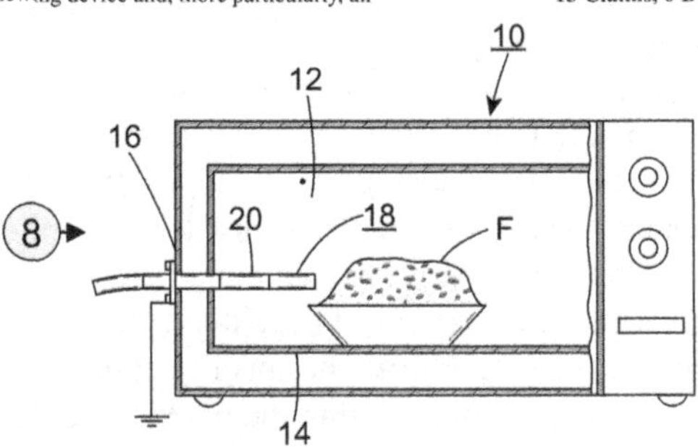

FIGURE 10.2 Front page of example US patent.

Figure 10.2 provides some means to look at food that is cooking—the drawing at the bottom is only one example—one *embodiment*—of that invention. Patents are broader than any of the embodiments they present; the invention is defined in the claims, as we will discuss shortly.

10.3.3 Specification

The specification explains the invention in detail. It is typically made up of three parts: the background, the summary of the invention, and the detailed specification. The first two sections are usually short, perhaps a few paragraphs. Together they usually explain the patent in enough detail for you to understand whether it is relevant to your project. The detailed specification is usually many pages long.

Bear in mind that the names of these sections vary. "Background of the Invention" may be called "Discussion of Prior Art" and "Summary of the Invention" may be "Object" or "Purpose of the Invention." Sometimes the sections may be merged together. In most cases, the information below will be presented in the order shown, but there is variation between patents.

Background of the Invention

The background of the invention explains what the world was like before this device or process was invented. It may be that people wanted to do something and they couldn't. It may be they could do it, but it was difficult, expensive, or unsafe, or they couldn't do it very well. The background helps establishes the value of the patent. It also tells a great deal of what the invention was intended to do. I find the background a quick way to get my bearings when reading a patent.

As an example, the background of Patent 4,711,979 (or just '979) starts with "The present invention relates to a microwave viewing device and, more particularly, relates to an arrangement facilitating the internal viewing of a microwave oven during the preparation of a food product or the like in a microwaving process." It then goes on to explain how difficult it is to see food cooking in a microwave oven and why that's a problem. Finally, it references other patents that attempted to solve a similar problem and what they don't do that '979 does.

Summary of the Invention

The summary of the invention gives on overview of what the invention does and perhaps why it's better than the prior art. In our example, the summary is several paragraphs long, but the key phrase occurs in the first paragraph. After mentioning a few shortcomings of the current methods, '979 states, this invention "...provides for a viewing device incorporating a tubular support which is extendable through an opening formed in a side wall of a microwave oven."

Detailed Specification

The detailed specification, as the name suggests, gives a great deal of detail about how the invention works. Normally, a large amount of the specification is written around the drawings, explaining each component: what it is, how it works, and its interdependencies with other components.

Attorneys often avoid using common terms so the patent will have broader meeting. Here are a couple sentences from '979 to illustrate two related points:

1. The inventor can define terms within the specification. This definition is then used to interpret the claims. Here the inventor defines "telescopable." Note that the component numbers (here, 20, 20a, etc.) are used to provide clarity—you can refer to the drawing at the bottom of Figure 10.2 to see 20.

Thus, the term "telescopable" refers to the changeability in the overall length of the tubular member 20 extending into the microwave oven 20, through the juncture of any desired number of tubular sections 20a, 20b, 20c et seq. which extend into the microwave oven 10

2. The inventor gives a specific example of using threaded brass tubes. However, the phrase "Generally, although not necessary…" makes it clear that invention can rely on other materials and other methods of joining the tubes:

Generally, although not necessarily, each tubular section may be constituted of a brass material…equipped with female screw threads at one end and male screw threads at the other end thereof so as to be able to threadingly engage with an adjoining tubular section.

One of the most important uses of the specification is that it defines all the numbered elements in the drawings. So, if you don't understand a component, you can search the detailed specification for that element number. (This is much easier when using an electronic version of the patent, which we'll discuss later). A search for "20" finds a description of that component showing that it holds an optical tube:

Located in a coaxial and concentric position within the sections of the tubular member 20 is an optical viewing device, such as an endoscope…

The full specification usually makes for heavy reading. Normally, you'll read the full specification in a minority of cases—when a patent seems close to the product being developed or when a method explained in the patent is particularly interesting.

10.3.4 Claims

Claims define the invention that the patent protects. A claim starts with a preamble like "A steering control system for an automobile comprising…" so you know the patent is not talking about a coffee maker or a submarine. The claim then deconstructs the invention into its elements. There is no limit on the number of elements, but typically there are five or 10 and, in most cases, the last element is the one that distinguishes the invention over the prior art.

For most patents (those with "comprising" claims) if and only if a product has all those elements does it infringe the patent. A product that adds extra features to the claim elements still infringes; however, removing just one of the listed elements prevents infringement. If someone finds a patent early in a development project, often they can "design around" the patent, which is to say, search for ways to eliminate at least one of the elements in a claim being in the product. However, if there are many claims in a patent, a product that infringes just one claim infringes the patent.

A patent cannot claim a natural law, a natural phenomenon, or an abstract idea. The abstract-ideas exception is currently important in the area of software and business patents. Many patents have been invalidated recently because they were found to claim only abstract ideas implemented with computers, with no particular limitations about hardware or specific algorithms. The natural phenomenon limitation is important in the field of biotechnology, and recently led to an important US Supreme Court case finding that segments of human DNA cannot be patented as such.

Claims are either *independent* or *dependent.* We'll focus on independent claims; these are the most basic. Dependent claims add more elements; they begin with phrases like "the system of Claim 1 and…." Dependent claims are written in case the independent claim has a defect and is invalidated in court. In that case, the patent owner may be able to assert the dependent claims. But when you're searching through patents, reviewing the independent claims is normally sufficient. Most patents have a handful of independent claims. Claim 1 is always independent.

An example claim from '979 is Claim 1, the only independent claim in this patent. The language is legally precise, which can make for difficult reading. With some effort, the elements are actually straightforward (refer to Table 10.1):

- A preamble explains this is for viewing food cooking in a microwave oven.
- Element 1: a telescoping structure penetrating the wall of the oven and able to extend to a position near the food.
- Element 2: an optical tube inside the telescope.
- Element 3: discs to support the optical tube in the telescope.
- Element 4: a mesh to reduce leakage radiation from the microwave oven.

The precise meaning of a claim is often difficult to understand. Claims are interpreted by reading the specification and sometimes the specification will use words in ways that are uncommon. However, a layman can often tell when the

TABLE 10.1 Claim 1, the Only Independent Claim of '979

Preamble	**1. An arrangement for the internal viewing of the preparation of a food product or the like in a microwave oven; comprising**
Element 1	**a telescopable tubular member** insertable through an aperture in a wall of said microwave oven so as to position the leading end of said tubular member in proximity to the food product in said microwave oven;
Element 2	a generally **cylindrical optical viewing device** being inserted in said tubular member in coaxially concentric relationship therewith, said viewing device having an outer diameter which is smaller than the inner diameter of said tubular member to form an annular space there between;
Element 3	**at least one ring-shaped apertured disc** extending transversely in the space between said viewing device and said tubular member **for radially supporting said device** within said tubular member;
Element 4	and **a mesh structure** being positioned in said tubular member to extend transversely in front of the leading end of said viewing device in said microwave oven so as to form a radiation screen **inhibiting the leakage of microwave radiation there through.**

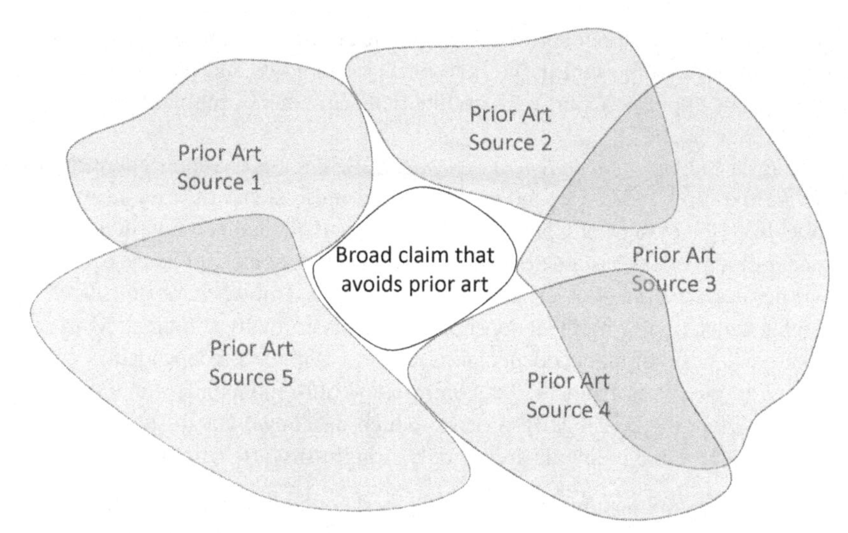

FIGURE 10.3 A broad claim can be valid only if it avoids the prior art.

claim is close to the product they are working on. In this case, if we were searching for a means of viewing cooking food in an electric oven (from Chapter 9), it is unlikely this patent would apply because element 4 specifically mentions a mesh that reduces microwave radiation. However, as with all issues in patent law, review your findings with a patent attorney or a patent agent.

Patents often have a combination of narrow and broad claims.

- Narrow claims include a great deal of detail. They are typically easier to validate but they offer less protection because they are easier to "work around."
- Broad claims seek to protect a much wider swathe of art, so they can offer much more protection than narrow claims. However, they are also more risky. If a claim is so broad that it includes prior art, it will probably be invalid.

Accordingly, patent attorneys seek to create claims that are as broad as possible but still avoid the prior art. As shown in Figure 10.3, patent claims often carve out a space among many pieces of prior art. Strong claims fill up the available space without crossing over into areas that are already known.

Another issue with broad claims is *enablement*. The specification of a patent is supposed to enable, which is to say, teach how to practice the invention in a manner commensurate in scope with the claims. A broad claim may cover many ways to solve a problem, but if the specification does not teach all of them, that claim may be invalidated.

10.4 SEARCHING PATENTS

If your company wants to patent an invention or be more confident they are working on something that is not patented by another company, it will probably be necessary to search one or more patent databases. There are several databases for international patents that are freely available and the tools to search patents are easy to use. So, you may decide you or the technical experts on the team would benefit from doing a preliminary search. Also, there are professional searchers; they are commonly engaged to support the patent application process. Fortunately you don't need to choose between the two. Instead you can use a two-stage approach:

1. The inventor performs a preliminary search.
2. Take any relevant search results to an attorney and discuss where a professional searcher might be needed.

The benefits of the preliminary search are:

- An inventor will have a deep understanding of the invention and will often be able to recognize related technology better than a professional searcher. By comparison, professional searchers are more skilled in the tools and have more knowledge of patent law than most inventors.
- You can reduce costs, especially as your technical experts become more skilled in preliminary searches. As stated by David Hitchcock concerning a professional search:

While [it is] very effective, the process is also very expensive. Instead of starting with this approach, you can save yourself some money by performing a preliminary search [4].

- You can save time. A professional search may take weeks (including the time to get approval to fund the search); a preliminary search might take just a few hours.
- Should you elect to engage a patent professional, the prior art found in the preliminary search can aid the professional searcher, speeding the process and improving the quality of the end result.

However, when a team member searches there are some legal risks and the inventor must disclose to an attorney anything of relevance. Information withheld from or misrepresented to the patent office can lead to a patent being held unenforceable for *inequitable conduct*. As PM, ensure team members are aware of this. There are other issues—talk with your attorney before you or your team members start searching.

The search methods I've used over the years have changed as the tools for searching have improved. When I started, I drove 4 hours to have a day or two in the patent office; there I flipped through hundreds of paper copies to search a topic. Today, multiple databases are freely available with powerful search tools. The method I use today is shown in Figure 10.4. The remainder of this section will review that process in detail; as an example, we'll search the oven with remote monitoring described in Section 9.3.3. Bear in mind that there are many search strategies and this is just one example; see suggested reading at the end of this chapter for texts with alternative approaches. Also, this approach is most appropriate for electrical and mechanical inventions—chemical and biotechnology use different strategies [5].

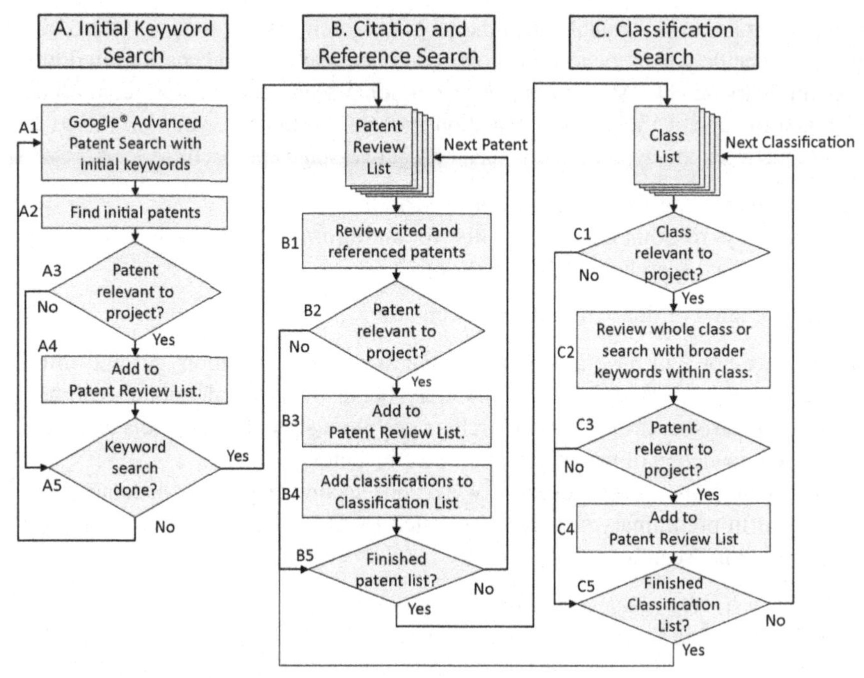

FIGURE 10.4 An example of a preliminary patent search.

10.4.1 A. Initial Keyword Search

The first patent in a new area of technology can be the most difficult one to find. You probably don't know the classifications to focus on and you may have a difficult time finding keywords that are broad enough to find patents but not so broad that they return an unmanageable number of candidates. You can study the classification systems to find a relevant class, but that can be time consuming—for example, the Cooperative Patent Classification system has 250,000 classes [6]. I rely on keyword searches to get started and use classifications to improve the search. Others start with a classification search, for example, the Seven Step Strategy suggested by the USPTO [7] (the University of Central Florida Libraries has a short, helpful video on a related method [8]). We'll discuss classifications further in Section 10.4.3.

A1. Google® Advanced Patent Search with Initial Keywords

The Google® Advanced Patent Search screen is shown in Figure 10.5 [9]. Here you can specify keywords; for this example, I specified that "oven" and "camera" be in the patent.

A2. Find Initial Patents

Google® then returns a list of patents, a portion of which are shown in Figure 10.6.

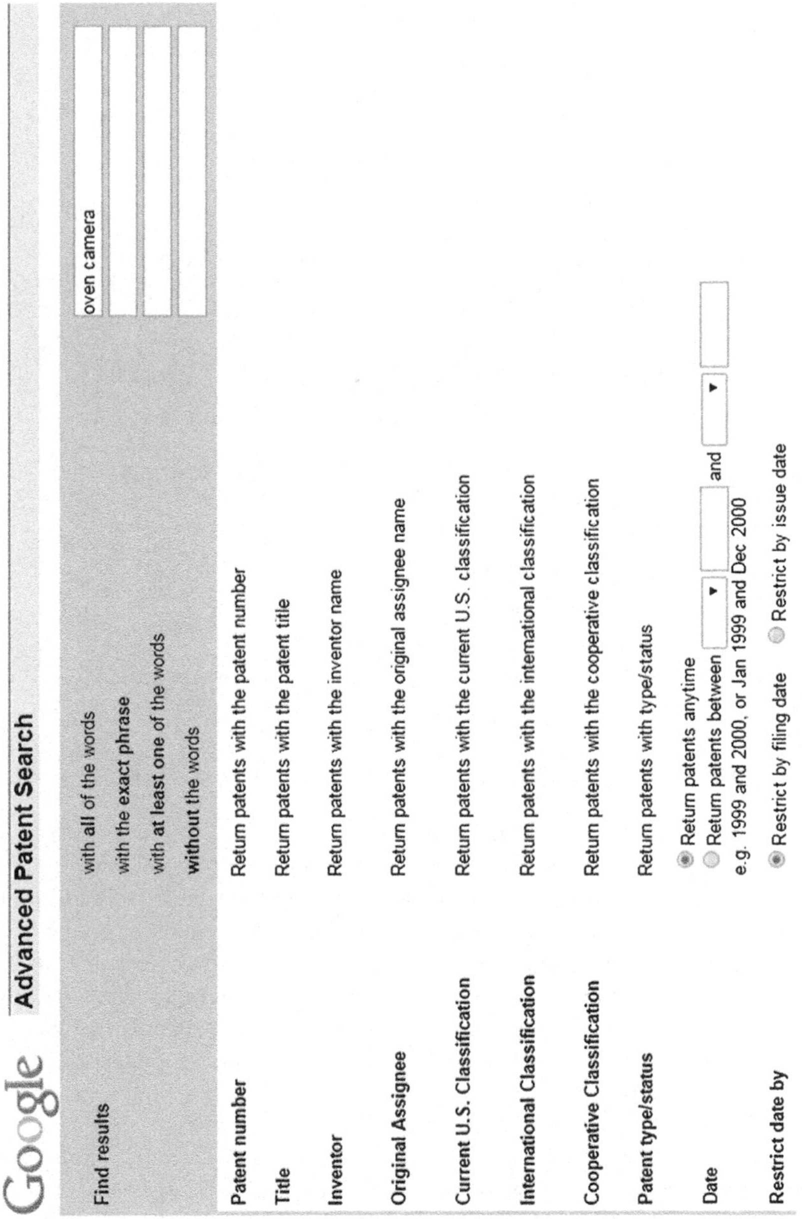

Google **Advanced Patent Search**

Find results	with **all** of the words	oven camera
	with the **exact phrase**	
	with **at least one** of the words	
	without the words	
Patent number	Return patents with the patent number	
Title	Return patents with the patent title	
Inventor	Return patents with the inventor name	
Original Assignee	Return patents with the original assignee name	
Current U.S. Classification	Return patents with the current U.S. classification	
International Classification	Return patents with the international classification	
Cooperative Classification	Return patents with the cooperative classification	
Patent type/status	Return patents with type/status	
Date	Return patents anytime Return patents between ▼ and ▼ e.g. 1999 and 2000, or Jan 1999 and Dec 2000	
Restrict date by	Restrict by filing date Restrict by issue date	

FIGURE 10.5 Google® Advanced Patent Search using just keywords.

Oven conveyor alignment system apparatus and method

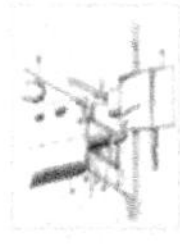

www.google.com/patents/US7131529
Grant - Filed Jul 1, 2003 - Issued Nov 7, 2006 - Ronald Meade - Casa Herrera, Inc.
An **oven** conveyor alignment system for maintaining a conveyor belt centered on ... The system comprises a **camera** that generates an image of ...
Overview - Related - Discuss

An oven detecting presence and/or position of tray

image not available

www.google.com/patents/EP2638327A2?cl=en
App. - Filed Nov 11, 2011 - Published Sep 18, 2013 - Mustafa Kuntay Karaaslan - Arçelik Anonim Sirketi
characterized by at least one indicator (11), disposed in the **oven** cavity (6) with features visually discernible by the **camera** (10) and at least ...
Overview - Related - Discuss

Coke oven machinery spotting system

www.google.com/patents/US4196471
Grant - Filed Apr 20, 1978 - Issued Apr 1, 1980 - Lawrence M. McClure - Koppers Company, Inc.
A coke **oven** machinery positioning and spotting system comprises system ... (b) said visual sensing means is a digital line scan **camera**, ...
Overview - Related - Discuss

Baking oven door and baking oven

image not available

www.google.com/patents/WO2012146523A1?cl=en
App. - Filed Apr 19, 2012 - Published Nov 1, 2012 - Florian Ruther - Electrolux Home Products Corporation N. V.
The invention is directed to a baking **oven** door 2 and baking **oven** 1. A **camera** 10 is mounted inside the door 2 and coupled to a heat sink 8 ...
Overview - Related - Discuss

FIGURE 10.6 Selection of returned list of patents from keyword search.

A3. Is Patent Relevant to your Project?

Review the patents from the search results. Many will be unrelated to the products being developed in your project. In this example, an industrial conveyor belt oven that uses a camera to count cookies might be on the list; if the purpose of the camera is unrelated to the cooking process, it's likely the patent is not relevant to this search.[4] From the list in Figure 10.6, the 4th item looked particularly relevant for this search. When opening that patent (or in this case, patent application[5]) in Google Patents, a summary is given as shown in Figure 10.7.

4. Seek legal advice when determining relevance. The invention is defined by the claims, which are interpreted through the specification; sometimes the patent abstract and title can describe a narrower invention than is protected by the patent.

5. This is a US patent application, the document published by the USPTO before the patent is awarded. This is evident because this reference has an 11-digit number, the first four of which are the year of the application. Both patents and patent applications are relevant in prior art searches.

Baking oven door and baking oven
US 20140048055 A1

ABSTRACT

The invention is directed to a baking oven door **2** and baking oven **1**. A camera **10** is mounted inside the door **2** and coupled to a heat sink **8** constituting an outer cover of the door **2**.

Publication number	US20140048055 A1
Publication type	Application
Application number	US 13/978,413
PCT number	PCT/EP2012/057118
Publication date	Feb 20, 2014
Filing date	Apr 19, 2012
Priority date ⓘ	Apr 29, 2011
Also published as	CN103501618A, EP2520169A1, WO2012146523A1
Inventors	Florian Ruther
Original Assignee	Electrolux Home Products Corporation N.V.
Export Citation	BiBTeX, EndNote, RefMan

Patent Citations (5), Classifications (6), Legal Events (1)

External Links: USPTO, USPTO Assignment, Espacenet

IMAGES (3)

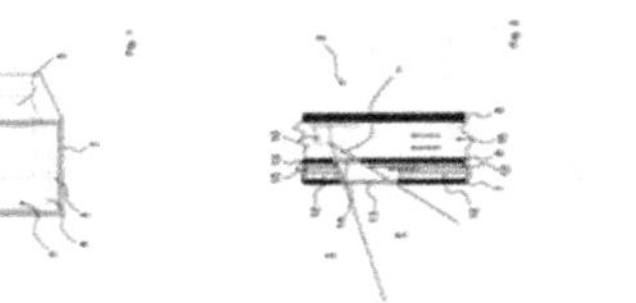

FIGURE 10.7 Example patent application returned in initial search.

The phrase in the abstract "A camera is mounted inside the door" indicates a deeper look is needed. Then, this was found in the background section (emphasis added):

As users in general want to observe the baking process within the muffle of the baking oven, optical systems and ***even camera systems have been proposed in order to visualize the baking chamber****...*

A4. Add Relevant Patents to the Patent Review List
This patent is certainly relevant and so was added to the review list.

A5. Keyword Search done?
Had I been working on a project for such an oven, I might have kept searching keywords. However, in this case a clearly relevant reference was found, so I chose to move on to the next section: B. Citation and Reference Search.

10.4.2 B. Citation and Reference Search

For each patent in the Patent Review List, execute a filtering and searching process.

B1. Review cited and referenced patents
Using Google Patents, it's easy to find citations (every other patent that this patent cites) and references (every other patent that references this patent). Citations are shown in Figure 10.8; references in Google Patents are in a similar form but missing here because this patent application had not been referenced at the time of this search.

PATENT CITATIONS

Cited Patent	Filing date
DE4333443A1	Sep 30, 1993
DE7934764U1	Dec 11, 1979
DE20103517U1	Feb 28, 2001
DE102008043722A1	Nov 13, 2008
DE202008000135U1	Jan 3, 2008
FR2693538A1 *	

FIGURE 10.8 Citations from patent found in keyword search.

B2. Are Cited or Referenced Patents Relevant to your Project?

Review each of the cited and referencing patents to see if they are relevant to the search.

B3. Add to Patent Review List

Add those patents that are related to your project work to the Patent Review List.

B4. Add Classifications to Classification List

For each patent of interest collect the classifications and add them to the Classifications List. This patent application has six different classifications (see Figure 10.9): two in the US system (e.g., 126/198), one in the international system, and three in the European–US cooperative classification.

- Finished Patent Review List?
  Repeat this process on all patents in the Patent Review List. When complete, move to the classification search.

10.4.3 C. Classification Search

Patents are usually given several classifications. Partly that's because the classes are sometimes so close, an invention belongs in both. Partly it's because one invention may use two different technologies. For example, is the oven-with-camera invention a modified oven or a special camera?

For our example, the patent has these classifications:

US 126/198	Oven Doors, Ventilating
US 126/190	Stove doors and windows
International F24C15/02	Doors specially adapted for stoves or ranges
Cooperative F24C15/02	Doors specially adapted for stoves or ranges
Cooperative F247/08	Arrangement or mounting of control or safety devices
Cooperative A21B3/02	Baker's ovens/doors and flap gates

Classifications divide patents in small groups of related topics. There are several patent classification systems including the USPTO (USPC [10]) system, the International Patent Classification (IPC [11]) system from the World Intellectual

CLASSIFICATIONS

U.S. Classification	126/198, 126/190
International Classification	F24C15/02
Cooperative Classification	F24C15/02, F24C7/08, A21B3/02

FIGURE 10.9 Classifications for patent found in keyword search.

Property Organization (WIPO), and the Cooperative Patent Classification (CPC [12]). The CPC is described as:

> *[...] a joint partnership between the United States Patent and Trademark Office (USPTO) and the European Patent Office (EPO) where the Offices have agreed to harmonize their existing classification systems (European Classification (ECLA) and United States Patent Classification (USPC) respectively) and migrate towards a common classification scheme [13].*

Using Google® Patents, the classifications are normally provided as hot links; for example, "F24C15/02" in Figure 10.9. Since this is an IPC class, clicking on this link goes to the WIPO site for details on this classification as shown in Figure 10.10.

At this point, you've probably searched numerous patents and have a list of classifications from each. Similar to the patent list, you can go through each classification in the Classification List with a filter/search process:

C1. Is the Classification Relevant to your Project?
If so, then search the classification. Otherwise, go to the next classification in the list.

C2. Review the Whole Class or Search the Class with Broader Keywords
Each relevant classification can be searched, the benefit being that there are a limited number of patents in any one classification, usually ranging from the 10s to the 100's. If there are a small number of patents in a classification (say, 25), you may elect to review all of them. If the classification is larger, you can filter with a keyword search. Of course, the keywords used here can be much broader than when searching a full database, which will contain millions of patents.

For our example, the initial keyword search was "oven camera". But if we're searching a classification specifically for ovens, there is probably no need to include the word "oven." Searching classifications helps with a primary shortcoming of the initial keyword search—the likelihood the inventor will use an unexpected word to describe the invention. For example, instead of oven an inventor might use the terms *range* or *cooking chamber*.

Because there are so few patents in one class, we can broaden the keywords for camera to *video*, *image*, or *optical*. Since any class contains a tiny fraction of the entire database, the likelihood that an irrelevant patent will be returned is much lower; so, using a broader set of search keywords is more practical.

C3. Is a Patent Found in the Classification Search Relevant to your Project?
Review each patent for relevance to the product being developed in your project.

C4. Add to Patent List
Add all relevant patents to the Patent Review List.

C5. Finished Classification List?
Are you finished searching the Classification List. If not, review the next classification. If so, return to the Citation and Reference search and review any relevant patents added to the Patent Review List from the Classification Search.

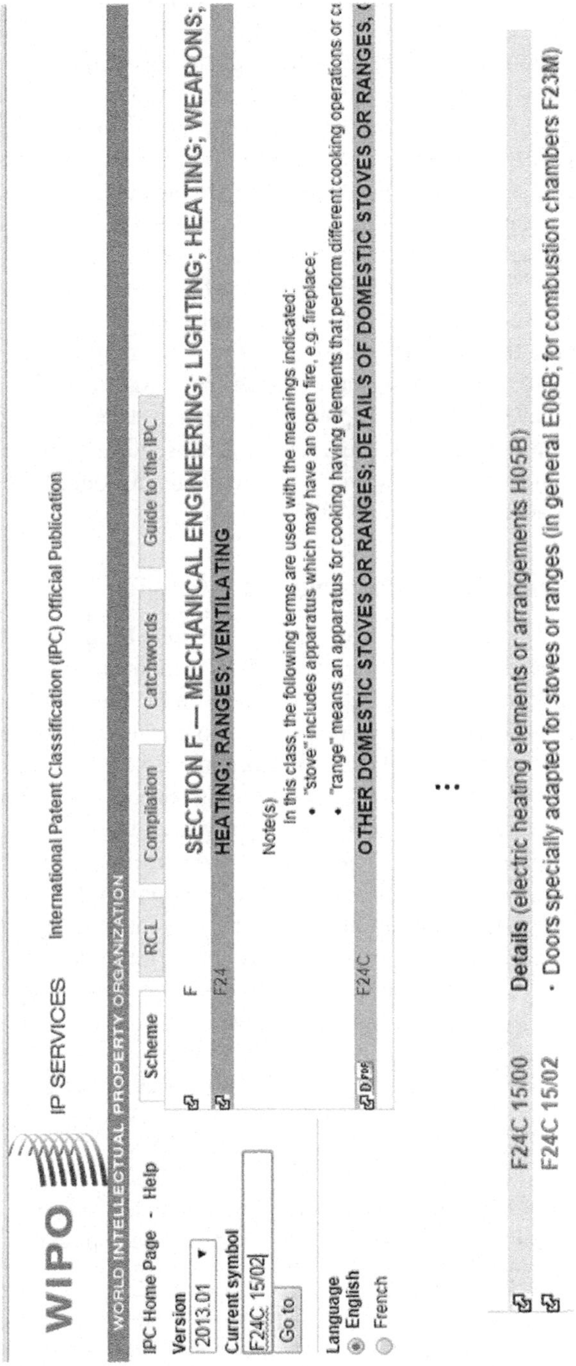

WIPO IP SERVICES International Patent Classification (IPC) Official Publication

WORLD INTELLECTUAL PROPERTY ORGANIZATION

IPC Home Page - Help

Scheme | RCL | Compilation | Catchwords | Guide to the IPC

Version
2013.01

Current symbol
F24C 15/02
Go to

Language
English
French

F SECTION F — MECHANICAL ENGINEERING; LIGHTING; HEATING; WEAPONS;

F24 HEATING; RANGES; VENTILATING

Note(s)
In this class, the following terms are used with the meanings indicated:
- "stove" includes apparatus which may have an open fire, e.g. fireplace;
- "range" means an apparatus for cooking having elements that perform different cooking operations or c

F24C OTHER DOMESTIC STOVES OR RANGES; DETAILS OF DOMESTIC STOVES OR RANGES, (

⋮

F24C 15/00 **Details** (electric heating elements or arrangements H05B)

F24C 15/02 · Doors specially adapted for stoves or ranges (in general E06B; for combustion chambers F23M)

FIGURE 10.10 Sample classification: International Class F24C15/02.

The search process of Figure 10.4 can be augmented with other search fields shown in Figure 10.5, especially the inventor and the assignee (normally the company the inventor worked for at the time of the invention). For example, if your search reveals an inventor is active in the area of interest, it would be wise to look over all patents awarded to that inventor; search engines support filtering by inventor. Also you may want to return to the keyword search, filter by companies that are active in the area, and then broaden the keywords similar to what was discussed just above for the classification search. For example, Figure 10.7 shows the assignee as Electrolux; a search for Electrolux patents with the keyword "camera" turned up over 100 results at the time of publication; such a list could be reviewed to improve the search results.

Keep repeating the process until your search stops finding relevant patents and classifications. The process can continue for some time, but in the end, you're likely to have a substantial portion of the art related to your invention. Bear in mind there is no perfect patent search. A seasoned expert can miss prior art and even the most complete search possible cannot account for unpublished patents applications that are in process.[6]

10.4.4 Freedom-of-Use versus Novelty Searches

There are two main purposes of searching for patents: *novelty* and *freedom-of-use*. Novelty searches are performed when there is an invention your company might want to patent. Here you are searching to find if the idea has been used before. Many things found in a patent search can bar patenting: an active patent, an expired patent, a published patent application that was rejected, or even an article in a magazine. In fact, if your invention has been used in a product that was sold or just offered for sale, you may be prevented from patenting it.

In a freedom-of-use search, the interest is to determine if the invention can be used without infringing an active patent. Freedom-of-use is a lower bar than novelty—normally only an active patent prevents freedom-of-use. So, in a freedom-of-use search, when you find your invention described in an old paper or an expired patent, it's good news; in a novelty search, the same result is unwelcome. In either search, finding the invention in an active patent is usually bad news. These rules are general; don't attempt to make a final decision on either freedom-of-use or novelty—that should be done with legal counsel.

10.5 THE PATENTING PROCESS

Suppose after a search, your company decides to pursue a patent. Here, you'll need to engage legal counsel. The first step is writing the application. Normally,

---

6. It's possible that a relevant patent could publish the day after your search is complete. You can mitigate this problem by using alerts on one or more patent search engines so patent applications that meet your search criteria are automatically sent to you when they are published.

the attorney (or agent) will ask the inventor to write a description of the invention in layman's terms. The attorney will use that information to write a description to meet the legal requirements of a patent application. It's important that all the relevant prior art known by the team is disclosed to the attorney. It allows the attorney to write an application more likely to lead to a patent grant and a patent that will be more likely to be validated if it's ever tested in court.

The application is then filed with the patent office. The process to award a patent is about 4 years in the United States at the time of this publication and about half of applications result in patent grants. The application is published typically 18 months after it's filed.

An alternative to a standard or *nonprovisional* patent application is the *provisional* application. This application has fewer requirements than a full application; these applications can be filed faster. They also have lower costs of initial filing and allow the inventor time (typically 1 year) to evaluate economic value before investing fully in the patent process. After the period of evaluation, the company can file a nonprovisional application or they can abandon the invention. There are downsides to provisional applications, for example, the total cost of filing a provisional patent and following with a nonprovisional patent is usually higher than just filing the nonprovisional patent. Discuss your situation with legal counsel.

10.5.1 Managing Information during the Innovation Process

If your company does elect to seek a patent, there are special responsibilities for managing information until the application is filed. As a PM, you need to be familiar with this issue because a team member or someone else in the company could disclose enough information to bar patentability. For example, if a sales document was published that described the invention before the application was filed, that document would become part of the prior art. Such a publication could put the invention in the public domain, making it impossible for you to patent while simultaneously making it available for anyone else to use freely.

The disclosure of information takes on many forms, some of which are subtle:

- Publication of documents anywhere in the world including the web.
  Publication of the invention before the application is filed can bar patentability, especially if you describe how it works in detail. This includes articles in other languages published in obscure magazines. It also includes blogs, press releases, presentations, sales literature, and product documentation. Any action that reveals your invention to the public is of concern.
- Revealing the invention to a third party.
  Revealing the invention to a third party before the application is filed without a legally binding agreement to keep the information confidential (a "nondisclosure agreement" or NDA) can constitute public disclosure. That

can apply to anyone outside your company—customers, suppliers, and even independent distributors who sell products for your company. Bear in mind that the protection an NDA offers is limited; if that third party violates the NDA and reveals the invention—even accidentally—it can bar patentability.

- Showing a working product/offering for sale.
  Showing a working product that uses the invention to a third party without an NDA before the application is filed can also bar patentability. This applies to almost anyone outside your company as discussed in the previous point. Including a customer in early testing or just demonstrating a product to a distributor—who may be selling the product in the near future—can be considered disclosure. To be clear, even if you don't explain the invention during the demonstration, just showing the product can be considered disclosure. Similarly, selling or even offering for sale a product that uses the invention can injure your right to patent.

In the United States, there is a 1-year grace period for many types of disclosure; most of the rest of the world does not offer such a grace period. There are many other rules about disclosure, which your attorney can inform you of. However, in general, before the application is filed, the less said about the invention to anyone, the better. You may want to limit disclosure even within your company to reduce the chances of accidental disclosure. As a PM, you'll need to manage this topic with the team to maximize the likelihood of a successful patent application process.

The PM may also want to ensure the team keeps permanent records of their development process. For example, the inventor could document when the idea was conceived, how it was tested, and what improvements were made over the course of the project. This is a good practice in case an accused infringer or rival applicant later claims that the invention was actually derived from the work of another.

10.5.2 Naming Inventors

Many patents will have a single inventor—one person who conceived the initial idea and worked through the issues to perfect it. However, often there are multiple inventors—perhaps one person thought of the basic approach but a second person was required to put in place all the pieces that made the invention work. In almost all cases, others will work on the patent, for example, building and testing prototypes or integrating the invention into company products. When filing a patent, the list of inventors must be complete and accurate. Every person who contributed to the conception of the invention must be listed; if an inventor is omitted, even unintentionally, the patent may be invalidated. At the same time, only those who contributed can be listed—those who acted at the direction of the inventors, for example, simply carrying out a test, cannot be included. The PM will need to communicate these rules; it can be sensitive because of the prestige associated with being named on a patent.

10.5.3 America Invents Act

The America Invents Act (AIA) of 2011 is one of the largest recent changes to US patent law [14]. It has many provisions, but the most important is probably the change in how *priority* is established. *Priority* determines who is entitled to patent when multiple people develop the same invention independently. Before AIA, priority in the United States was established by who developed the invention first; after AIA, it's determined by who filed first. This gives more reason to file patents quickly during the project. Some people feel the AIA creates a "race to the patent office," which gives advantage to large companies because they are more likely to have the resources to win that race. Others feel it gives advantage to small companies and individuals by reducing the fees they pay [15]. This provision of the AIA does bring the United States in line with most other countries where "first to file" has been the rule for some time.

If you're familiar with patent law from the past, also be aware that the US long-time1-year grace period is in a new and, according to many, weaker form in the AIA. The 1-year grace period used to allow US inventors to publish their invention 12months before filing an application; this allowed inventors time to better understand the value of their invention before investing in filing for a patent. In the AIA, for the United States the 1-year grace period exists but there are more exceptions and this has led some attorneys to recommend not trusting the grace period completely.

10.6 RESOURCES

Here are several websites you may find helpful as you learn about patents [16]:

- The US Patent and Trademark Office site offers a great deal of documentation for laymen and attorneys: www.uspto.gov
- Invent Now, Inc. is an organization that encourages innovation in children and adults: http://InventNow.Org
- Google Patents offers a powerful search engine, providing electronic and PDF copies of patents: http://www.google.com/advanced_patent_search
- Pat2Pdf provides free PDF versions of US patents: http://pat2pdf.org/
- Inventors Digest offers articles on trends and news in intellectual property: http://www.inventorsdigest.com/
- Free Patents On-line also offers articles and news related to patents, international patent searches, alerts for patents, and easily retrievable PDFs: http://www.freepatentsonline.com/
- There are several subscription services that offer services for intellectual property including Innography (http://www.innography.com/), Intellectual Property Solutions by Thomson Reuters (http://ip.thomsonreuters.com/), and ip.com (http://ip.com/).

10.7 RECOMMENDED READING

There are a great number of books about patent law written for laymen. NOLO publishes several books on patents that are readable; two examples are below. When choosing books, be aware that patent law changes frequently, so it's wise to get the most recent edition of any book on the topic.

Charmasson H, Buchaca J. Patents, copyrights, and trademarks for dummies. For Dummies (Wiley); 2008.

Durham AL. Patent law essentials: a concise guide. Praeger; 2013.

Hitchcock D. Patent searching made easy: how to do patent searches on the internet and in the library. NOLO; 2013.

Hunt D, Nguyen L, Rodgers M. Patent searching: tools & techniques. Wiley; 2007.

Pressman D. Patent it yourself: your step-by-step guide to filing at the U.S. Patent Office. NOLO; 2012.

10.8 REFERENCES

[1] Shinra S. Engineering project management for the global high-technology industry. McGraw Hill; 2014. p. 88.

[2] Becoming a Practitioner, US Patent and Trademark Office. Available at: http://www.uspto.gov/learning-and-resources/ip-policy/becoming-practitioner.

[3] Durham AL. Patent law essentials: a concise guide. Praeger; 2013. p. 9.

[4] Hitchcock D. Patent searching made easy: how to do patent searches on the internet and in the library. NOLO; 2013.

[5] Hunt D. Patent searching: tools and techniques. Wiley; 2010. p. 16.

[6] Espacenet Patent Search, http://ep.espacenet.com/help?locale=en_EP&method=handleHelpTopic&topic=cpc.

[7] Seven Step Strategy, US Patent and Trademark Office. Available at: http://www.uspto.gov/learning-and-resources/support-centers/patent-and-trademark-resource-centers-ptrc/resources/seven.

[8] How to Do a Patent Search. UCF Libraries; 2014. Available at: https://www.youtube.com/watch?v=vr5aMjUTVOc.

[9] Google® Advanced Patent Search, http://www.google.com/advanced_patent_search.

[10] Patent Classification System, US Patent and Trademark Office, http://www.uspto.gov/web/patents/classification/selectbynum.htm.

[11] International Patent Classification (IPC), World Intellectual Property Organization (WIPO), http://www.wipo.int/classifications/ipc/en/.

[12] Cooperative Patent Classification (CPC), European Patent Office, http://www.epo.org/searching/essentials/classification/cpc.html.

[13] Classification Standards and Development, US Patent and Trademark Office, http://www.uspto.gov/patents-application-process/patent-search/classification-standards-and-development.

[14] US Patent and Trademark Office. America Invents Act: Effective Dates. 2011. Available at: http://www.uspto.gov/sites/default/files/aia_implementation/aia-effective-dates.pdf.

[15] Jordon R. The new patent law: end of entrepreneurship? Forbes; November 13, 2012. Available at: http://www.forbes.com/sites/robertjordan/2012/11/13/the-new-patent-law-end-of-entrepreneurship/.

[16] Blum J. How to protect your intellectual property rights. Entrepreneur; July 26, 2011. Available at: http://www.entrepreneur.com/article/220039.

Chapter 11

Reporting

This chapter will discuss reporting for the project manager (PM). The focus will be on reporting up—to the sponsor or a steering committee. Reporting up is a requirement for PMs so the company leadership can provide approvals and monitor project health; as a secondary effect, the preparation usually brings clarity to critical issues and so, while it is aimed at the sponsor/steering committee, reporting normally serves the entire team. The first section will focus on oral presentations given for project reviews and/or approvals. The remainder will discuss the use of quantification in project management, beginning with a broad discussion of metrics available to the PM. The chapter will then turn to the selection of the most important metrics for a given project, key performance indicators (KPIs); finally, we'll look at creating a project dashboard, a single picture that tells all stakeholders what is most important about the project. The discussion here builds up from the reporting material from Section 3.3 and builds on the metrics and visual management presented in Chapters 5–9.

11.1 MANAGEMENT PRESENTATION

Clear and accurate reporting enables managers to make fast decisions. Reinertsen, a well-known expert in lean product development, investigated the problem of slow decisions from management on many occasions and consistently found that information was missing:

> *Invariably, I discover that the cost and benefit of the decision are either unquantified or poorly quantified. In my experience, most managers make amazingly fast decisions when they are presented with compelling economic arguments [1].*

The management presentation is probably the most important single activity the PM will undertake; it can also be one of the most intimidating. For one thing, you're usually presenting to a smart group of people ready and willing to ask tough questions. Your colleagues may caricature the senior managers, but my experience is they are some of the most insightful people in the company. They know how to find the weaknesses in your work and they're not afraid to drill in when they have a concern. It's a great opportunity to learn but it can be painful.

As we discussed in Section 3.3, the stakes are often high at project reviews. A good review will pave the way for approvals to spend and for general support.

Project Management in Product Development. http://dx.doi.org/10.1016/B978-0-12-802322-8.00011-5

But if the work isn't going well, the presentation can result in the project being deprioritized, put on hold, or even canceled. So, when it's your turn to present, you'll want to be prepared.
This section will help you prepare in three ways:

1. Suggest a form for the presentation,
2. Guide you in developing the content, and
3. Give examples of tough questions you can expect during the presentation.

Being well prepared will provide the best results, but don't expect to emerge without problems being identified. Remember that the forum of project reviews gives a decided advantage to the management team: there's one of you and probably many of them. And, they can interrupt at any time and take the conversation down a path you could not have anticipated. Also, they have been to a lot more reviews than you have. Finally, reviews are by their nature a *management by exception* exercise, one that focuses on problems rather than successes; so, if you've done a 98% bang-up job, they may spend most of their energy on the other 2%. So come ready to learn and keep realistic expectations.

For your presentation, follow the writer's mantra: *write to your audience*. Their point of view will probably be different from yours; that's expected because your roles are different. When preparing and presenting for each topic, ask yourself: "Why does the management team care about this?" Their primary agenda at a review is to evaluate the health of a project and to identify hidden issues. Build your presentation so every slide helps them do that.

11.1.1 The Form of Your Presentation

While this section is targeted at project reviews, most of what's here applies to management presentations in general.

Be Open and Honest

As we've discussed throughout this book, be open and honest. Every time you present to management, you build your reputation. Don't hide bad news. Don't understate problems. Don't dance around questions. If the news is bad, they will likely find out eventually. It's better if you're the one that tells them. Better still if you volunteer it at the appropriate time.

There is a balance to strike—don't overreact to bad news. Similar to what was presented in Section 9.4.4, don't raise the alarm too soon and when you do raise it, be sure it's to the appropriate level. When you raise an alarm without verifying the information or when you raise too high an alarm, the results can bring a sense of chaos.

No Surprises

As has been mentioned, avoid surprises in meetings. Make sure your sponsor is aware of any news, especially bad news so she has time to prepare—it's only fair because she bears a level of responsibility for your project. Avoid surprising any stakeholder. And don't spring information on colleagues to pressure them for action you want. If you have bad news that affects others, treat them as you want to be treated—tell them ahead.

Show the Right Level of Detail

Keep the discussion focused on broader project goals such as updates for launch, revenue estimations, and risk exposure. The single most often made mistake in my experience is when PMs present too much technical information about either minor issues or issues that have been resolved. Normally, this is not why managers review projects. So give brief progress reports with little technical detail; if someone wants more detail, they'll ask.

Be Direct

Make your presentation direct. Start with a summary and work your way down to the detail. Avoid the temptation to tell a mystery story like "Kenneth saw this behavior in a test and thought it might be a problem with that but Kim got involved and had Brad repeat the test at high temperature and then Ethan realized the problem was really this and then we talked with Merix Supply and it turned out everything was okay." Instead, change the order: "It turned out everything was okay with the part from Merix Supply, but Kenneth saw…". There are two benefits to starting the discussion with the result:

1. First, the audience can weigh how much they need to hear the discussion. You might get 10 words in and have your sponsor wave you on to the next topic. If so, you've got more time to speak on subjects of more interest.
2. It alerts people to an appropriate level of concern. Imagine the amount of attention that would be attracted by starting a lengthy explanation with "This issue is likely to delay launch" versus ending with it.

Build your entire presentation from a summary level, working down to detail as necessary. As the example in Figure 11.1 shows, you can begin with a summary, so, from the start, the review team knows what the urgent issues are. Then you can work down to the next level of detail. So, the summary might simply say "We identified a serious risk with vibration. It may delay launch, but it's unlikely. This will be covered in detail in Slide 4."

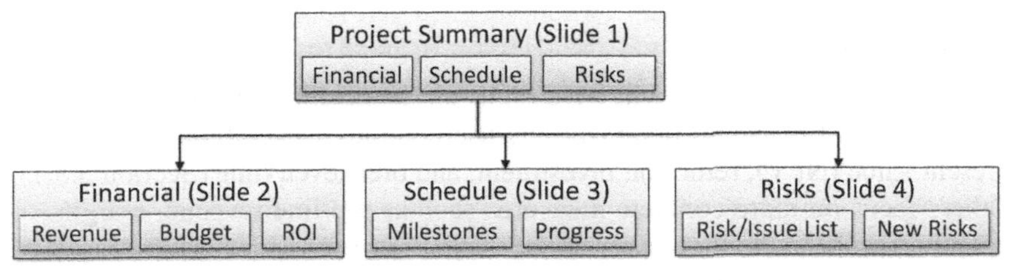

FIGURE 11.1 Present topics "top down."

TABLE 11.1 Sample Risk/Issue Balance Sheet

Risks/Issues	High Concern	Medium and Low Concern	Total Risk List
At start of month	4	16	20
Added this month	+1	+3	+4
Retired this month			
Planned	−2	−7	−9
Actual	−1	−7	−8
At end of month	4	12	16

likelihood of success in a general market. For example, you can speak to how well the value proposition has been validated (Section 9.3.3), early orders, or results of prototype testing. Happy customers create confidence in the project.

Organizational Issues

Does the team need help from anywhere in the organization? Does the test lab schedule threaten to delay the project? Does the team need sourcing to dedicate a resource for a few weeks so material can be ordered? Do you need time from the machine shop or assembly line to build prototypes? The steering committee is often in the best position to help with cross-department resourcing. Of course, no surprises—tell your colleagues in the other departments of your plans well ahead.

Approvals

Does the team need approvals for purchase orders, extra resources, capital requests, and milestone completion? Be ready to explain why the approval is needed. Also, be clear about when it's needed, for example, "we need $40k to upgrade Line 11; no delays in project anticipated if approval received by May 3."

11.1.3 15 Tough Questions

No matter how well you prepare, it's likely something will come up you're not ready for. Senior management is usually good at hitting you with tough questions that can hijack your presentation. I can recall several times in my career being stuck 30 min on the second slide because someone asked something I wasn't ready for.

After you've prepared thoroughly, you will want to think about the questions that might come up during your talk. Consider the following questions your managers might ask[1] [2].

1. These are adapted from Tom Gilb's 12 Tough Questions. See Ref. [2].

1. What data did you use to drive that decision? Can I see it?
2. How does this measurably affect project goals?
3. How do you know you're at root cause?
4. Who in the organization have you created consensus with on this? Why didn't you include [whoever you might have left out]?
5. What are the leading indicators that will tell us your new idea is working?
6. Does the team have the capability to handle this? The capacity? Why do you say that?
7. What commitments do you have from early customers? If you have some, how do we know they are committed? If you don't have any, how do we know this project is worth doing?
8. Why should I think that that customer is serious? Do they have a real project? Are they successful in the target market? Do they pay their bills?
9. How is this product better than the competition? What unmet needs are we now able to meet?
10. Did you follow the standard process for this? If not, how did you deviate? Why?
11. Why are we finding out about this now?
12. (After a subjective statement like "this is better") Can you quantify this? If not, why not?
13. What is the history of this issue?
14. How do we know this won't happen again? What processes did you improve to help those that follow you?
15. Do we have to pay the consultant/supplier even if the solution doesn't work? If so, why?

11.2 METRICS

A metric is simply a measure of activity or results related to project performance. Good metrics are powerful tools, but thousands can be contrived, collected, and published—which ones are worth measuring? Common advice in creating metrics is to start with the SMART [3] acronym:

- **S**pecific—a clear purpose for making the measurement.
- **M**easurable—built up from credible data with an understandable method.
- **A**chievable—the team believes their actions can reach the goal associated with the measurement.
- **R**elevant—it brings value to something important.
- **T**ime phased—the goals have clear timing.

Metrics are founded on the power of quantitative measurements. Quantification brings clarity and the accountability of data creates credibility. Unfortunately, qualitative discussions are easier and many PMs have a difficult time creating quantitative ones. In fact, every one of the areas in Section 11.1.2 can be presented quantitatively. As pointed out in Chapter 10, Tom Gilb [2] makes a

TABLE 11.2 Quantitative Presentation of Approvals

Request	Date of Request	Date Approval Needed to Prevent Delay	Likely Delay as of Today
Patent attorney for new vent structure	Feb 22	Mar 8	2 weeks
Capital expense for glass molds	Mar 14	Apr 26	0 weeks

compelling case that every topic can be quantified.[2] For example, let's consider the topic of approvals from Section 11.1.2:

- Qualitative discussion example
  "The team applied for approval of capital expense for glass molds several weeks ago. We also have a request for a patent attorney to accelerate an application for the new vent structure. Without these approvals, launch may be delayed.
- Quantitative discussion example
  Table 11.2 is a quantitative discussion of the same topic.

The quantitative discussion brings several advantages:

- Quantitative discussions can be tied more directly to steering committee goals. The example of Table 11.2 ties what the team needs (approval) to what the steering committee wants (on-time launch). Steering committees hear every day "I need more resources" or "slow approvals make work harder." Those are discussions centered around what you want. Like most other people, managers respond better when the discussion centers on what they want.
- Quantitative discussions are verifiable. If you give a rambling list of why the project needs something, the steering committee may not be convinced. Verifying qualitative arguments is difficult. By presenting quantitative reasoning, you open yourself to being fact-checked, and that gives the argument credibility.
- Quantitative discussions are easier to give context. The two most common types of context are historical (how this project is doing compared to the past) and comparative (how this project is doing versus other projects, either internal to your organization or across an industry or market). Consider if

2. This stands in contrast to a quote from W. Edward Deming, who is commonly misquoted as saying "If you can't measure it, you can't manage it." In fact, he said, "It is wrong to suppose that if you can't measure it, you can't manage it—a costly myth" (Deming, The New Economics, 1994, p. 35). So, readers may be circumspect in adopting Gilb's advice wholesale. Nevertheless, Gilb's arguments are compelling in that a great deal of value can be derived by quantifying more than comes naturally to most of us.

					Today						
Request	22-Feb	1-Mar	8-Mar	15-Mar	22-Mar	29-Mar	5-Apr	12-Apr	19-Apr	26-Apr	3-May
Capital for molds	*Requset*		**Start delay**	**Delay**	**Delay**						
Attorney for vent				*Request*						*Start delay*	

FIGURE 11.2 A history of the approval process showing project delay.

the discussion above concerning approvals and assume delays continued for weeks. Table 11.2 could be augmented to show a historical context as shown in Figure 11.2, which could be updated and presented weekly, adding credibility.

This is one example. Metrics can be applied to virtually any facet of project management from approvals to financial return to project team satisfaction. The power of quantitative presentations has led most companies to use a variety of metrics [4]. They have numerous purposes, which for product development projects include [5]:

- To support decisions on projects approvals.
- To support resource allocation decisions (expense and people).
- To estimate contribution to organizational goals, most notably project revenue.
- To design incentives.
- To aid in the learning of the PM, the project team, and all project stakeholders.
- To support the improvement of development processes.

Different metrics are used in different phases of a project. Typically a larger number are applied at the initial project approval. After that, it's common to use a smaller set of metrics, most commonly updates on revenue projection and team performance to schedule.

Metrics are often divided into two types: results and activity.

- Results metrics are lagging indicators—accurate, but slow. They show if the plan worked, but they don't usually tell what's needed to react quickly if the plan is not working. If someone was driving from Baltimore to Boston and expected to leave at 1:00 PM and arrive by 9:00 PM, the primary results metric would be arrival time.
- Activity metrics are used as leading indicators—fast, but unable to tell if the goal will be met. They show if the plan is being followed. Driving the 400 miles from Baltimore to Boston in 8 hours leads to an activity metric of 50 mph; an odometer could be checked each hour to see if the trip is on plan. If traffic in Philadelphia delayed the trip, the odometer would be a leading indicator of delay; using it, the driver might react by deciding not to stop for a sit-down lunch.

A results metric like "on time launch," "first year revenue," and "zero quality defects" will typically link closely to organization goals. Activity metrics like "% critical path complete," "estimated revenue," and "# significant risks closed" tell whether the team is on track to meet the results metrics. The two combine to give a better picture than either can alone.

11.2.1 Types of Metrics

There are a large number of metrics used in product development. Some of the most common are listed here.

Price and Cost Metrics

There are several metrics related to price and cost of the "average" product, several of which are shown in Figure 11.3.

- The average sales price (ASP) of the product line—including a range of models and accessories, and the effects of discounts to large-volume customers.
- The average product cost of the models used for the ASP calculations. These costs include direct expenses (labor and material cost) and overheadd. Depending on the organization, the preferred overhead may be *standard*, which includes all overhead; or it may be *variable*, which includes only the overhead costs that scale with the number of units built, such as factory line supervisors and costs of managing material. Variable overhead excludes fixed costs such as rent and heating.
- The difference between price and cost is called "margin" or "gross profit." If variable overhead is used, this produces variable margin if standard overhead is used, it produces standard margin.

For example, if the company built a security camera family that had a list price between $800 and $1200, but sold to several distributors for $500–$800, the ASP might be $600. If the cost (with overhead) was $400, the margin would be $200 on average. This is typically reported as a percentage of average price: 1 − $400/$600 = 33% margin.

Financial Performance Metrics

Financial performance metrics measure the return of the project as shown in Figure 11.4. For example, the average price and product cost from Figure 11.3 can be scaled by the number of units sold to estimate the total revenue and the *cost of goods sold* or COGS. The difference between revenue and COGS is the contribution to the company. The return is the difference of the contribution and the project development costs.

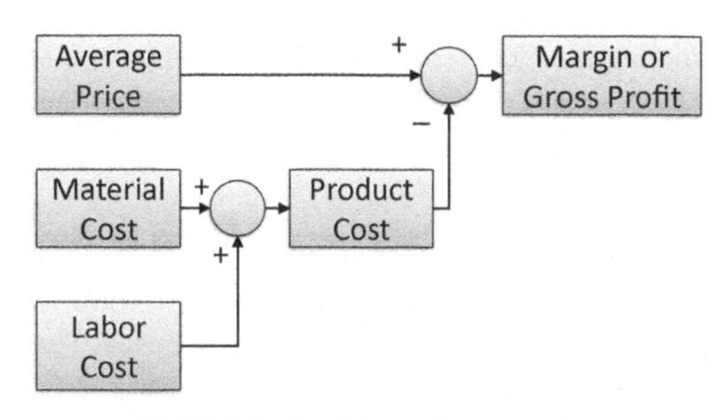

FIGURE 11.3 Price and cost metrics.

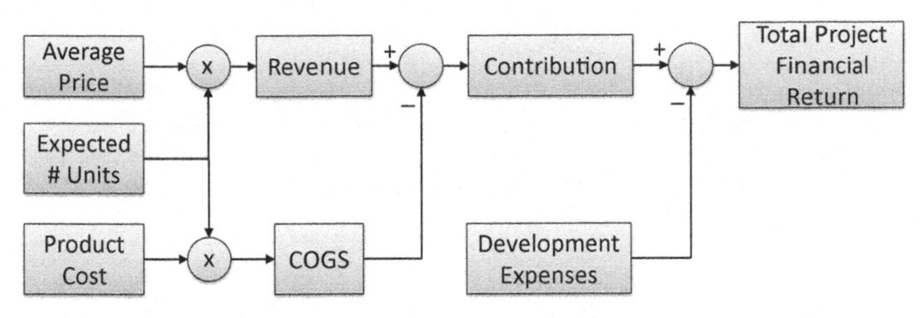

FIGURE 11.4 Financial analysis at project initiation.

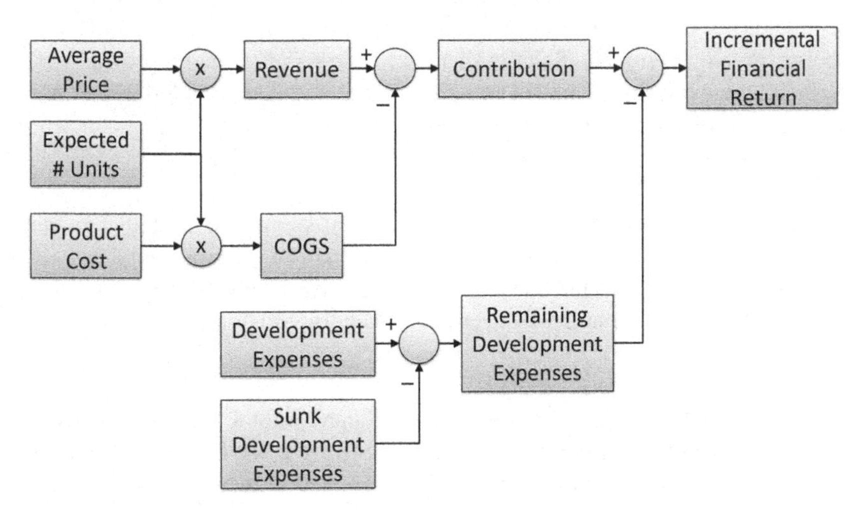

FIGURE 11.5 Financial analysis for project cancellation.

Unlike Figure 11.3, the block diagram of Figure 11.4 does not represent simple arithmetic. Because financial performance must be measured over several years for product development projects, the cost of capital must be accounted for. Today, a common cost of capital is 10%/year. For example, $500k 3 years from now has a present value of $500k/(1.10)^3 = $375k today. So, the financial performance must be calculated with formulas such as NPV, IRR, and break-even time (see Section 5.3.1).

Financial performance metrics are usually of the most concern at the initiation of the project. As the project proceeds, they are typically less critical because more and more of the estimated development expenses are spent or *sunk*. However, a common scenario when a project runs into financial problems—either due to reduced estimations of contribution or increased development expenses—is that cancellation may need to be considered. In these cases, the *incremental financial return* is usually calculated, which is the financial return ignoring the sunk costs as shown in Figure 11.5. This avoids the *sunk cost fallacy*: showing a benefit from cost reduction when costs are actually unrecoverable. According to Heerkens, when sunk cost is properly accounted for, few mature projects will be canceled:

For most projects that are more than 60 or 70 percent along in their life cycle, an unimaginably bad set of circumstances—or a sudden...shift in the project environment—would be needed to make termination the most appropriate choice [6].

Schedule Metrics

Metrics on schedule are among the most common for ongoing projects. Of those, projected days late to milestone completion or launch are probably seen most often. As has been discussed, these are typically lagging indicators—by the time the PM is certain enough to show a milestone slip, it's often too late to recover. The fever, run and Schmitt charts of Chapters 5, 6, and 8 provide leading indicators—an important feature since project delays are among the most common complaints regarding project management.

Spending Metrics

Spending metrics can include expenses or capital spent to date, either in currency or as a percentage of budget. Updated estimations of total spend can also be measured. Sometimes there can be metrics focusing on short-term spending, for example, expected purchases over the succeeding 30 or 90 days.

Resource Metrics

Similar to spending metrics, resource metrics can be total hours consumed to date, either in units of time or as a percentage of the total plan, and updated estimations of total resources required for the project. A metric for resources required versus allocated can also be tracked.

Other Metrics

Depending on the history and needs of an organization, there are many other metrics that can be used in projects. Metrics are often given as a backlog/flow pair. Some of these include:

- Number of *failure mode and effects analysis(FMEA)* issues with a Risk Priority Number (RPN) greater than 100 (backlog) and rate of closure (flow).
- Number of open issues from design reviews and rate of closure.
- Number of open action items from Kaizens or other events and rate of closure.
- Number of unreleased drawings remaining and rate of release.
- Number of components or suppliers remaining to be qualified and rate of qualification.
- Number of open risks items and rate of closure.
- Number of features complete and planned, and the rate of completion.
- Number and severity of changes accepted, in progress, and complete.
- Number and severity of open software bugs and rate of resolution.

- Team morale (via surveys) and change over time.
- Number of tests completed versus planned, and the rate at which tests are completed.
- Number of prototypes supplied and/or current prototype customers evaluating products.

The list continues. There are hundreds of metrics that can be defined and measured for a project. Metrics can bring focus and clarity; they can also require substantial time to report on weekly or monthly. As the list of metrics grows, the burden for reporting grows while the focus is blurred. Further, many metrics drive opposing actions—for example, the delay metric may be improved by hiring a few temporary employees to accelerate drawings; that same action will likely worsen the expense metric. So, to drive the right action, a strategy is needed to pick a modest set of metrics with clear goals for each. This set of metrics is called the key performance indicators or KPIs.

11.3 KPI: METRICS TO DRIVE IMPROVEMENT

KPIs [7] are the most important metrics applied to a business or part of a business. As shown in Figure 11.6, they are the metrics that bring focus to the organization. They are commonplace in modern business, used to improve such far flung areas as customer call satisfaction, revenue generation, emergency room wait time, and product cost. In this section, we'll talk about KPIs and how you can use them for better project reporting. Your company may already be using project KPIs that you're required to report on. If not, you may want to develop a few KPIs for your project. Either way, KPIs are powerful tools to lead. They create consensus and point the way to improvement.

KPIs create clarity by aggregating large amounts of data from multiple sources into a single measure. They display quantification of how the project is doing against what the organization expects. KPIs provide a history, showing trends, current performance, and gaps. Everyone involved uses the same data and sees the same result. KPIs bring focus by imposing a process where a few areas are selected for improvement at any one time and within those areas a few metrics are used to measure that improvement. Everyone wants to improve; KPIs make improvement attainable by defining where you'll get better, by when, how we'll measure it, and how much improvement is enough.

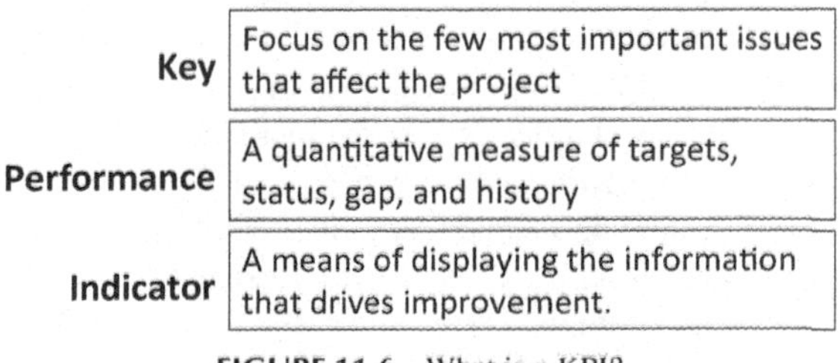

Key
Focus on the few most important issues that affect the project
Performance
A quantitative measure of targets, status, gap, and history
Indicator
A means of displaying the information that drives improvement.

FIGURE 11.6 What is a KPI?

11.3.1 Example KPI: Product-Cost Tracking KPIs

Since schedule is usually of primary importance in project metrics, KPIs that will bring benefit to most product development projects are variance to launch schedule (*results* KPI) and variance to critical path such as shown in the fever chart (*activity* KPI). But KPIs are agile—they can be developed for a department or even a single project. For example, consider a product-cost tracking KPI. Suppose the department has a history of starting projects with favorable product-cost estimates, but they rise over time; the problem is so severe that over the last few years, several products have launched with high costs that forced high pricing in order to generate barely-acceptable margins. Here, the leadership might decide the goal is for projects to maintain their initial cost estimates; if so, a KPI could be used to track product cost throughout the project.

Normally there should be at least two KPIs: one measuring results and the other measuring activity. For the product-cost goal, let's start with:

- Results KPI
  Estimated product cost built up from BOM costs and labor estimates.
- Activity KPI
  Results of reducing cost of individual items each month. This is a combination of reducing: component pricing, the number of components, factory labor time for any given step, and the number of steps. The idea is that each month, the team will select two or three items to cost reduce; this metric will capture the sum of that activity.

Now, it's time to choose targets. Since the historical problem has been maintaining cost over the project life, the target can be fixed as the initial estimate. For this product, let's assume the starting cost is $75. This creates a *maintaining* KPI—a measure meant to ensure the team holds performance at current levels. We could have chosen an *improving KPI*—for example, reducing cost 1%/month.

For the cost-reduction KPI, some historical data is required. Suppose that looking over the history of projects over the past 3 years, the average cost increase was 30% over a 1-year average project life, about 2.5%/month. The goal for an activity KPI could be simply to reduce costs 2.5%/month, but that assumes that in the past, teams sat by idly while costs increased. More likely, the teams made some effort to hold cost, but they were not successful. It's doubtful there's much data on this since it is something that is not commonly measured. So let's estimate the team was 50% successful in the past: costs rose 5%/month and the team's partially-effective response lowered that to 2.5%/month. A KPI goal that takes that into account would set the activity goal at 5% cost reductions each month, about $3/month.

It's common for activity KPIs to be new measures, so the need for some guesswork isn't a surprise. It shouldn't cause alarm–the activity KPIs are directional; the primary need for accuracy is with the results KPI. Results KPIs can be monitors and activity KPIs adjusted accordingly. For example, if it turns out that after 6 months a 4%/month reduction meets the results KPI, just adjust the activity KPI to 4%.

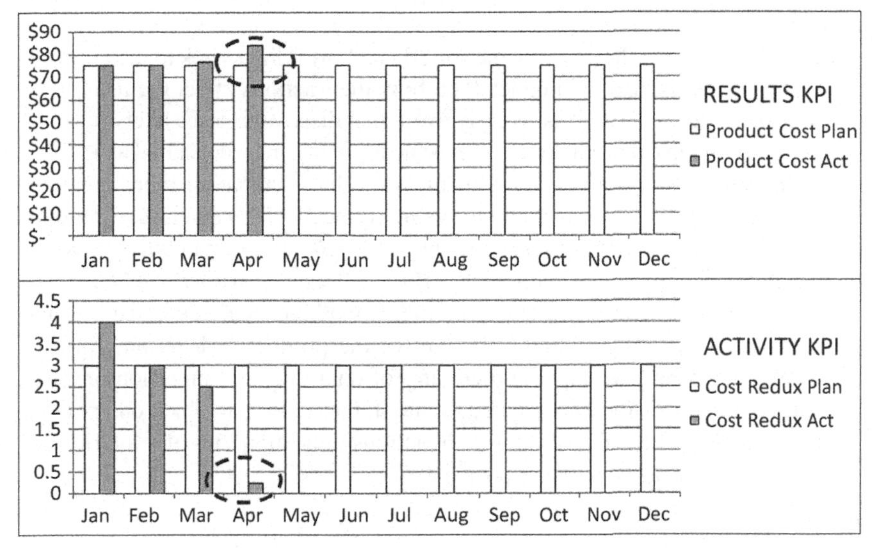

FIGURE 11.7 Results KPI (above) and activity KPI.

Now the KPIs are set. Let's simulate how the KPI might guide improvement. Suppose data is collected for 4 months as shown in Figure 11.7. Over the first 3 months, the activity KPI shows the team found a total of about $10 in cost reductions to offset what must have been $10 in increases (we can see the offset because the results KPI is almost flat over that time period). But by the fourth month, there's a problem. Costs have crept up $3 and the activity KPI shows virtually no cost reductions. Now the work to counter measure begins: Did we stop expending effort? Did the effort continue without success? Do we need more focus? Do we lack capability? It's time for the hard work of counter measuring to begin, but the KPIs have served their purpose for the month: identifying a gap that needs to be addressed.

The more the project KPIs align with organization goals, the more likely the KPIs will be sustained over time. Don't choose a topic where the leadership of the company doesn't have focus. If the management team is focused on quality, pick KPIs that will help the team improve quality.

11.3.2 Process to Create a KPI

It is easy to create KPIs, but creating a set of KPIs that leads to improvement is challenging. The characteristics of a strong set of KPIs are:

- Purpose
  KPIs must have a clear purpose that the project stakeholders are in consensus with. The purpose must have value in their eyes and it must be attainable. Wrapping KPIs around problems people don't believe need solving or don't believe can be solved is unlikely to drive good behavior. If you have a vision for your KPI that others don't see, use your leadership skills to convey that vision.

- Intuitive
  It should be clear what is being measured and how it links back to the purpose. Everyone involved should understand how their actions affect the result. For example, a design FMEA is an early event that discovers issues likely to cause quality problems in production. Each issue is given a measure called an RPN—RPNs range from 1 to 1000, with higher values indicating more concern; issues with an RPM above a certain value (typically 100) demand resolution. If the purpose of a KPI is to better manage FMEA issues, an intuitive measure might be "# open issues with RPN > 100" or "days to resolve RPNs > 100." A counterintuitive measure would include a mathematical manipulation such as "sum of squares of RPNs" or "sum of the product of RPN and days to resolve that RPN." Complex processing of data may add precision or other value in the eyes of the people that create it, but it also can block others from understanding the link between their actions and improving the metric. So, keep your KPIs as concrete as possible.
- Credible
  People will not act on KPIs they don't believe. A credible measure for a cost target includes cost of a BOM that the team understands and has access to. Barriers to credibility include hidden sources of data (the BOM is only on the PM's laptop), dubious measures (the PM and technical lead score "cost efficiency" on a scale of 1 to 10), and data that is inaccurately or incompletely recorded (the BOM is a patchwork no one can reconstruct).
- Focused
  KPIs must focus behavior because the team cannot address every problem at one time. Use lean thinking here. There are many things that can be improved—pick the two or three that need attention now. Spend several months improving in those areas; after success, come back and choose other areas.
- Balanced
  A KPI set must be balanced to avoid "gaming," the undesirable behavior of intentionally taking actions to improve the KPI in a way that brings little value. For example, a cost KPI might be balanced with KPIs for quality or timely completion. Also, use results KPIs to balance activity KPIs, which are usually easier to game. In the example of Section 11.3.1, the KPI for cost reduction activity could easily be gamed by allowing unnecessarily large cost increases (for example, using a quote with low quantities) so the obvious reduction strategies (using proper quantities) will count towards the activity goal.

As much as possible, create the KPI set as a team. Team engagement is required to build in the proper focus, balance, and credibility. And, the more the team participates in the process, the better they will understand the KPIs and own the results.

Building a KPI in Seven Steps

A step-by-step process to create a KPI is shown in Figure 11.8. A PM will likely focus on KPIs for their projects; the company leadership usually sets KPIs for

FIGURE 11.8 Creating a KPI.

the organization. However, whether you're developing a KPI specific to a project or a broad measure for the department, the process is essentially the same. This section will go through that process using the example of Section 11.3.1 to demonstrate the steps.

1. The first step in creating a KPI is to define the purpose or the objective the KPI should support [8]. The purpose must be one the team understands and that is in line with company strategy and the project objectives. It must be within the team's scope and capacity.
   - Example:
     Maintain original product costs estimates throughout the project.
2. Write out a list of actions the KPI should drive if the metric misses the target. These actions must be in proportion to the problem being resolved and they must be within the authority of the responsible team.
   - Examples:
     Requote major cost contributors with more suppliers.
     Hold cross-functional Kaizen on internal designs.
     Review functional specification to identify and validate line items that drive cost.
     Hire consultant to design lower-cost circuits.
3. Define the reporting for the KPI. First, ask who the primary audience is. Is it the core team? The sponsor? The company senior managers? Then, what forum is the intended for the primary presentation: a chart on the wall in the team area, a slide in the weekly team meeting, or a part of management project reviews? Knowing to whom and how you will present will guide you in creating the right visualization.
   - Example:
     A chart on the wall in the team's primary meeting room.
4. Define the visualization—the charts, action lists, and bowlers needed to drive the defined actions. Focus on the "view" now and the "data" later so you can avoid measuring what's easy or familiar. Ask, what visual will it take to drive the actions above?
   - Example:
     A cost-reduction action list together with the charts of Figure 11.7.
5. Define the targets for the KPIs. They should be achievable but also push the team to improve. Hone your transformational leadership skills: get buy-in from the team at the outset and they will work harder to achieve the targets later.

   Targets require context: sometimes the context is simple, such as a known cost. Other times the context must be built up over time. If your team is just learning to use FMEAs, they probably can only guess at how many open issues are acceptable or how long they should take to resolve. In these cases, start with the best estimation possible and be willing to adjust as the team gains experience.

- Example:

  Hold cost constant at the starting estimate through the project life.

6. Define the raw data and its source. The data must be credible, accessible, and be able to be updated often and with few errors.
   - Examples:

     (Early in project): Use Excel costed BOM stored in central site.

     (Later in project): Put BOM in MRP system where costs reports can be pulled as needed.
7. Finally, define the data collection: who owns it, how often it will be done, and what method will be used. Automate the method as much as possible for two benefits:
   a. Reduce time, which saves that time for other activities. A more subtle benefit is, if it's easy, the data will be more likely to be collected on time every time; this leads to more trust of the data.
   b. Reduce errors, which also leads to more trust of the data. Let the team spend 2 hours addressing an issue that it turns out was generated by a data transcription error, and you'll have a hard time keeping them engaged in the future.
   - Example:

     MRP report pulled weekly and pasted into Excel sheet.

11.4 DASHBOARD

A project dashboard is a simple display that gives a balanced view of the entire project. Modeled after an automobile dashboard, it tells at a glance everything someone needs to know for a quick check on the project. Of course, a car dashboard doesn't replace the need to have the car serviced or the need to have a new noise checked even if the dashboard reports no problems. Similarly, a project dashboard seeks to identify the most common issues, understanding that some things will be revealed only with closer review. A project dashboard need only tell the sponsor and the rest of the leadership team most of what they need to know, but it should require only a glance: Are we meeting project goals? Are there new known risks? Are approvals required? The right dashboard will depend on the culture at your company, the type of project, and the issues that have affected your organization in the past.

11.4.1 What Goes in a Dashboard?

The first thing to consider is the content of the dashboard. Focus on issues most relevant to your organization. For almost all projects, schedule will occupy a fair amount of space; so will risks. These two are common in almost all product development projects. Of course, you'll want to include any company-wide

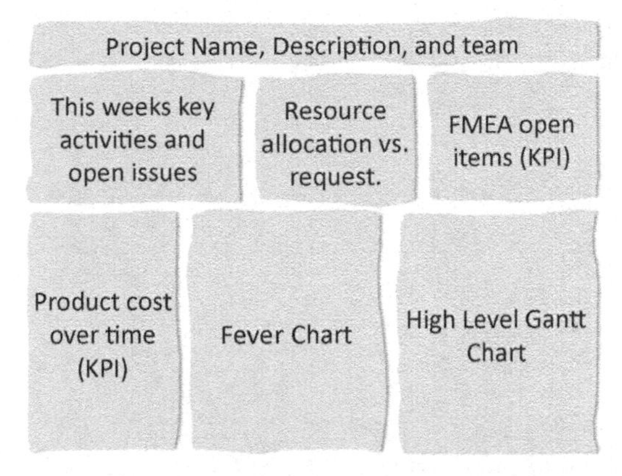

FIGURE 11.9 Sketch of a project-level dashboard.

KPIs you are managing for the project. After that, look at issues in your department. For example, if your department is often constrained by internal resources but getting approvals for expenses is less of an issue, place a resource measurement on your dashboard and skip the expenses. The issues for a dashboard can include any of the metrics listed in Section 11.2.1. But you'll need to prioritize—everything needs to fit easily on one page.

11.4.2 Sketch an Outline

Now that you've chosen your dashboard content, outline the dashboard appearance [9]. For example, Figure 11.9 lays out seven sections in the dashboard including two for schedule and two KPIs.

11.4.3 Build the Dashboard

Build up the dashboard section by section. Often a spreadsheet works, creating the dashboard as a summary tab. Then, use other tabs for raw data that build up the dashboard. Except for static items such as project name, resist hard coding data—it leads to stale data and transcription errors. For example, pull the FMEA open items directly from a copy of the FMEA action list. That way, you can update the dashboard with a single cut and paste. If you attempt to copy data cell by-cell, you're sure to eventually encounter transcription errors. Errors in your dashboard will injure trust—work hard to avoid them. The sketch from Figure 11.9 is built into a dashboard in Figure 11.10.

11.4.4 Use the Dashboard

Now that you've built your dashboard, start using it with a lean mentality—let the new processes stabilize before making changes. Expect several weeks just to smooth the process of collecting the data and eliminating errors in calculations

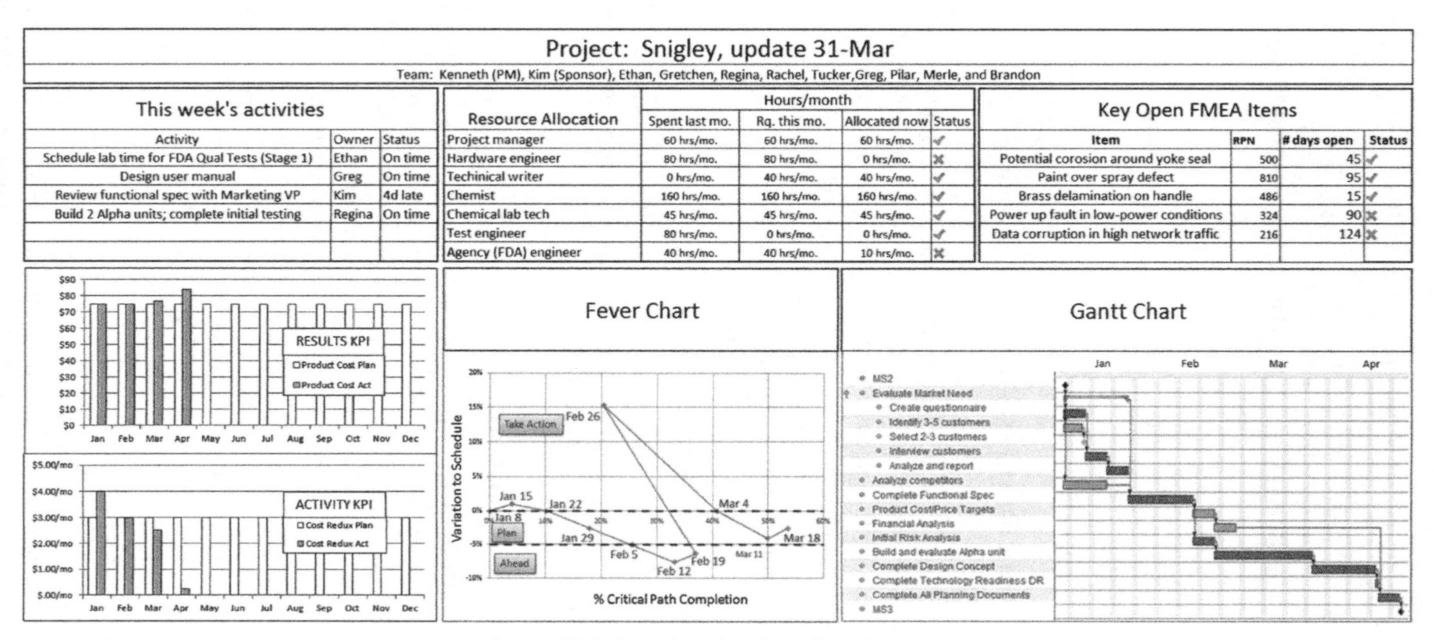

FIGURE 11.10 Sample dashboard.

and graphs. But after a month or so, you should start to see clarity around key issues. The use of a dashboard will probably represent a culture change—depending on your organization, that change may be substantial. Initially, people may be guarded; over time, the resistance usually subsides, especially if you are able to deliver results.

Sustain the dashboard. If the dashboard often becomes outdated or has a history of errors, don't expect much action from it. Apply the principles of continuous improvement to your dashboard. Keep what works, get rid of what doesn't. Over time, your dashboard can become a powerful tool for understanding the issues in your project, building consensus with the team, and creating convincing reasoning for your sponsor.

11.5 RECOMMENDED READING

Keyte C. Developing meaningful key performance indicators. Intrafocus; 2014.

Loch C, Kavadias S, editors. Handbook of new product development management. (Oxford, UK): Butterworth-Heinemann; 2008.

Tufte ER. The visual display of quantitative information. Graphics Press; 2001.

11.6 REFERENCES

[1] Reinertsen DG. The principles of product development flow: second generation lean product development. Celeritas Publishing; 2009. p. 51.

[2] Gilb T. Quantify the un-quantifiable: Tom Gilb at TEDxTrondheim. TEDx Talks. Publ; November 3, 2013. Available at: https://www.youtube.com/watch?v=kOfK6rSLVTA, about 16 minutes in.

[3] Gaddis G. 5 ways to make the most out of having a mentor. Forbes; November 5, 2013. Available at: http://www.forbes.com/sites/gaygaddis/2013/11/05/5-ways-to-make-the-most-out-of-having-a-mentor/.

[4] Goldense B, Gilmore JB. Measuring product development. Machine Design July 6, 2001.

[5] Tatikonda MV. Product development process measurement. In: Loch, Kavadias, editors. Handbook of new product development management. (Oxford, UK): Butterworth-Heinemann; 2008. p. 199–216.

[6] Heerkens GR. The sunk-cost dilemma. Project Management Institute; January 1, 2013. Available at: http://www.pmi.org/learning/business-project-sunk-cost-dilemma-2676.

[7] Glossary, The KPI Institute, Available at: http://www.performancemagazine.org/glossary/.

[8] Keyte C. Developing meaningful key performance indicators. Kindle ed. Intrafocus; 2014. Location 81.

[9] Duggirala P. Project management dashboard. October 6, 2009. Available at: http://chandoo.org/wp/2009/10/06/project-status-dashboard/.

Appendix A

Certifying Agencies for Project Managers

Certifying organizations for project management include [1]:

Australia:	Australian Institute of Project Management (AIPM) 139 Macquarie Street Sydney NSW, 2000 Web: www.aipm.com.au Email: info@aipm.com.au
United Kingdom:	Association for Project Management (APM) Ibis House, Regent Park Summerleys Road Princes Risborough Bucks HP27 9LE Web: www.apm.org.uk
Europe/Rest of world:	International Project Management Association (IPMA) About 30 addresses for various countries on the website Web: ipma.ch Email: web form at http://ipma.ch/about/contact/
USA:	Project Management Institute (PMI) Global Operations Center 14 Campus Boulevard Newtown Square, PA 19073-3299 Email: customercare@pmi.org Web: www.pmi.org

USA:	American Society for the Advancement of Project Management (ASAPM) 6547 N. Academy, #404 Colorado Springs, CO 80918 Email: web form at http://www.asapm.org/about-us/contact-us Web: www.asapm.org

REFERENCE

[1] Cagle RB. Your successful project management career. American Management Association; 2005. p. 14f.

Appendix B

Sorting Problems People Express with Their Jobs

This appendix sorts the list from Section 4.3.3, which is a set of reasons people express as to why they are dissatisfied with their jobs. The reason given is listed in the first column. The second column lists one or more possible root causes and then whether those causes are most likely transactional (xA) versus transformational (xF), and whether the issue is most focused on people issues (P) or company goals and objectives (O). The list of root causes is not meant to be exhaustive—there are many reasons a person might say "we are underresourced." It might be company policy that consistently underresources projects, it might be due to a colleague being out for extended periods due to unexpected reasons, or it might be the project manager (PM) has significantly underreported the needed effort. The focus here is on root causes that are within the scope of the PM to act on.

Reason Given	Possible Root Cause	xA/xF	P/O
The work is boring/I'm not challenged	This is below my capability; it should be done by a more junior person.	xF	P
	I don't believe in the vision for this project. It doesn't seem important.	xF	O
	We don't automate repetitive tasks. There are too many mundane activities in my daily work.	xA	O
I don't understand the business value of what I'm doing	I have not heard the vision for this project.	xF	P
	I don't agree with the vision for this project.	xF	P
My function here is unclear	The project plan is broken down so poorly that I don't understand what I should do.	xA	O
	There is conflict in the team so it's not clear who is doing what.	xF	P
	The PM can't make decisions so I don't know what do to.	xF	P

Continued

—cont'd

Reason Given	Possible Root Cause	xA/xF	P/O
I have no say in how we do things here	My company doesn't allow people in my position to have the appropriate responsibility.	xA	P
	My PM doesn't understand that I can take on more responsibility.	xF	P
My career is not going where I want	My boss isn't coaching me or giving me opportunities.	xF	P
	My company lacks a consistent approach to career growth.	xA	P
I don't like my teammates	Team members are allowed to act out. This creates a poor environment for collaboration.	xA	P
	Our project team is constantly blaming one another.	xF	P
There is too much red tape here	Company processes are inefficient.	xA	O
	Our team is not empowered to make decisions.	xF	P
I get in trouble when I deliver bad news	My PM blames the team when things don't go well.	xF	P
I'm not capable of doing this task	The complexity of this task has been underestimated.	xF	O
	The PM has overestimated my capability.	xF	P
	The company does not allow me to grow in a needed area.	xA	P
My team is not capable of doing this job	The complexity of this project has been underestimated.	xF	O
	The PM has overestimated the capability of the team.	xF	P
	The company does not have processes needed to properly support project teams with needed resources.	xA	O
We are underresourced	My company consistently underresources projects.	xA	O
	My PM has underestimated the complexity of this task.	xF	O
Our PM has not been successful	*This is challenging to categorize because the PM could have a poor history for any number of reasons.*		
The goals of this project are not realistic	The PM doesn't understand what's required.	xF	O
	The PM has not openly explained what's required.	xF	P

Reason Given	Possible Root Cause	xA/xF	P/O
This project is too complex	The project has not been properly planned and resourced.	xA	O
The politics around here are against this project	My company has policies that tolerate and even encourage politics.	xA	P
Our company isn't good at this type of project	My company lacks the processes and discipline needed to finish this type of project.	xA	O
	My company does not have the ability to deliver the products we design to market.	xA	O
Our sponsor is weak	Our company doesn't have a clearly defined role for a sponsor.	xA	O
	Our sponsor is disengaged or not in a position to help us.	xF	P
The customer is not engaged	Our company doesn't know how to get needed commitments from customers.	xA	O
	Our PM doesn't understand our customer.	xF	O
Our suppliers are not capable	We are using suppliers outside our qualification processes.	xA	O
	Our qualification processes need to be updated to include the kinds of issues we run into.	xA	O
Our meetings are ineffective	We don't have agendas or goals for meetings; we don't track decisions from meetings.	xA	O
	Our PM doesn't listen during meetings.	xF	P
	We have team members that act out during meetings. The meeting tone is negative.	xF	P
We don't resolve issues—they just linger on	Our PM doesn't follow up on identified issues in a reliable way.	xA	O
The support system around here is poor	My company doesn't create the support structure (e.g., labs) that are needed to be successful.	xF	O
The workplace environment is bad	My company tolerates politics and finger-pointing on a regular basis.	xA	P
	My company tolerates unacceptable behavior from my colleagues.	xA	P
	My PM isn't getting us the tools we need because she doesn't understand what's required.	xF	O
My boss/PM is bad	My boss/PM does not display the character I want in a leader.	xF	P
	My boss/PM does not show respect for me.	xF	P
	I don't have a good relationship with my boss/PM; she doesn't understand me.	xF	P

Continued

—cont'd

Reason Given	Possible Root Cause	xA/xF	P/O
My salary is too low	My company doesn't compensate us at the market rate.	xA	P
	My boss undervalues the associate or values them correctly but does not communicate it well.	xF	P
My job is not secure	I don't trust my company.	xA	P
	My company has an unsure future.	xF	O
I'm not valued here	My boss undervalues me and/or my team.	xF	P
	My organization consistently undervalues associates.	xA	P
My values are not the same as the my company's	My company has poor values.	xA	P
Problems in my personal life	My company is too inflexible to allow a person to deal with ordinary personal issues.	xA	P
	My boss doesn't understand what's going on in my life.	xF	P

Glossary

Activity metrics Measures of activities toward accomplishing a goal. Well-designed activity metrics are the basis for leading indicators. Also called *process* metrics. See *Results metrics*.

Adoption Accepting new processes and practices and integrating them into standard work.

Agile project management A family of product development methods oriented to low-cost-of-iteration projects, most commonly software development. Includes Scrum, Scrumban, and eXtreme Programming.

Average sales price (ASP) The average price of a family of products sold in varying regions and volumes.

Automatic test equipment (ATE) Factory equipment that tests products prior to shipment.

Breakeven time The point in time where the net present value of a project is zero.

Burn-down (or burn-up) charts Progress measurement charts for Agile projects similar to the Schmidt chart for CPM except they measure total progress, not differentiating between critical and noncritical paths.

Claims (From patents) A list of the required elements of an invention which, for comprising claims, must be present to constitute infringement.

Critical chain project management (CCPM) A project management method derived from the critical path method that was created in large part to improve the on-time delivery of WBS-based projects.

Cost of capital The cost for an organization to obtain money. See also *Net present value*.

Concurrent engineering (CE) Cross-functional teams working together for the full length of the project. CE has largely replaced serial engineering in product development.

Continuous improvement The mindset of lean manufacturing to constantly be searching for opportunities to improve, creating process to generate the improvement (typically Kaizen events), and then creating standard work to permanently adopt the improvement.

Copyright Protection offer by a government for the expression of an idea.

Critical path The path through the work breakdown structure that is longer than all other paths.

Critical path method (CPM) A method of planning and managing a project developed first in the 1950s that focuses on maintaining the schedule of the critical path, even when tasks off the critical path are delayed.

Cross-functional Activities that simultaneously use people with a wide range of capabilities, for example, a project with a team that included design engineers, manufacturing engineers, product marketers, and accountants.

Current state The ability of an organization to plan and execute projects, commonly used as a reference point when embarking on a path of change.

Daily management Detailed work for project managers such as following up on tasks and updating project documents.

Dashboard A single display that provides the most important information about a project or a complex problem. Typically includes project KPIs. See also *Visual workflow management.*

Deliverables Items a team is responsible to produce for a project. They may come from the product specification, from standard work in the company (e.g., part of a Phase–Gate process), or commitments to stakeholders.

Deming cycle See *Plan-do-check-act.*

Design for manufacturability (DFM) Creating products that are easy to manufacture with high quality.

Design review A project team event to review a design in detail, typically lasting from a few hours to several days.

Design specification See *Functional specification.*

DFMEA Design FMEA, a FMEA targeted at product design.

Disclosure The act of passing information to another person. Disclosure takes on many forms including publication, demonstration, or explanation. In the case of patents, selling or even offering a product for sale that includes a new invention can qualify as disclosure.

Error proofing (From lean manufacturing) Designing a product or process so errors are prevented or they are so obvious they will be corrected rather than becoming defects. The seatbelt is often given as an example. Also called *poke-yoke.*

eXtreme Programming A set of programming processes frequently added to Agile Scrum.

Extrinsic motivators Motivational factors provided by others such as monetary reward.

Fever chart A plot of variance to project schedule (buffer burn in CCPM) in the vertical against critical path complete in the horizontal. Is a leading indicator of progress in WBS project management methods like CPM and CCPM. See also *Run chart.*

Fishbone diagram A means of organizing a large number of details in a hierarchical manner so that when drawn, the appearance often resembles a fish skeleton.

FMEA (Pronounced fee-mah) Failure modes and effect analysis, a cross-functional event that reviews the details of a process or design attempting to identify potential failure modes and determine likelihood of occurrence and the severity. See also *RPN.*

Flow From LPD, the movement of tasks through a process as they mature toward completion. A common goal in LPD is for smooth flow without accumulation of WIP.

Full kit From CCPM, the state of a task or project that is ready for execution.

Functional managers Department leaders such as vice president of engineering or director of quality.

Functional specification A document defining the needs a product or product family must fulfill including such things as features, performance, size, regulatory requirements, and price. The functional specification defines what the product must do; the design specification defines how the team will create a product that does those things.

Gantt chart A method of displaying a WBS where task names and other details are shown in text on the left and the duration, completeness, and predecessor relationships are shown graphically as part of a calendar to the right. See *Pert chart.*

Gates See *Phase–Gate.*

Gemba The place where the action is. Lean thinking recommends a person "go to Gemba" when solving a problem, by which is meant go see the problem in person.

Global team A project team distributed in diverse regions, typically with strong cultural and language differences.

Gold plating The tendency for developers to continue adding features after all the features that add value are complete. See also *Parkinson's law.*

Granularity The size a task is divided down to in a work breakdown structure. For example, one-day granularity is common for upcoming tasks when the project team meets weekly.

Hard skills Skills required for a project manager to ensure the team follows process such as keeping documents organized, following up on progress, and managing meetings. Contrasted with soft skills, hard skills are easier to quantify.

Handoff The process of transferring responsibility for a task from one person or department to another. Handoffs have a poor history due to loss of knowledge and ownership for the task across the handoff. Concurrent engineering reduces the need for handoffs in product development projects.

Innovation The use inventions to solve an unmet customer or market need.

Inspiration The ability to lead people in a way that they willingly apply themselves beyond a defined minimum.

Intellection property (IP) The knowledge a company owns. For project managers, the most important IP is patents, trade secrets, copyrights, and trademarks.

Internal rate of return (IRR) The effective interest rate of a project taking into account all expenses and revenue over an extended period of time.

Intrinsic motivators Motivational factors internal to a person such as personal advancement or a sense of belonging to a team.

Iron Triangle A representation that a project team will be given (1) time and (2) resources (people, expenses, and support) in exchange for delivering (3) a product that meets the specification.

Kaizen A cross-functional multiday event targeted at identifying specific ways to make the next increment of continuous improvement. Kaizens can focus on problems on the factory floor, customer service, sales, design, finance, HR, logistics, or any other part of a business.

Kanban project management A project management method that relies primarily on sticky notes and a white board (or the software equivalent) to flow tasks from left to right, towards completion. Targeted at simpler projects because work breakdown structures with predecessor/successor relationships are difficult to represent.

Key performance indicators (KPIs) A focused set of metrics that are selected to provide a balanced measure of project/process health or effectiveness.

Kick-off meeting A project team meeting called at the start of a project or, in a Phase–Gate project, to start a new phase.

Lagging indicators Indicators that reliably measure progress toward a defined goal; however, they are produced so slowly, they are difficult to use to improve behavior and so are usually complemented with leading indicators.

Leading indicators Indicators that are generated rapidly and so support continuous improvement; however, they only indirectly relate to goals and so are usually accompanied by lagging indicators.

Lean product development (LPD) A mindset borrowed from lean manufacturing and applied to product development.

Low balling The practice of using misleadingly, overly conservative estimates in the planning phase of projects to take pressure off during the execution phase.

Margin The difference between the price of a product and the cost to produce it. See also *Standard margin* and *Variable margin*.

Metric A means of quantifying the effectiveness of an organization or group, especially as it relates to the execution of a process.

Milestones See *Phase–Gate*.

Minimal batch size The smallest number of items that can be processed economically at one time. In lean thinking, smaller batch sizes reduce waste. See *Single-piece flow.*

Minimal viable product The variant of product with just enough features and performance to deliver real value, even if only to a narrow slice of customers. It is a fast way to get real-world feedback and so provides learning during the development cycle.

Net present value (NPV) A method of summing expenses and revenue over long periods of time taking into account the time-dependent value of money. See also *Cost of capital.*

Obeya (or Obeye or Oobeye) room A large room used for permanent display of information important to the project. Sometimes many of the team members will have their offices located in the Obeya room.

Parkinson's law The tendency for tasks to expand so they fill the time originally allotted for them. See also *Gold plating.*

Patent A temporary monopoly offered by a government for an invention. *Utility patents* protect the function or method of a device or process; *design patents* protect ornamental designs.

Patent agent A nonattorney that can advise a client regarding patent law for many issues related to filing and prosecuting a patent application.

Patent attorney An attorney specializing in patent law. Can be licensed to represent a person regarding all aspects of a patent.

Patent infringement The intentional or unintentional act of using an invention protected by an active patent without the permission of the patent assignee.

Patent search A process to find patents or patent applications that disclose a method that is similar to a method under study. Novelty searches seek to establish if the method under study can be patented; freedom-of-use searches seek to establish if the method under study is freely available for use whether or not it can be patented.

Payback period A simple financial return method that divides the annual sales some years after launch divided by the total investment.

Pert chart A method of displaying a WBS where tasks are displayed in blocks with duration and start/stop dates, and then blocks are interconnected to highlight the predecessor/successor relationships of the tasks. See *Gantt chart.*

PFMEA Process FMEA, a FMEA targeted at production processes.

Phase–Gate An addition to WBS-based projects that divides projects into defined sections (called phases or stages) separated by approval cycles (milestones, gates or toll gates). Phase–Gate processes are usually defined the same for all projects in a company's portfolio and so simplify the management of the portfolio.

Plan-do-check-act (PDCA) The foundational process of continuous improvement where improvement is planned and executed, then checked to see if the goals were achieved. If not, improvement is attempted again; if so, changes are standardized and the next area for improvement is planned. Also called the *Deming cycle.*

Poke-Yoke See *Error proofing.*

Portfolio management Managing all projects in an organization as a group, applying consistent metrics, approval cycles, and review processes.

Portfolio progress charts Charts for progress such as the run chart or fever chart to plot progress of many projects on the same graph.

Predecessor tasks The set of tasks that must be complete before a task can begin.

Process flow chart A visual representation of a process used to make the process easier to understand.

Process metrics See *Activity metrics.*

Prosecution (of a patent application) The processing of a patent application between the time when the application is filed and the patent is granted or the application is abandoned.

Project A sequence of activities undertaken to accomplish a specified outcome at a defined time using a defined set of resources.

Project Management Body of Knowledge (PMBOK®) A text from the Project Management Institute for general project management.

Project Management Institute (PMI) A large certifying body of project managers.

Pull From lean manufacturing, a process based on customers starting the flow of value by purchasing product. The removal of the product signals the next upstream process to produce. That then signals the next upstream process to produce and so on through the entire production process.

RACI matrix A technique used to map responsibilities against roles in a project. The name comes from a mnemonic for the categorizations of responsibilities: Responsible, Accountable, Consulted, and Informed.

Refactoring Rewriting code to improve the way a function is implemented without changing the function. Typically used to make the code easier to maintain and extend.

Regression test Typically software tests that exercise functions to ensure they execute properly on newer versions of firmware (in other words, to ensure the software did not "regress").

Relay racer From CCPM, the mentality of increasing efficiency by working on something with total focus or not working on it at all it. The allusion to relay racing is that the runner is either waiting for the baton or running as fast as possible.

Resources Typically people on a project team but can also refer to budget for project expenses.

Results metrics Direct measurement toward achieving a goal. Results metrics normally create lagging indicators. See also, *Activity metrics*.

Risk A problem that may or may not occur. The term *risk* is often substituted for *issue*, a problem is certain or nearly certain to occur.

Risk funnel A list of all risks and issues that are at least partially likely to affect a project. Also called *risk register*.

Risk mitigation Actions taken to reduce the effects of a risk or the likelihood it might occur.

RPN Risk Priority Number, the rating of risk identified in an FMEA that ranges from 1 to 1000. Risks with RPN above a certain number (for example, 100) normally need to be addressed after the FMEA.

Run chart A plot of variance to project schedule in the vertical against time in the horizontal. Is a leading indicator of progress in WBS project management methods like CPM. See also *Fever chart*.

Scrum An Agile project management method that specifies team interactions (e.g., daily Scrum meeting, retrospectives, sprint planning) and roles (Scrum Master, Product Owner). Most commonly applied to software projects.

Scrumban From Agile, a simplified, lower-overhead variant of Scrum.

Schmidt chart A plot of critical path progress against time. Is a leading indicator of progress in CPM projects. See also *Fever chart* and *Run chart*.

Single-piece flow A process with a minimal batch size of one, the ideal size.

SMART metrics Metrics that have five characteristics that form the mnemonic SMART (Specific, Measurable, Achievable, Relevant, and Time phased).

Scope creep The phenomenon where features are added throughout the project. Those changes often appear at a slow but steady pace (hence "creep") and so the project scope increases slowly, so slowly it may not be noticed for some time.

Soft skills The skills needed to inspire and/or build relationships with others. For example, the abilities to create common purpose and build trust.

Sprint From Agile scum, a fixed period of time (typically 2–4 weeks) over which one iteration of product development is completed.

Sponsor A coach for the project manager and project team; a champion for the project at the highest levels of the organization.

Stage See *Phase–Gate*.

Stakeholder Anyone that has a substantial interest in the project including the project team, the sponsor, and the company management.

Standard margin The difference between the price of a product and the standard cost to produce the product. Standard cost includes fixed costs such as rent and utilities.

Steering committee The group of senior leaders that manages the project portfolio, for example, allocating resources, approving expenses, or passing a project to the next phase. In many cases the sponsor will be on the steering committee.

Story points From Agile, a method of estimating the length of time to complete a task. One story point is the smallest task the team takes on. Larger tasks are then estimated in comparison. Story points are thought by many to give more accurate estimates than estimating in hours or days.

Student syndrome From CCPM, the way people procrastinate when a deadline is used as the primary driver for completion.

Sunk costs Investments in projects that cannot be recovered if project plans are changed, for example, if the project is canceled.

Sunk cost fallacy Changing project plans with the intention of realizing a cost reduction when the costs are actually unrecoverable.

Supply chain Your suppliers and all the suppliers they rely on.

SWOT analysis A method of comparing products and organization according to four measures that form the mnemonic SWOT (strengths, weaknesses, opportunities, and threats).

Tollgates See *Phase–Gate*.

Trade secrets Intellectual property that is protected by limiting disclosure. Trade secrets can include process details, customer lists, and design calculations.

Trademarks Indications of the source of goods and services. Trademarks can be protected to prevent others from copying them or using similar markings.

Traditional project management Alternative term for *critical path* project management with or without Phase–Gate.

Transformational leadership Leading a team through tasks that demand a change in how things are normally done. "Thinking outside the box." Requires extensive use of "soft skills."

Transactional leadership Leading a team through tasks that are repeated regularly. Requires extensive use of "hard skills."

Trystorm A combination of creating ideas (borrowed from brainstorming) and then building a simple embodiment to demonstrate how well the idea works.

United States Patent and Trademark Office (USPTO) The governmental agency in the United States responsible for issuing protection of intellectual property.

Use case The interactions between a single type of user and a product to accomplish a single goal. Most products have many use cases, all of which sum to the total functionality of the product. A narrow use case for a phone is placing a call to a busy number within the user's area code.

Value In LPD, the features or performance in a product that a customer is willing to pay for.

Value add In LPD, any action that increases the value of a product.

Value stream mapping The technique of analyzing a process by identifying and displaying each increment of value that flows to the customer.

Variable margin The difference between the price of a product and the variable cost to produce the product. Variable cost counts only those costs that are expended due to producing the product. It excludes fixed costs such as rent and utilities.

Velocity The rate at which projects or tasks within projects are completed.

Virtual team A team where all or nearly all team members are in separate locations.

Vision The ability to identify a worthy goal normally well before others.

Visual workflow management The process of creating simple, credibility views of complex goals or issues that brings consensus and drives good action.

Waste Steps in a process that do not add value or support other steps that add value.

Waterfall project A project using *critical path* or critical chain project management with or without Phase–Gate. Usually used for high-cost of iteration projects. Stands in contrast to Agile projects.

Work in progress (WIP) From LPD, the undesirable accumulation of partially completed tasks or projects.

Work breakdown structure (WBS) A structured list of tasks that together define all work required for the project. The WBS is organized into a predecessor/successor structure. Often displayed with Gantt and Pert charts. See *Predecessor tasks*.

Working project manager One whose primary role is to complete project work and secondary role is to execute the modest amount of project management. Typically reserved for small projects.

Index

Note: Page numbers followed by "f" and "t" denote figures and tables, respectively.

D

H

I

J

K

L

U

V

W

X